Telecommunication Technologies
Voice, Data & Fiber-Optic Applications

Telecommunication Technologies
Voice, Data & Fiber-Optic Applications

By John Ross

PROMPT®
PUBLICATIONS

©2001 by Sams Technical Publishing

PROMPT® Publications is an imprint of Sams Technical Publishing, 5436 W. 78th St., Indianapolis, IN 46268.

All rights reserved. No part of this book shall be reproduced, stored in a retrieval system, or transmitted by any means, electronic, mechanical, photocopying, recording, or otherwise, without written permission from the publisher. No patent liability is assumed with respect to the use of the information contained herein. While every precaution has been taken in the preparation of this book, the author, the publisher or seller assumes no responsibility for errors or omissions. Neither is any liability assumed for damages resulting from the use of information contained herein.

International Standard Book Number: 0-7906-1225-9
Library of Congress Catalog Card Number: 00-109377

Acquisitions Editor: Alice J. Tripp
Editor: Will Gurdian
Assistant Editor: Kim Heusel
Typesetting: Will Gurdian
Proofreader: Kim Heusel
Cover Design: Christy Pierce
Graphics Conversion: Christy Pierce and Phil Velikan
Illustrations: Courtesy the author and 3COM Corporation, 3COM/U.S. Robotics, ADC Telecommunications, Inc., Advanced Micro Devices, Inc., Cisco Systems, Fluke Corporation, Hewlett-Packard Company, Hughes Network Systems, IBM, and MilesTek, Inc.

Trademark Acknowledgments:
All product illustrations, product names and logos are trademarks of their respective manufacturers. All terms in this book that are known or suspected to be trademarks or services have been appropriately capitalized. PROMPT® Publications and Sams Technical Publishing cannot attest to the accuracy of this information. Use of an illustration, term or logo in this book should not be regarded as affecting the validity of any trademark or service mark.

PRINTED IN THE UNITED STATES OF AMERICA

9 8 7 6 5 4 3 2 1

Dedication

I thank God for granting me the skills and patience required for completing this project. All things are possible through Him. My parents—John C. and Larraine N. Ross—have been wonderfully supportive at all times and have always given great advice. As work progressed with this book, my mother burned midnight oil while assisting with the production of the figure drawings. Many of the staff at Forsyth Library gave positive support and also assisted with locating needed research materials. Representatives of some of the top industry leaders also provided detailed information that adds to the uniqueness of this book. I express the deepest appreciation to my friends at Sams Technical Publishing—Alice Tripp, Will Gurdian, and Christy Pierce—for their assistance, diligence, and positive attitude.

Acknowledgments

I would like to thank the following companies and manufacturers for allowing me to use these images. The following companies and manufacturers do not state or imply any certification or approval of the material covered in this book.

3COM Corporation: Figures 4-13 and 4-15.

3COM/U.S. Robotics: Figure 6-10.

ADC Telecommunications, Inc.: Figures 5-9, 7-2, 7-4, 7-8, and 8-6.

Advanced Micro Devices, Inc: Figures 4-7, 4-10, 4-27, 6-8, 9-4A, 9-4B, 9-5, and 10-1.

Cisco Systems: Figures 4-11, 4-26, 5-1, 5-13, 8-2, 8-4, 8-5, 8-11, 8-12, 8-14, 8-15, 9-9, 9-11, 9-15, 9-16, 9-17, 11-1, 11-2, 12-1, 12-2, and 12-6.

Fluke Corporation: Figure 3-27.

Hewlett-Packard Company: Figure 4-12.

Hughes Network Systems: Figure 12-11.

IBM: Figure 4-28.

MilesTek, Inc.: Figures 3-21, 3-22, 3-23, 3-24, and 3-27.

Contents

Preface .... *xi*

About the Author .... *xvi*

1 Voice, Video, and Data 1
Signals 2
Signal Bands 2
Analog Signals 4
Digital Signals 6
Binary and BCD Number Systems 6
What Is Bandwidth? 8
Compression 10
Noise 10
Networks 11
Baseband and Broadband Communications 12
Network Standards 13
Transmitting a Signal Across the Network 14
Multiplexing 14
Demultiplexing 15
Network Components 15
Network Categories 16
Metropolitan-Area and Wide-Area Networking Basics 16
Internetworking 17
Network Quality-of-Service 18
Network Management 19

2 Layers, Protocols, and Reference Models 23
Network Architecture 24
Protocol Suites 26
Communication Between Layers 28
The OSI Reference Model 28
The Physical Layer 29
The Data Link Layer 31
The Network Layer 36
The Transport Layer 37
The Session Layer 39
The Presentation Layer 40
The Application Layer 40
The TCP/IP Reference Model 42

3 Cabling, Connectors, and Cabling Systems 49
Factors that Affect Cabling 50
Twisted-Pair Cabling 54

Telecommunication Technologies

Modular Jacks and Plugs 60
Terminating UTP Cables 65
Coaxial Cable 66
Coaxial Cable Connectors 68
Fiber-Optic Cables 71
Structured Cabling Systems 74
Testing the Cable Installation 81

4 Working with LANs 87
LAN Protocols 87
LAN Transmission Methods 87
Network Topologies 88
Interconnections for Local-Area Networks 93
Ethernet Networks 107
100VG-AnyLAN 116
Token Ring Networks 122
Thin Client Networks 126
Virtual LANs 127

5 Wide-Area Networking, Internetworking, and Telephony 129
Internetworking 129
Internetwork Addressing 132
Internetworking Devices 136
Single-Protocol Backbone 139
Multiprotocol Routing Backbone 140
Internetwork Transmission Services 144
Point-to-Point Services 144
WAN Dial-Up Services 144
T1 and T3 Dedicated Circuits 145
Switched Services 145
WAN Virtual Circuits 149
Examples of Switched Service Technologies 150
Telephone Services 151
Telephone Systems 151
POTS 155
Dialing 156
Switching 157
Digital Telephony 158
LECs, LATAs, and Carriers 159
The Internet 161
The World Wide Web 162

6 Analog Modems 165
Analog-to-Digital Conversion 165
Modulation 165
Demodulation 169

Serial Communications	170
Looking Inside a Modem	178
Data Flow Control	180
Frequency Response and Line Speeds	181
Modem Standards	184
Analog Modem Operation	187

7 ISDN and Frame Relay 193

ISDN Services	194
Dial-on-Demand Routing	208
Frame Relay	208
Frame Relay Virtual Circuits	209
Virtual Remote Nodes	212
ISDN Providers and Customers	213
ISDN Dial Backup	213
ISDN Security	214
Limiting ISDN Costs	214
Using SNMP	216
ISDN and Internetworking	216
The ISDN Marketplace	216

8 ATM Networks, Frame Relay, and SMDS 219

Looking Inside ATM Networks	220
ATM Layers	221
Switched Multimegabit Data Service	226
ATM Devices	234
ATM Connections	234
ATM Services	240
ATM Virtual Connections	240
ATM Service Classes	241
ATM Switching Operations	242
ATM Addressing	243
ATM LAN Emulation	245
ATM and Internetworking	249
Network Internetworking/Service Internetworking	252

9 Optical Networks 255

Synchronous Optical Network (SONET)	255
SONET Signals	257
SONET Frame Format Structure	259
SONET Multiplexing	264
SONET Network Components	265
SONET Network Configurations	269
Synchronous Digital Hierarchy	271
ATM Networks and SONET/SDH	271
Fiber Distributed Data Interface	271

10 Cable Modems and Hybrid Fiber/Coax Networks 281

Cable Modem Operation 281
Quadrature Phase Shift Keying Modulation 282
Quadrature Amplitude Modulation 282
Cable System Architecture 284
Cable Modems at the Home 287
Cable Data Systems and the OSI Model 287
Cable Data Network Topology 289
Hybrid Fiber/Coax Networks 291
Hybrid Fiber/Coax Architecture 291
Hybrid Fiber/Coax Topology 292
Switched Hybrid Fiber/Coax Networks 295

11 DSL Technologies 297

Advantages of xDSL 298
Differences Between xDSL Technologies 300
Implementing DSL Technologies 301
Asymmetric Digital Subscriber Line 301
An ADSL Network 303
Typical ADSL Chip Set Operation 307
Symmetric Digital Subscriber Line 309
Very-High-Data-Rate Digital Subscriber Line 309
Looking Inside VDSL 311
Rate-Adaptive Digital Subscriber Line 314
High-Bit-Rate Digital Subscriber Line 315

12 Wireless Networks and Satellite Distribution 319

Planning for the Wireless System 319
A Typical Broadband Wireless System 320
Multipath Distortion 320
Antennas 328
Interference 331
Broadband Wireless Transmission Frequency Bands 334
MMDS/WCS Internet Access 336
Licensed and Unlicensed Frequencies 336
Receiving the Signal 337
Connecting to the Internet with a Satellite System 339
Satellite Distribution of Data Services 339
Satellite Networks 340
Very Small Aperture Terminal Networks 346
Receiving Data Services with DirecPC 348
Installing and Running DirecPC with Windows 95 or 98 351
Using DirecPC on a Network 353

Index 355

Preface

The right data communications, voice communications, or telecommunications solution increases productivity and saves time and money. This book provides the information needed to develop a complete understanding of the technologies used within telephony, data, and telecommunications networks. As you progress through *Telecommunication Technologies: Voice, Data & Fiber-Optic Applications*, you will have the opportunity to not only define but also learn about the integration and benefits of these technologies.

In addition, this book and its straightforward style will increase your technical vocabulary by defining terminology and fully explaining functions. It will provide information so that you or your business can select, install, maintain, and service networks that fit your needs. To do this, *Telecommunication Technologies* approaches difficult-to-understand technologies in everyday language, while its photographs, line art, schematics, and other graphics aid your understanding. This book also:

- Explains basic characteristics of networking technologies
- Illustrates the capabilities and limitations of modern communications networks and the possibilities for adapting the technologies to changing needs
- Defines the hardware used within telephony, data-communications, and telecommunications networks
- Shows methods for installing different network technologies
- Describes common network applications
- Describes how broadband technologies will benefit small and large companies
- Considers voice- and data-communications standards and the impact of those standards on existing and future communications networks

From the opening to the closing chapters, *Telecommunication Technologies* accomplishes several key goals. In the opening chapters, it gives you the opportunity to learn how network systems function, comparing digital transmission techniques to analog methods. With this, you can begin to develop a network strategy that's based on need and includes the best combination of services and features.

Telecommunication Technologies

From beginning to end, *Telecommunication Technologies* relies on a building-block approach. Each section of every chapter begins with basic definitions and then moves into a concise yet complete overview of entire systems. While Chapter 1 establishes the book's purpose and provides an overview of network technologies and applications, Chapter 2 continues to lay the groundwork for the remainder of the text with a thorough discussion of layers, protocols, and reference models.

The layers, protocols, and reference models illustrated in Chapter 2 serve as a framework for the functionality and operation of a network. With respect to actual network operation, two key points need to be emphasized: (1) Data communications and telecommunications networks use all or portions of the models. (2) As this book works through various types of network technologies, it continually emphasizes the relationships between each technology and the OSI model.

Chapter 3 introduces transmission standards, defines different transmission media used in networking, and describes the factors that affect various cabling systems used in local- and wide-area networks. In addition, Chapter 3 defines and compares the characteristics of twisted-pair cabling, coaxial cabling, and fiber-optic cabling. The descriptions of each cabling type also include an overview of the jacks, plugs, and connectors that make up a cable's connection to network devices. Chapter 3 also outlines the requirements for structured cabling systems, highlighting major components such as racks, conduit, and patch panels. With this approach, the chapter demonstrates that the telecommunications closet and wire-management system must integrate into the entire building plan. Chapter 3 concludes with a discussion of cable testers and test methods used to ensure the proper operation of connected networks.

Chapter 4 provides an overview of popular local-area networking architectures and describes applications of Ethernet, Fast Ethernet, and Gigabit Ethernet. As a result, Chapter 4 builds the foundation for discussions on internetworking found in the following chapters. Just as importantly, Chapter 4 also introduces the hardware used to build local-area networks (LANs), Switched Multimegabit Data Services (SMDS), and internetworks. With this, the chapter covers network-interface cards (NICs), hubs, bridges, routers, and switches.

Preface

Discussions on wide-area networking and internetworking make up Chapter 5. As with other chapters, Chapter 5 uses the OSI reference model to illustrate specific characteristics and functions. This leads to definitions of protocol routing and descriptions of router operation. Like Chapter 4, Chapter 5 describes a wide variety of network devices such as CSUs, DSUs, Multiplexers, and Gateways. Chapter 5 also moves into the world of telephony by illustrating private branch-exchange technologies and applications. An overview of these technologies begins with a comparison of point-to-point and switched services. From there, Chapter 5 progresses to definitions of dial-up services, dedicated services, and switched services. Each of those definitions leads to broader discussions of packet-switched and circuit-switched networks. All this creates a better understanding of the telephone-system technologies described in the latter portions of the chapter. While Chapter 5 defines uses for trunks, it also provides detailed information about telephony services. In addition, the chapter discusses digital telephony and sets the tone for later chapters.

Chapter 6 introduces analog-modem technology and begins with a discussion of the principles of heterodyning, modulation, and demodulation. In addition, the chapter covers the internal operation of modems, briefly describing simple diode-detector operation and showing the pin connections for DTE and DCE devices. Chapter 6 builds on this discussion with a detailed look at UART ICs. It also considers the differences between internal and external modems, discussing modem-to-modem communication in a WAN environment. Although Chapter 6 doesn't cover digital modems, the overall consideration of analog modems lays the foundation for other chapters in this book. Digital subscriber line (DSL) and cable-transmission technologies rely on digital modems. And digital modems rely on the same principles discussed in Chapter 6.

Chapter 7 provides information on implementing these applications by describing ISDN services. The chapter considers components such as terminals, terminal adapters, network-termination devices, line-termination equipment, and non-ISDN terminals. The overview of these components and the ISDN methodology leads to a definition of in-channel signaling and common-channel signaling. The chapter then progresses to a detailed description of the Signaling System Number 7 standard and the protocols used to control network signaling. Chapter 7 concludes with a discussion of Frame Relay technologies. Besides illustrating uses for Frame Relay virtual circuits, the chapter also describes Frame Relay error checking and management and the Frame Relay frame format. Information on Frame Relay leads to a larger discussion of ISDN and internetworking.

This book uses Chapter 8 to introduce asynchronous transfer-mode networks as a platform for enlarging ISDN strategies. While comparing the ATM network model with the OSI model, the chapter provides information on ATM cell switching, multiplexing, and layers. In addition, Chapter 8 describes equipment used to implement the technology—ATM concentrators, interconnects, and switches. It also considers SMDS. Within that discussion, the chapter explains ATM-addressing and LAN-emulation techniques. Chapter 8 concludes with a discussion of ATM as an internetworking technology that connects to ISDN and Frame Relay networks.

Chapter 9 provides a detailed look at optical networks such as SONET and FDDI. The consideration of the SONET/SDH infrastructure begins with a definition of optical carrier levels and continues with a brief discussion of wave-division multiplexing. The chapter builds from that discussion to an overview of SONET signals and techniques for multiplexing those signals. As the chapter describes the signals, it also provides information about the SONET signal hierarchy and frame-format structure. Chapter 9 also defines a number of devices used to carry SONET signals such as optical receivers and cross-connect switches. This leads to an overview of FDDI solutions. Chapter 9 first compares the FDDI reference model with the OSI reference model and then progresses to the operation of the FDDI networking scheme.

From there, Chapter 10 continues the examination of broadband technologies, referring back to earlier chapters as it provides information on cable-modem and hybrid fiber/coaxial networks. As the chapter describes these technologies, it provides a detailed overview of cable-modem operation and different modulation techniques. It also describes cable-system architecture and defines the operation of headends, distribution hubs, superhubs, and cable-data centers. Moreover, Chapter 10 discusses physical and network-layer devices used in the cable-system network such as routers, switches, and servers. All this leads to a comparison of cable-system implementation methods and the layers that make up the OSI reference model. The chapter also defines topologies for cable-modem and hybrid fiber/coaxial networks. The chapter concludes by defining hybrid fiber/coaxial characteristics and network architectures, as well as the benefits of combining broadband services with hybrid fiber/coaxial and cable-modem technologies.

As the book continues, Chapter 11 discusses another up-and-coming broadband service—digital subscriber line. While the chapter begins with a generic look at the benefits gained through implementing xDSL technologies, it quickly moves to a technical examination of ADSL, RADSL, VDSL, HDSL, and SDSL. It also illustrates the differences among the xDSL family. Throughout the chapter, the focus remains on devices used to build xDSL networks. The chapter describes ADSL modems, as well as the multiplexing schemes used within these devices. In addition, the chapter provides information on the chip sets used within the ADSL line interfaces. Chapter 11 concludes with a detailed description of VDSL and HDSL2 technologies and the design behind these.

Chapter 12 brings the discussions full circle as it describes wireless technologies and satellite-communication technologies. As the chapter opens, it emphasizes planning for the wireless system and outlines potential problems that can prevent the system from operating normally. The chapter considers site selection and regulations concerning the placement of a tower and transmitter. Potential problems for wireless systems include multipath distortion, wind loading, obstructions, and interference. Chapter 12 also provides information on antenna construction, polarization, and gain while describing key antenna specifications such as beam width and aperture illumination. The chapter moves from hardware issues to a description of broadband frequency bands and wireless Internet access. From there, Chapter 12 changes tempo by considering the use of satellite technologies to access data networks and the Internet. The analysis of satellite technologies includes a look at VSAT technologies and DirecPC options.

Each of the chapters in *Telecommunication Technologies* brings you closer to understanding the factors that continue to drive technology convergence. For example, the capabilities of ATM technologies push other technologies such as ISDN and Frame Relay to the forefront. Moreover, the gap between voice communications, data communications, and telecommunications has vanished. Modern technologies allow voice transmissions over IP networks and have encouraged the downloading of audio and video information across a variety of platforms.

About the Author

John Ross has authored four books on electronics and communications technologies, including *DVD Player Fundamentals* and *Guide to Satellite TV Technology,* both from PROMPT®. In addition, he has served Fort Hays State University in a variety of capacities since 1989. Past responsibilities for FHSU include working as the manager of microcomputer services, director of the University Card Center, and director of special academic projects. As assistant to the provost, he worked with university accreditation and assessment issues while providing project management for a number of diverse initiatives. In his current role for the university, he works as the director of Forsyth Library. Ross holds a bachelor's degree in English and a master's in political science from FHSU.

Chapter 1
Voice, Video, and Data

Introduction

If we use consumer expectations as a measuring stick, data communications, telecommunications, and video services have become as common as any household tool. As voice services have become a common necessity in the office, home, or automobile, consumers become frustrated and angry if their telephone signal lacks clarity. Educational services have multiplied as a result of video teleconferencing. Now consumers ask for rock-steady transmissions and the perception of eye-to-eye contact. In our information-intensive world, consumers who were once satisfied with 9,600-baud modems now seem disappointed with anything less than full-bandwidth capabilities when sending data over long distances.

Despite the commonplace nature of telecommunications, none of the technologies—voice, video, or data—would have widespread use without standards. A quick look at the simple definitions for each type of transmission makes the needs for standards evident:

1) Voice transmissions involve any frequency within the audio frequency range of a human voice.
2) Video signals represent data displayed as an image.
3) Data is information in an analog or digital format.

Specific terms—transmissions, frequency, range, data, display, information, and format—in each definition point toward some of the standards that dominate telecommunications.

Chapter 1 is a foundation for understanding the technologies examined in the remainder of this book. No matter how brilliant, one cannot understand the definitions of telecommunications technologies without first gaining

In our information-intensive world, consumers who were once satisfied with 9,600-baud modems now seem disappointed with anything less than full-bandwidth capabilities when sending data over long distances.

some understanding of the standards and fundamental properties that affect functions and operations. This level of understanding also goes far when considering the value of some technology for customers, employees, or an organization. Whether listening to customer expectations, managing a project team made up of technical and nontechnical professionals, or attempting to negotiate a vendor contract, an awareness of basic standards is a necessity.

Signals

We can define a signal as a voltage or current that has deliberately induced, time-varying characteristics. A signal voltage or current is different than a source voltage or current for several reasons. Every electrical signal has a distinctive shape described in terms of the time domain (or the amplitude) of the signal as a function of time and in terms of the frequency domain (or the magnitude and relative phase) of the energy.

Signal Bands

Radio frequency (RF) signals occupy the frequency range between 10^4 Hz and 10^{11} Hz in the frequency spectrum. As with the low-frequency and dc energy seen with supply voltages, we can transmit RF energy over wires. Unlike the low-frequency energy, however, we can also transmit RF energy through space in the form of electromagnetic waves.

Conventional RF transmissions break down into three fundamental categories. While we can define any portion of a radio wave that travels along the surface of the Earth as a ground wave, we call the portion of the wave radiated at an angle greater than horizontal the sky wave. Radio waves that travel from one antenna to another without the affect of the ground or the upper atmosphere are called direct waves.

The Radio Frequency Spectrum

As shown in *Table 1-1*, the RF spectrum divides into eight categories. When considering the uses of each band, how the frequencies differ in wavelength

Whether listening to customer expectations, managing a project team made up of technical and nontechnical professionals, or attempting to negotiate a vendor contract, an awareness of basic standards is a necessity.

Designation	Abbreviation	Frequency Range (kHz)
Extremely High Frequency	EHF	30,000,000 to 300,000,000
Super High Frequency	SHF	3,000,000 to 30,000,000
Ultra High Frequency	UHF	300,000 to 3,000,000
Very High Frequency	VHF	30,000 to 300,000
High Frequency	HF	3,000 to 30,000
Medium Frequency	MF	300 to 3,000
Low Frequency	LF	30 to 300
Very Low Frequency	VLF	10 to 30

Table 1-1.
Radio frequency bands.

is important. For example, we could transmit frequencies within the EHF band across very long distances, because of the extremely long wavelength seen with those frequencies. In practice, the earth would absorb much of the power from the frequency waves. Even with those power losses, EHF transmitters still have a great range.

Frequencies in the LF band are even more prone to losses due to earth absorption. Yet the shorter wavelengths seen with those frequencies allow the use of highly efficient antennas. When considering the MF band, note that the commercial AM broadcast band—533 kHz to 1605 kHz—lies with those frequencies. Because of the wavelength of those frequencies, most AM radio transmissions take the form of ground waves.

The HF band covers frequencies used by foreign broadcast stations and amateur radio stations. Given the frequency band, ground wave transmissions have a limited range of 10 to 20 miles. Sky wave transmissions between 30 MHz and 60 MHz, however, have a much greater range.

Frequencies found in the VHF band cover the VHF television bands, or 54 MHz to 72 MHz, 76 MHz to 88 MHz, and 174 MHz to 216 MHz. In addition, the VHF band includes frequencies used for commercial FM broadcast transmissions. Most conventional VHF transmissions involve direct wave transmissions. The UHF band covers UHF television band frequencies, or 470 MHz to 890 MHz. Conventional UHF transmissions also involve direct wave transmissions. The last band of frequencies, the SHF band, has extremely short wavelengths and is recognized as microwave frequencies.

Analog Signals

Signals can be found at transmission points, may be generated within electronic systems, and may take different forms. In brief, an analog system operates with continuous waveforms over a given range, such as the sinusoidal wave shown in *Figure 1-1A*. By definition, an analog signal is a quantity that may vary over a continuous range of values. One of the most basic analog signal waveforms is the sinusoidal wave, or sine wave. Other basic analog waveform types include the rectangular wave, the triangle waveform, the sawtooth waveform, and rectangular pulses.

Sine Waves. The sine wave shown in Figure 1-1A represents a mathematical relationship of an alternating voltage or current produced by an alternator, inverter, or oscillator. When a sine wave goes above and below the zero line twice, each combination of maximum positive and negative values equals one cycle. In turn, each cycle subdivides into two alternations, or one-half cycle. We can view an alternation as the rise and fall of voltage or current in one direction.

Rectangular Waves. Any waveform consisting of high and low dc voltages has a pulse width and space width and is defined as a rectangular waveform. The pulse width is a measure of the time spent in

Figure 1-1A.
A sinusoidal (sine) wave, which represents a mathematical relationship of an alternating voltage or current produced by an alternator, inverter, or oscillator.

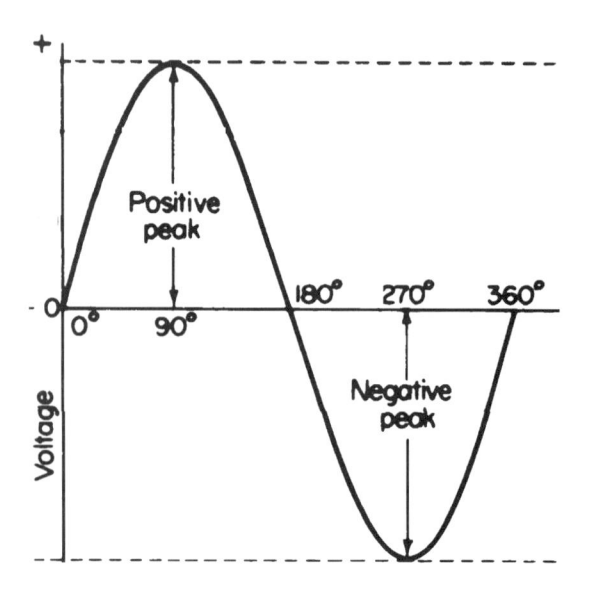

the high dc voltage state, while the space width is the measure of the time spent in the low dc voltage state. Adding the pulse width and space width together gives the cycle time or duration of the waveform. The values for the pulse width and space width are always taken at the halfway points of the waveform.

As shown in *Figure 1-1B*, a square wave is a rectangular waveform that has equal pulse-width and space-width values. While the pulse width and space width for a square wave equal one another, the opposite is true for pulses. We can define a pulse as a fast change from the reference level of a voltage or a current to a temporary level and then an equally fast change back to the original level. As shown in *Figure 1-1C*, pulses usually appear in a series called a train and are measured in terms of pulse-repetition rate and repetition time period.

Ramp, Triangular, and Sawtooth Waves. Although square waves and pulses provide the basis for modern digital communications, other types of waveforms also exist. Illustrated in *Figure 1-1D*, a ramp waveform has a slow linear rise and a rapid linear fall. In contrast, the triangular waveform shown in *Figure 1-1E* rises and falls at a constant rate and has a symmetrical shape. The sawtooth waveform shown in *Figure 1-1F* appears similar to the triangular wave, but has a longer rise time and a shorter fall time.

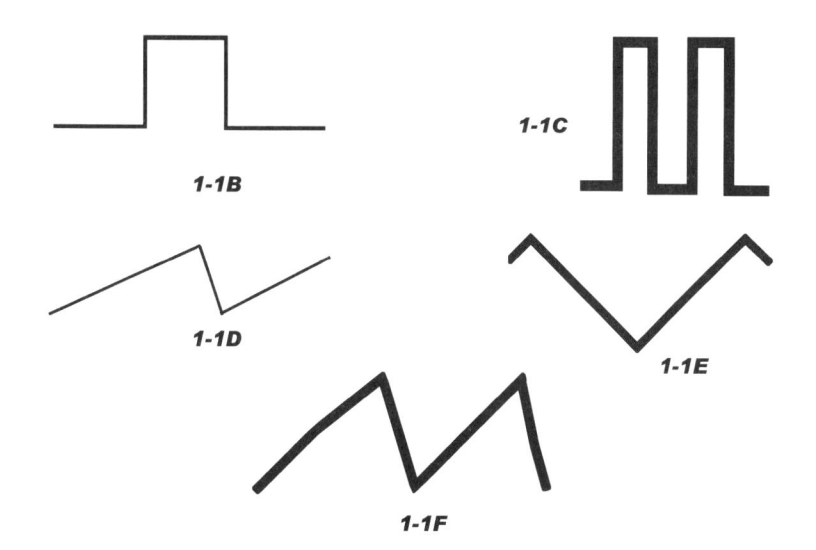

Figures 1-1B, 1-1C, 1-1D, 1-1E, and 1-1F. A square wave; a pulse; a ramp waveform; a triangular waveform; a sawtooth waveform.

Digital Signals

Digital information is data that has only certain, discrete values. In contrast to analog information, the digital information shown in *Figure 1-2* offers precise values at specific times and changes step by step. When considering electronic systems, this preciseness translates into immunity against electrical noise distortion and variations in component values. Digital information consists of data held in one of two states: low and high. Noise and component variations do not cause the low and high states of a digital signal to appear as opposite values.

Binary and BCD Number Systems

Computer systems work with the binary number system, a system that allows only two values—0 and 1. The use of binary numbers in those systems breaks information down into elementary levels. In addition, because the binary number system only relies on 0s and 1s, it provides a very basic method for counting and accumulating values.

If we look at the digits of a binary number, each value in the columns equals a value based on the powers of 2. For example, a binary number represented by 111 has 2^0=1 (or 1x1) for the rightmost column, 2^1=4 (or 2x1) for the middle column, and 2^2=8 (or 4x1) for the leftmost column. In the decimal system, this value would equal 1x1+1x2+1x4 (or 7). A binary number of 1101 equals 1x1+0x2+1x4+1x8 (or a decimal equivalent of 13).

Figure 1-2.
Digital information, which offers precise values at specific times and changes step by step.

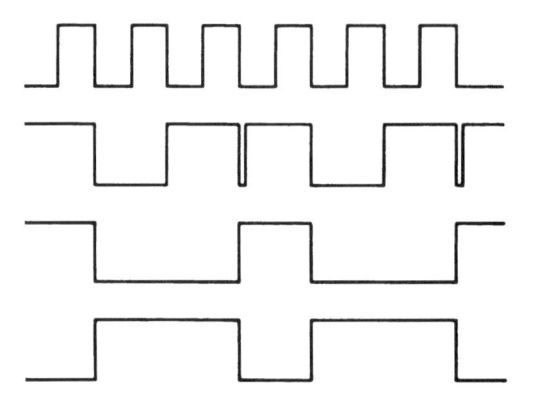

As mentioned earlier, an electronic system uses high and low signals to represent binary numbers. Each high and low signal is separated by an area of voltage that has no binary meaning. While a high signal has a value of 3V dc to 5V dc, a low signal has a value of 0V dc to 1V dc.

Bytes and Bits

The simplicity of the binary system allows computer systems to move numbers from one part of a system to another and to work with large numbers. Each binary position is called a bit, while a group of eight bits is called a byte. The sum of bits provides a method for assigning a value. As a result, the number of bits required to complete a task depends on the magnitude of the number. With each bit existing as either a 1 or 0, the value of each successive bit can increase by a maximum value of 2. As a result, the individual bits of a binary number translate to the following decimal values.

32	16	8	4	2	1	1/2	1/4	1/8	1/16
2^5	2^4	2^3	2^2	2^1	2^0	2^{-1}	2^{-2}	2^{-3}	2^{-4}

Going back to the addition of binary numbers, the sum of true values in a byte equals a decimal value—that is, a byte that appears as 11001101 has an equivalent decimal value of 206.

Binary-Coded Decimal

Another counting system—called binary-coded decimal (BCD)—combines the efficiency of the binary system with the familiarity of the decimal system. Keyboards, LED displays, and switches rely on BCD. The BCD system uses four bits to represent each digit of a decimal number. For example, decimal number 759 uses 0111 for the 7, 0101 for the 5, and 1001 for the 9 and appears as 0111 0101 1001.

Boolean Algebra

All of the operations performed through the use of digital signals originate with a concept called Boolean algebra. Named after mathematician George Boole, Boolean algebra is a mathematics system designed to test logical

The simplicity of the binary system allows computer systems to move numbers from one part of a system to another.

statements and show if the result of those statements is true or false. An example of a logical statement may appear as: "If it is raining and I have my raincoat, I will stay dry." Boolean algebra will test the validity of the statement by asking if the person is wet or dry because of rain or no rain and if the person has a raincoat.

Logical statements divide into two possible values: 0 and 1. A 0 represents either false, negation, disable, inhibit, or no. A 1 represents either true, assertion, enable, or yes. Truth tables describe how the output of a logic circuit depends on the logic levels present at the input of the circuit.

In electronic equipment, voltage and current levels represent the two possible logic states. For example, a +5V dc level may represent a logic 1 or true statement, while a –5V dc level may represent a logic 0 or false statement. In another example, the presence or absence of current could represent the logic 1 and 0 states. Most of the time, however, voltage levels express logic conditions.

What Is Bandwidth?

Looking at *Table 1-2*, we find a list of the frequencies that make up the electromagnetic, radio, infrared, and visible light spectrums. Every frequency category breaks down into frequency ranges measured in units called a hertz (Hz) and its multiples of 1,000. Each frequency range has a usable range of frequencies called the bandwidth. Ranging from hertz to terahertz (trillions), bandwidth is calculated by finding the difference between the highest and lowest frequencies of the range. Whether the transmission media used for a network consists of cabling or radio signals, bandwidth is a standard measure of the rate that a network can carry information between two points.

Every type of transmission media has different capacities for bandwidth. While copper cables were traditionally limited to lower bandwidth applications within the 300 Hz to 3,000 Hz ranges, new technologies such as digital subscriber line technologies have increased the usable range of copper cabling. By comparison, fiber-optic cabling has a larger bandwidth in the 30 to 400-terahertz range. Thus, the same length of fiber-optic cable can carry thousands of times the traffic as that carried by a copper cable.

Whether the transmission media used for a network consists of cabling or radio signals, bandwidth is a standard measure of the rate that a network can carry information between two points.

Frequency (Hz)	Frequency Range	Service
10^2	ELF Extremely Low Frequency	Telegraph
10^3	VF Human Voice Frequency	Telephone
10^4	VLF Very Low Frequency	High Fidelity
10^5	LF Low Frequency	Maritime Mobile Radio, Navigational Broadcast Radio
10^6	MF Medium Frequency	Land and Maritime Mobile Radio, Radio Broadcast
10^7	HF High Frequency	Amateur Radio, Mobile Maritime and Aeronautical Radio, Radio Broadcast
10^8	VHF Very High Frequency	Television Broadcast, Radio Navigation, Amateur Radio, Mobile Maritime and Aeronautical Radio
10^9	UHF Ultra High Frequency	Television, Military
10^{10}	SHF Super High Frequency	Space and Satellite Communication, Microwave Communication
10^{11}	EHF Extremely High Frequency	Radar, Radio Astronomy Infrared
10^{12}	Far Infrared Region	
10^{13}	Mid-infrared Region	
10^{14}	Near Infrared Region	Optical Fibers
	Visible Light	

Table 1-2.
The electromagnetic, radio, infrared, and visible light spectrums.

Microwave radio signals can range from 2 to 25 gigahertz. But the bandwidth is split into usable, lower frequency channels, which, in turn, are split again. For example, a radio signal operating at the high frequency of 30 megahertz can be split into 6,000 lower frequency channels that accommodate the range of the human voice.

Telecommunication Technologies

Compression

A direct relationship exists between the number of bits making up data and the transmission speed for that specific set of information. For example, a full-page typewritten memo with no special fonts consists of approximately 15,000 bits. If a user sends that memo to someone else with a modem, the transmission will take from two to four seconds before completion. If the same user adds different fonts, shading, and higher resolution to the same memo, the number of bits could grow to almost 100,000 and the transmission time increases to about 30 seconds with the same modem.

Two to 30 seconds does not seem like a long time. But modern business transmissions often include anything from multipage e-mail messages to full-motion video clips. Each type of transmission requires more data bits and longer transmission times unless some type of data compression is employed. Data compression reduces the number of bits required for encoding a digital transmission by sending bursts of data, rather than one bit at a time. Since data compression cuts transmission times and the need for high transmission speeds, telecommunications has become more cost effective and efficient.

Noise

Modern electronic devices include circuitry for cutting the impact of noise on received signals. Noise is an unwanted signal—such as a transient, spike, or high- voltage electrical transmission—that causes signal distortion. When considering telecommunications applications, noise can affect the quality of a transmission and appear as a bit error. Although bit errors may seem insignificant on the surface, one bit error in a banking network could change the value of an electronic funds transfer.

Telecommunications networks measure distortion as bit-error rates and include methods for detecting and correcting errors. A bit-error rate of one error per million bits is considered unacceptable. To further address noise problems, network designs specify different cable types for noise reduction. While copper phone lines are very susceptible to noise, fiber-optic lines feature no noise-related distortion.

Since data compression cuts transmission times and the need for high transmission speeds, telecommunications has become more cost effective and efficient.

Networks

A network consists of a set of devices that share a resource, can access each other, and have specific addresses. As shown in *Figure 1-3*, different types of components make up any network. Nodes serve as access points to the network. Depending on the network function, devices such as microcomputers, automated teller machines, barcode readers, point-of-sale devices, mainframe computers, and telephones serve as nodes. While transmission links include cables, microwave relays, and satellite stations, a node is the common connecting point for the transmission media and the computers attached to the network. Transmission methods and network architectures define the operating standards of the networks.

Network designs fit within five main categories. These categories are:

- Public networks
- Private networks
- Local-area networks
- Metropolitan-area networks
- Wide-area networks

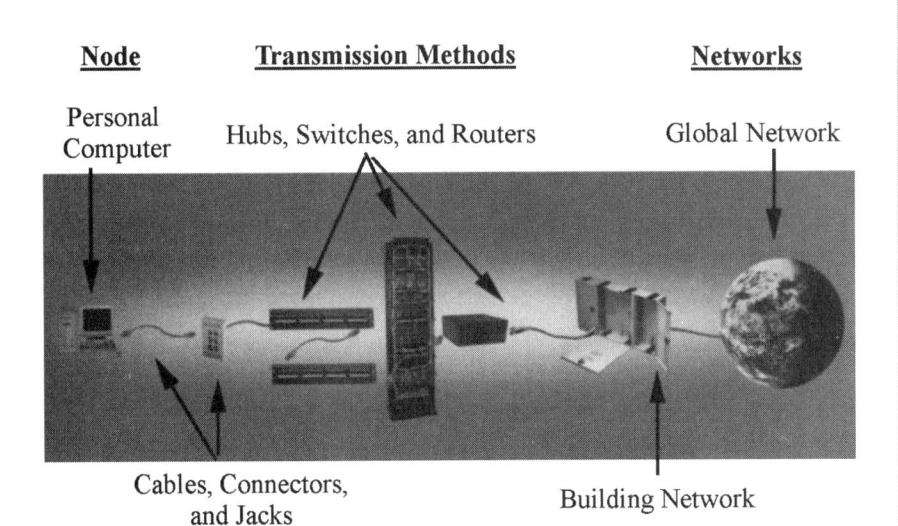

Figure 1-3.
Components of a network. A network consists of a set of devices that share a resource, can access each other, and have specific addresses.

The means of access and geographic coverage make up the main differences between the network designs. In most cases, users must pay for services when accessing networks that span regional, national, and international borders. Smaller networks that connect offices or buildings remain open to intra-organizational sharing. Most organizations use a combination of these services when conducting business.

Baseband and Broadband Communications

The transmission of data through a network occurs in either a broadband or baseband format. With broadband transmission, a single wire can carry several channels at once. For example, all cable television transmissions occur through broadband transmission. On the other hand, the baseband transmission format allows the transmission of only one signal at a time. Most communications between computers—including most local-area networks—involve the use of baseband communications.

When we discuss data communications, the term "channel" describes a communications path between two computers or devices. In this general sense, a channel may refer to wiring or to a set of properties that separates one transmission band from another. For example, the use of television channels provides a method for separating frequency bands. With this, transmissions on channel two do not interfere with transmissions on channel four.

With baseband transmission, a signal flows in a bidirectional manner. Broadband transmission also differs from baseband transmission in the direction of signal flow. With broadband transmission, a signal flows in a unidirectional manner. Rather than the signal flowing in both directions along the cable, a unidirectional signal moves in only one direction along the cable. As a result, broadband signals require two paths for the flow of data so that the data may reach all devices in the network. Broadband systems rely on either "single cable" or "dual cable" configurations.

In a single cable configuration, the cable carries transmissions over two channels with each channel set to a different frequency range. The system uses one channel to transmit signals, while the other allows data to flow to the receiver. During transmission, the signal travels to a head-end device that includes a frequency converter that changes the frequency of the signal. The system then retransmits the signal in the opposite direction along the same cable so that all devices attached to the cable receive the signal.

In a dual cable configuration, two cables connect to each station, with one cable carrying transmitted signals and the other carrying received signals. During transmission, the signal reaches the headend and passes through a connector to the other cable. No frequency change occurs. Devices attached to the network receive the signal as it passes along the second cable.

Network Standards

More than a thousand standards ensure that different networks will have compatibility. Specifications govern everything from equipment and cable interfaces to message formats, signal transmission, physical structures of networks, and network architectures. When considering standards, two significant terms surface: (1) Proprietary standards are specified by individual vendors and may not be universally accepted. (2) Open standards set up procedures that remain independent of individual vendors or manufacturers. To ensure compatibility and the strength of the marketplace, telecommunications providers, manufacturers, and suppliers join together to accept standards.

Three major international groups develop and set standards for the telecommunications industry. The International Standards Organization (ISO) developed a network architecture standard designated as the Open Systems Interconnection (OSI) standard. Designers, manufacturers, and suppliers adopted OSI standards to guarantee that information could move between two different systems.

Another group, the Consultative Committee on International Telegraphy and Telephony (CCITT) also develops and sets standards for the telecommunications industry. Two examples of CCITT standards include the transmit/receive standards for modems and the open systems standards for digital networks that carry voice, video, and data on the same phone line.

When considering standards, two significant terms surface: (1) Proprietary standards are specified by individual vendors and may not be universally accepted. (2) Open standards set up procedures that remain independent of individual vendors or manufacturers.

The Institute of Electrical and Electronic Engineers (IEEE) also sets specifications for telecommunications usage. For example, the IEEE 802.3 10BaseT specification sets standards for local-area networks using unshielded twisted-pair cabling.

Transmitting a Signal Across the Network

Some networking standards define how information flows across a network. For example, full-duplex transmission allows signals to flow in both directions simultaneously on the same line. Half-duplex transmission allows signals to flow in only one direction at a time. With asynchronous transmissions, no regular timing relationship exists between the transmitting and receiving terminals of a network. Each set of data bits also includes a control bit that indicates the beginning or end of a transmission for the receiving terminal. Synchronous transmissions use time as a standard to synchronize the transmitting and receiving terminals. Since time intervals separate the data and control bits, the receiving terminal recognizes that a specific interval length represents one or the other.

Multiplexing

Multiplexing allows more than one signal to coexist in a channel by developing a scheme to share space, frequency, or time. As a result, many signals utilize unused channel capacity through the sharing of existing channel bandwidth. The method of multiplexing depends on available bandwidth, distance, number and type of signals, cost and complexity, and reliability. One benefit of multiplexing becomes apparent when considering a communications system where users send messages between the same points. Rather than install additional physical cabling between the users or establish a new transmitter/receiver pair, a network design applies multiplexing so that different types of signals can share the same channel.

Space-division multiplexing (SDM) establishes multiple physical paths by installing new lines alongside existing installed lines. Local telephone systems rely on space-division multiplexing. With frequency-division multiplexing (FDM), each user signal modulates a different carrier frequency in the available

Full-duplex transmission allows signals to flow in both directions simultaneously on the same line. Half-duplex transmission allows signals to flow in only one direction at a time.

bandwidth. Along with broadcast radio and television stations, telemetry and telephone systems use frequency-division multiplexing. Time-division multiplexing assigns a time interval or slice to each signal. With this, the signal sequentially uses a channel link and frequencies. Digital transmission systems, wireless transmission systems, and satellite transmission systems take advantage of time-division multiplexing.

Demultiplexing

Demultiplexing circuitry separates and recovers multiplexed signals. With this, demultiplexing circuitry handles a composite signal that includes distortion, undesired mixing between signals, unknown variations in timing, and a wide range of signal levels. In addition, demultiplexing circuitry sends the recovered signals to appropriate circuits for processing.

Network Components

PC-based networks feature clients attached to a server. While the client devices run applications software, the devices rely on the storage and management capabilities of the server computers. Generally, servers have either large or multiple hard disks, have backup capabilities, and run multitasking software. Network-interface cards installed in microcomputers, servers, and printers allow the parts of the system to communicate over transmission media or cabling.

The transmission media used to carry voice, video, and data communications range from copper telephone lines and coaxial cable used for television signals to fiber optics and wireless systems. Each type of transmission medium varies in terms of capacity and carrying distance. *Table 1-3* lists data-transfer rates that describe the transmission speed of data from one device to another.

Measurement	Abbreviation	Quantity vs. Time
Kilobytes per second	Kbps	Thousands of bytes per second
Megabits per second	Mbps	Millions of bits per second
Megabytes per second	MBps	Millions of bytes per second
Gigabits per second	Gbps	Trillions of bits per second

Table 1-3.
Data-transfer rate measurements.

In client/server networks, the cables attached to the network-interface card (NIC) connect to a wiring hub in a central location. Wiring hubs are available with ports for any type of cable used in the network and contain logic circuitry that accommodates different network-interface cards. For more sophisticated networks that require network management, some wiring hubs have circuitry designed to isolate malfunctioning segments of the network and run network-management software.

Network Categories

In the past, clear distinctions existed between local-area, metropolitan-area, and wide-area networks. Today, however, networking technologies continue to come together due to several reasons. First, efficiency and effectiveness require compatibility between systems—whether the systems are located in one building or separated by 50 miles. The desire to remain competitive and to cut costs also pushes organizations to integrate technologies as much as possible. With many of the standards once associated with only local-area networks or only wide-area networks merging, networks are increasingly defined as global networks.

A local-area network (LAN) is a data communications system housed in an office or campus environment. With the installation of a local-area network, employees can share information, software, storage devices, modems, and printers. Local-area networks appear in different configurations called topologies and feature different standards for the transmission of data. But each type of network has client computers that are terminals on the network, server computers that allow resource sharing, network-interface cards, some type of transmission media, and interconnection hardware.

Metropolitan-Area and Wide-Area Networking Basics

Sending data between buildings requires another type of networking technology called the metropolitan-area network (MAN). With this approach, local-area networks in individual buildings can interconnect with LANs in other buildings through routers and gateways. A metropolitan-area network covers an area within a 30-mile radius.

Wide-area networks (WANs) serve much larger geographical areas and may carry voice, video, or data communications. While connecting to many different types of devices, WANs also carry many types of traffic such as e-mail, voice, and videoconferencing. Originally developed to accommodate long-distance telephone networking needs, wide-area networks use different technologies than those seen with local- and metropolitan-area networks. Yet growing private and public sector uses for telecommunications networks have increased the interdependence between the technologies. Both MANs and WANs use a higher bandwidth than local-area networks.

Internetworking

A network has limits in terms of how far it can extend, how many stations can connect to it, how fast data can transmit between stations, and how much traffic the network can support. If an organization needs to go beyond those limits and link more stations than a LAN can support, it must install another LAN and connect the two together in an internetwork. As illustrated in *Figure 1-4*, the term "internetworking" refers to the linking of individual LANs together to form a single internetwork.

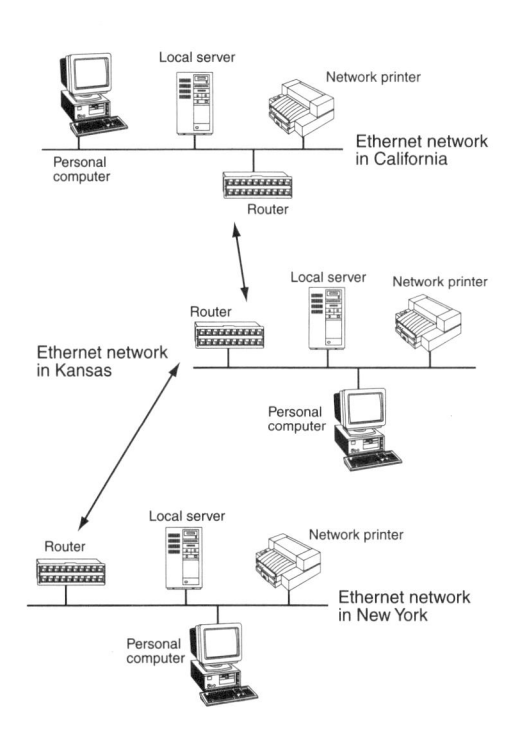

Figure 1-4.
Diagram of an internetwork, which is formed by linking together individual LANs.

An internetwork is sometimes called an enterprise network, because it interconnects all of the computer networks throughout the entire enterprise. As a result, internetworking extends the geographic coverage of the network beyond what a single network can support. This coverage may extend to multiple floors in a building, to nearby buildings, or to remote sites. Internetworking also allows multiple networks to share traffic loads. If the load increases beyond the carrying capacity of a single network, users will see the results of data congestion and lose much of the productivity achieved by installing the network.

Network Quality of Service

"Quality of service" (QoS) refers to the capability of a network to provide better service to selected network traffic over networks utilizing broadband technologies. The primary goals of QoS include dedicated bandwidth, controlled jitter and latency required by some real-time and interactive traffic, and improved loss characteristics. Quality-of-service technologies establish the building blocks used for future business applications in networks.

Software used for quality-of-service applications allows complex networks to control and predictably service a variety of networked applications and traffic types. Almost any small or large network can gain efficiency through QoS through the use of network analysis, management, and accounting tools that show network usage trends. In addition to enhancing network efficiency, QoS allows control over resources such as bandwidth, equipment, and wide-area facilities. This type of control ensures that applications will receive required bandwidth and that minimum delays will not interfere with network traffic. Quality of service also provides a method for allowing Internet service providers (ISPs) to offer different grades of service for customers.

As shown in *Figure 1-5*, ISPs can provide three basic levels of end-to-end QoS across a heterogeneous network. While best-effort service establishes basic connectivity with no guarantees, differentiated service provides faster handling, additional bandwidth, and a lower loss rate for some network traffic. Guaranteed service reserves network resources for specific traffic.

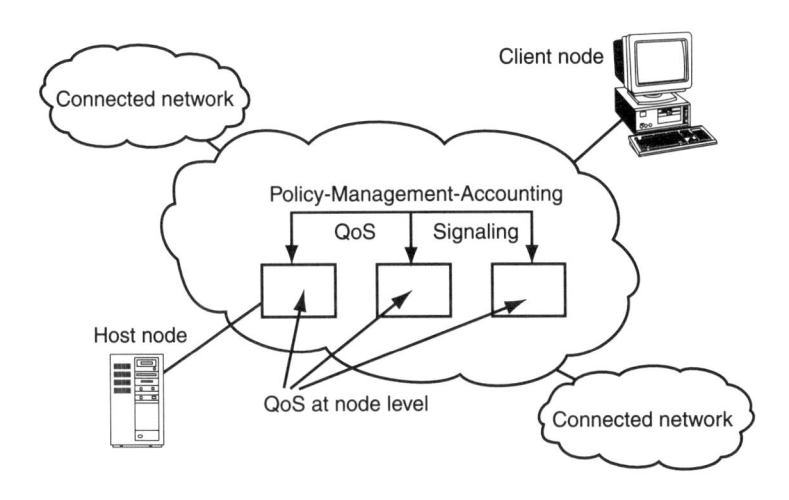

Figure 1-5.
Three levels of end-to-end quality of service (QoS).

Network Management

Network management may involve a network consultant monitoring network activity with protocol analyzer or the use of a distributed database, the auto-polling of network devices, and the use of high-end workstations that generate real-time graphical views of network topology changes and traffic. A very basic definition of network management employs a variety of tools, applications, and devices to assist human network managers in monitoring and maintaining networks.

Depicted in *Figure 1-6*, most network management architectures use the same basic structure and set of relationships. Managed devices such as computer systems and other network devices run software that allows the devices to send alerts when problems occur. When these alerts occur, management entities use software programs to execute one, several, or a group of actions, including operator notification, event logging, system shutdown, and automatic attempts at system repair. Management entities can also poll end stations to check the values of certain variables.

Polling can occur through either automatic or user-initiated methods with agents in the managed devices responding to all polls. Agents consist of software modules that compile information about the managed devices in which they reside, store this information in a management database, and

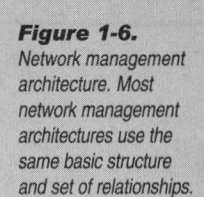

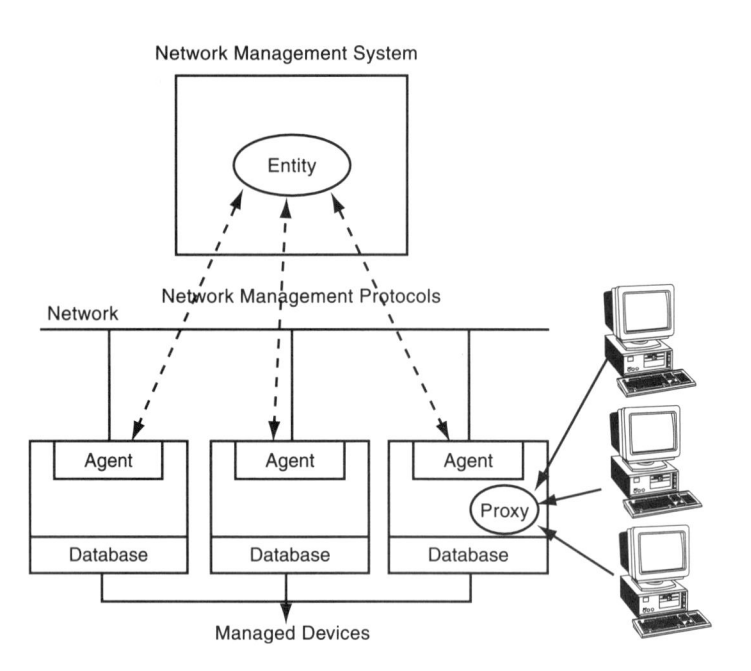

Figure 1-6.
Network management architecture. Most network management architectures use the same basic structure and set of relationships.

finally provide the information to management entities within network management systems (NMSs) through a network management protocol such as the Simple Network Management Protocol and Common Management Information Protocol. Management proxies provide management information on behalf of other entities.

Performance Management

Performance management measures and provides network throughput, user response times, and line utilization so that internetwork performance remains at an acceptable level. Performance management involves the gathering of performance data about variables of interest to network administrators, the analysis of data to determine normal (or baseline) levels, and the determination of appropriate performance thresholds for each important variable so that exceeding these thresholds indicates a network problem worthy of attention.

Management entities continually monitor performance variables and work on either a reactive or proactive basis. When a network operation exceeds a performance threshold, the managed devices generate an alert and send the alert to the network management system. With the proactive method, a network simulation projects the effect of network growth on performance metrics. With this data in hand, a network administrator can act to prevent possible problems.

Configuration Management

Configuration management monitors network and system configuration information so that an administrator can track and manage the effects on network operation of various versions of hardware and software elements. Configuration management subsystems store version information in a database for easy access. When a problem occurs, an administrator can search the database for possible solutions.

Accounting Management

Accounting management measures network-utilization parameters for the efficient regulation of individual or group uses on the network. Such regulation minimizes network problems through the allocation of resources on a capacity basis and maximizes the fairness of network access across all users. As with performance management, accounting management measures utilization of all important network resources. Analysis of the results shows current usage patterns.

Fault Management

Fault management detects, logs, notifies users of, and automatically repairs some network problems to keep the network running effectively. Since faults can cause downtime or unacceptable network degradation, fault management remains the most widely implemented network management element. Fault management involves determining problem symptoms, isolating the problem, repairing the problem, testing the solution on all important subsystems, and recording the detection and resolution of the problem.

Security Management

Security management controls access to network resources according to administrative levels and prevents the access of sensitive information by anyone who does not own the proper authorization. A security management subsystem can monitor users logging on to a network resource and refuse access to those who enter inappropriate access codes.

Security management subsystems work by partitioning network resources into authorized and unauthorized areas. For some users, access to any network resource cannot occur. Other network users can gain access to information held in specific areas but not to the entire network. To accomplish these tasks, security management subsystems identify sensitive network resources such as systems, files, and other entities, determine mappings between sensitive network resources and user sets, monitor access points to sensitive network resources, and log inappropriate access to sensitive network resources.

Summary

Many successful coaches preach "fundamentals, fundamentals!" when working with all levels of players. Successful organizations also need a firm grip on fundamental definitions, standards, and procedures. Members of an organization have little chance to innovate, produce valued products, and improve any methods without having a definite sense of the fundamentals. While this first chapter employed a very brief and basic approach toward telecommunications technologies, it also laid the groundwork for the next chapters.

Chapter 2 takes another step toward learning the fundamentals with its emphasis on layers, protocols, and standards. Although many topics may seem more exciting than protocols and standards, those concepts provide the basis for all data communications and telecommunications technologies. As a result, having knowledge about layers, protocols, and standards eases the understanding of why specific technologies take certain forms and how those technologies operate. Given that vantage point, it becomes easier to see the long-term value of the technologies.

Although many topics may seem more exciting than protocols and standards, those concepts provide the basis for all data communications and telecommunications technologies.

Chapter 2
Layers, Protocols, and Reference Models

Introduction

Networks involve workgroup computing, or a collection of individuals working together on a task. In a very limited sense, workgroup computing occurs when all the individuals have computers connected to a network that allows them to send e-mail to one another, share data files, and schedule meetings. Each device attached to the network is called a workstation. The following characteristics differentiate one network from another.

- Protocols: The rules and encoding specifications for sending data. The protocols also determine whether the network uses a peer-to-peer or client/server architecture.

- Media: Devices can be connected by twisted-pair wire, coaxial cables, or fiber-optic cables. Some networks do without connecting media altogether, communicating instead via radio waves.

- Topology: The geometric arrangement of devices on the network. For example, devices can be arranged in a ring, star, or in a straight line.

In some ways, telecommunications networks face the same problems encountered by world travelers. Along with accepting new cultural attitudes, travelers may also face language barriers. Both the cultural attitudes and the language barriers work as analogies for compatibility issues seen with

In some ways, telecommunications networks face the same problems encountered by world travelers. Along with accepting new cultural attitudes, travelers may also face language barriers.

In the most basic sense, the use of a protocol guarantees that each device connected to the network uses the same language.

networks. With networks, protocols—or a consistent set of operating standards—establish compatibility between communications systems.

Protocols may vary with the manufacturer of the network and often yield differing standards of reliability, speed, and ease of use. During the early years of data communications and telecommunications, manufacturers worked with different sets of protocols. As a result, networks produced by one company did not communicate with networks manufactured by another company. Due to the need to have compatibility and to link networks together, the International Standards Organization created the concept of layered protocols seen with the Open Systems Interconnection (OSI) model.

Network Architecture

The complexity of network designs requires an organizing method called layers. When comparing networks, the number of layers, the names of the layers, the contents of each layer, and the function of each layer differ from network to network.

Each layer consists of a program or a set of programs that provide services for the next higher layer or set of layers, with the highest layer providing services to the user. Since a layer provides services only to the layer above and uses services only from the layer below it, any change to any given layer affects only the layer above. The concept of layering breaks a single large program into parts according to function. As a result, the management and editing of the program becomes easier.

Protocols describe a set of rules and methods used by devices to communicate over a network. In the most basic sense, the use of a protocol guarantees that each device connected to the network uses the same language. A very simple protocol may specify that numeric values indicating message length constitute the first three characters of each message. The remainder of the message would consist of the actual information. When the receiving program accepts the message, it recognizes that the first three characters indicate length and do not show as part of the message.

Layers of Protocols

Networking relies on different sets of protocols for different layers. Protocols also use layers as a format for communication and providing services. A protocol establishes: (1) the type of error checking used for data compression, (2) the rules for data transfer, (3) the method for showing that a message is complete, and (5) the method for showing that the message is received.

Since different networks may use different error-detecting and error-correcting codes, protocols produce an agreement about the type of error control used between communicating networks. Data transfer covers the direction or directions that data can travel, the number of channels allocated per connection, and the route selected for the data transfer. The protocols also determine the type of control information used to signify the end of the message and the receipt of the message. Headers and trailers contain the control information. Illustrated in *Figure 2-1*, a five-layer network uses different sets of protocols at each layer to establish

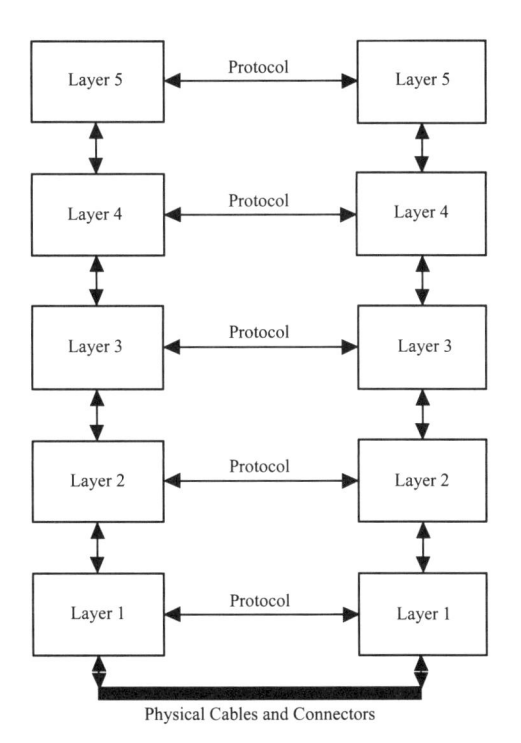

Figure 2-1.
A five-layer network model uses different sets of protocols at each layer to establish communications between different computers.

communications between different computers, or peers. Rather than having data pass from one layer of one computer to the corresponding layer of another computer, the layers of each system pass information downward to the lowest layer. Service access points (SAPs) within each layer establish points within the layers where the next layer can access services. From there, communication occurs through some type of physical medium.

Interfaces between the layers define the operations of the layers and services offered from a lower layer to the next upper layer. Sets of rules govern each interface and the passing of interface data units (IDUs) that contain service data units (SDUs) and control information from layer to layer. While the SDU envelopes information that passes across the network from peer to peer and then through the receiving computer layers, the control information establishes tasks for the lower layer.

Services define the operations provided from a lower layer to the next upper layer. With this, the lower layer becomes the service provider and the upper layer becomes the service user. Although each service defines the operation that the layer should perform, a service does not define the implementation method. Entities within each layer perform the operation.

Network design at the layer level considers the number of layers, the type of interface between layers, and the specific functions of each layer. Following detailed guidelines at this level allows a network to function consistently despite changes in physical transport media. The set of layers and protocols used within a network also defines the network architecture. Specifying the architecture in this way allows the writing of a software program or the building of a hardware device that takes advantage of the services provided by particular layers and establishes protocols for the software or hardware. The matching of one protocol per layer for use by a system creates a protocol stack.

Protocol Suites

We can refer to layers of protocols as either protocol stacks or protocol suites. The lowest layer of a suite operates at the bit level, while the highest layer operates at the program level. At the lowest level, streams of bits flow across some type of physical medium between two points. Higher layers of a suite communicate with peers through other nodes and—as a result—through an indirect link. Communication at the highest level includes routing information that allows data to flow across more than one link and enables large-scale network functions.

During actual operation, programs found at the highest layer work with more complicated data routines. Protocols found at the layers below the highest layer work with implementation of the network operation and associated tasks. If we use the routing of web pages across the Internet as an example, lower levels of the protocol suite route data packets across the Web. Higher levels of the protocol suite involve the operation of the server or the browser. As a result, the data flows only through the highest necessary level of the suite.

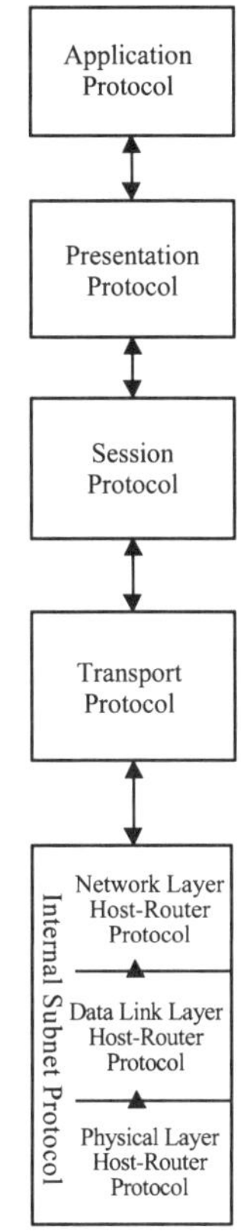

Stacking Protocols

A typical network transaction involves multiple protocols transmitting information between two applications. As shown in *Figure 2-2*, the protocols arrange in stacks that correspond to the transfer of the information. Using the example of sending web pages across the Internet, the efficient sending of data across the network may also involve the dividing of long messages into shorter segments.

When this occurs, the receiving node must have a method for finding the number of transmitted segments and the assembly order. To do this, the transmitting node encapsulates the data with headers and trailers. The header contains a field that specifies the length of the encapsulated data and at least one field that provides information about the data. If the data is a segment of a larger message, then the header also specifies the relative position of the segment in the complete message, along with the number of segments.

Figure 2-2.
Protocols communicating between applications. The protocols arrange in stacks that correspond to the transfer of the information.

Unstacking Protocols

The receiving node removes the added information after processing the data. Since this occurs before the node passes the data on to the remainder of the suite, the higher layers do not see the encapsulating information. The removal of the protocol information at each layer unstacks the protocol suite.

Communication Between Layers

Layers communicate through a basic form of messaging. Each message includes a header that identifies the type of message through control, size, and timing information. While all messages receive a header, some also receive trailer information. In most cases, layer four of a network architecture will not have a size limit for the message, while layer three will establish a size limit. As a result, the transmission of messages from layer to layer may result in the reformatting of the message into packets. In turn, each packet receives a header.

Protocol data units (PDUs) form the packets and carry peer-process information used to perform peer protocols. Headers contained within the PDUs identify the PDUs that contain data and the PDUs that contain control information. The PDUs also provide the sequence and count information so that the receiving system layers can reassemble the fragmented data back into the original order. As the packets pass from a lower layer to the next upper layer in the receiving computer, the upper layer strips away the corresponding header information before passing the packet to the next layer. Again referring to Figure 2-2, we see that peer processes include horizontal transactions between layers. Hardware devices such as network-interface cards or routers contain software instructions embedded as firmware within microcontrollers, microprocessors, or read-only memory devices. The software instructions allow the devices to implement the lower layers of the network architecture.

The OSI Reference Model

Illustrated in *Figure 2-3*, the Open Systems Interconnection model consists of seven layers of protocols that cover the communication of raw data and the networking of applications. Looking at the model, the layers range from the Physical Layer that covers the transportation of data to the Application Layer that contains user applications. With this, the OSI model establishes a reference for layered networking architectures. The model offers "open" standards, because the interconnection between networks occurs without specifying any type of hardware. Instead, the communications software adheres to given standards.

To better understand the functionality of the OSI model, we can place actual network functions alongside each layer. In brief review, the bottom four layers of the model cover the communication of raw data, while the top three layers cover the networking of applications. *Figure 2-4* supplements the OSI model with a diagram of network functions.

The Physical Layer

The Physical Layer resides at the lowest layer of the OSI model and provides the service of transferring bits of data across some type of physical link. As we will see in Chapter 3, physical links may consist of twisted-pair cabling, coaxial cable, or fiber-optic cabling. In addition, the physical links also include wireless radio signals and satellite transmissions.

Given the definitions of the Physical Layer and physical links, we can begin to see a much clearer picture of network operation. Every network consists of nodes—that is, the devices that connect to a network and have communications capabilities. The personal computers that attach to a network function as nodes. Other personal computers that have more processing power, memory, and storage capabilities work as servers on the network. Along with the network operating system, the server provides print sharing, file sharing, and application sharing capabilities.

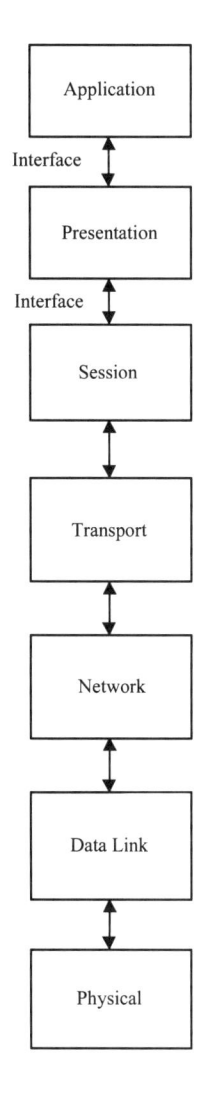

Figure 2-3.
The Open Systems Interconnection model consists of seven layers of protocols that cover the communication of raw data and the networking of applications.

29

Figure 2-4.
The OSI model and network functions.

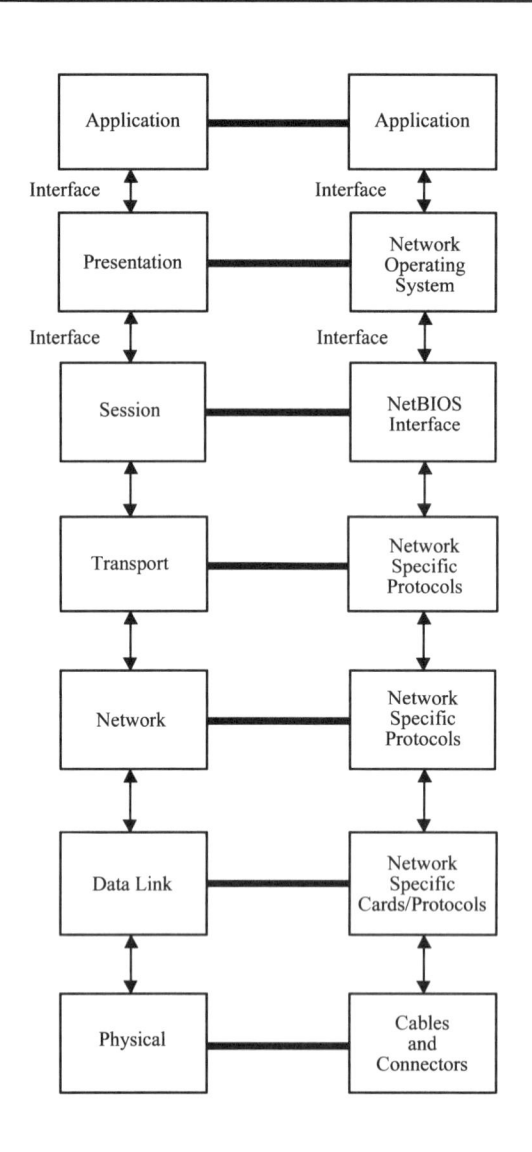

The Subnet Protocol

Within the Physical Layer, the subnet—or transportation—protocol varies according to the type of physical link used within the network. The subnet protocol establishes the method for representing bits of data. In addition, the subnet protocol provides a method for showing when a transmission begins and ends. Along with those functions, the subnet protocol indicates whether bits can flow in only one direction or both directions simultaneously.

All this depends on the type of physical link used within the network. For example, the subnet protocol for twisted-pair cabling has a different set of standards than the subnet protocol used for fiber-optic cabling. Examples of subnet protocols include the Ethernet and Token Ring networking standards, as well as the RS-232 transmission standard.

Theory to Function: The Physical Layer

All cables, connections, and equipment such as hubs, routers, and switches occupy the Physical Layer. In turn, the specifications and categories that cover cabling, relays, repeaters, and cabling interfaces fit the requirements of the OSI model. For example, the twisted-pair cabling used for local-area networks offers certain bandwidth and data-carrying capabilities. Fiber-optic cables also fit into the Physical Layer, but have greater bandwidth and data-carrying capabilities.

The Data Link Layer

The different layers of the OSI model work with data and information contained in different units. At the Physical Layer, a network transmits data in the form of bits, while the Data Link Layer moves data in the form of frames. A frame consists of data placed in a specific sequence and encapsulated by a header and a trailer that contain addressing and messaging information.

At the Network Layer, networks transmit datagrams or packets as units of information that may consist of a message or a segment of information. Each packet includes a header that contains the address of the destination node.

Again looking at the OSI model shown in Figure 2-3, we see that the Data Link Layer covers the transmission of frames, rather than the transmission of bits of data as seen with the Physical Layer. Given the purpose of the Data Link Layer, it functions in tandem with the Physical Layer and the Network Layer. Within this function, the Data Link Layer may provide either connectionless transmission that requires no acknowledgment or connection-oriented transmission that requires an acknowledgment of service.

The Data Link Layer covers the transmission of frames, rather than the transmission of bits of data as seen with the Physical Layer. Given the purpose of the Data Link Layer, it functions in tandem with the Physical Layer and the Network Layer.

The Data Link Layer protects against errors by regulating the speed of the transmission and by requesting that the transmitting peer retransmit any data that shows errors.

Connectionless service works well if the Data Link Layer provides high reliability, and if the upper layers of the OSI model provide error correction. With connectionless service, a protocol designates each individual data packet as an individual unit, and the network ensures that the packet reaches the proper destination.

Since each packet exists as an individual unit, the network does not establish and maintain a connection before sending and receiving the packet. During connectionless operation, the Data Link Layer breaks the stream of data received from the upper layers into frames. Referring to the Data Link Layer only, connectionless operation does not have a great amount of complexity.

Connection-oriented network operation uses a protocol in which the network establishes and maintains a connection throughout the data transaction. The stations involved in the transaction remain in contact with one another during the sending and receiving of the packets. During connection-oriented operation, the data link layer:

- Establishes a connection
- Accepts packets of data from the Network Layer
- Divides the packets into frames
- Passes the frames to the Physical Layer for transmission

Within this framework, the Data Link Layer becomes active at both the transmitting and the receiving ends of the network. At the transmitting end of the network, the Data Link Layer accepts the address of the node that contains a Physical Layer link so that the network can transmit data to the node. Then the Data Link Layer accepts data packets from the Network Layer. Once the Data Link Layer accepts the packets, it begins to "handshake" with the peer to ensure the correct reception of the data. We define handshaking as the initial setup process that occurs between two computers attached to a network, establishing the communication parameters between the two nodes.

At the receiving end of the network, handshaking begins the process of passing data packets back to the Network Layer. During this process, the

Data Link Layer ensures that each frame retains the proper sequencing information. In addition, the Data Link Layer applies error detection and correction codes to the frames. With this, the receiving process can generate a report to show the location of an error.

The error-detection, location, and correction functions found within the Data Link Layer may exist as the most important functions provided by this layer. At the receiving end of the transmission, the handshaking information added by the Data Link Layer works with the peer process to correct problems that may occur. Examples of network errors include corrupted, lost, delayed, and duplicate data. In addition, the data may arrive at the destination node out of sequence. The Data Link Layer protects against errors by regulating the speed of the transmission and by requesting that the transmitting peer retransmit any data that shows errors. All this combines with the error detection and correction offered in higher layers of the OSI model.

Theory to Function: The Data Link Layer

The variations in matching the OSI layers to generic networking functions become more apparent as we continue to move up the model. *Figures 2-5A and 2-5B* compare the OSI models for a Novell network and an AppleTalk network. In Figure 2-5A, the Data Link Layer has a path to Ethernet and Token Ring protocols, just as the Data Link Layer in Figure 2-5B has a path to LocalTalk, EtherTalk, and TokenTalk protocols.

Regardless of the network type, the Data Link Layer serves the primary medium for the network. In comparison to the protocols seen with the upper layers of the model, the Ethernet, Token Ring, EtherTalk, and TokenTalk protocols provide the rules for transporting data across the network and—more specifically—the Physical Layer. In practice, the Data Link Layer also affects the connection of different medium. For example, a computer that includes a Token Ring network-interface card will not attach to an Ethernet network. Within Ethernet, TokenRing, and AppleTalk networks, different types of network cabling and styles of network-interface card connections exist. All types and styles, however, have the same purpose as set by the Data Link Layer standards.

Figures 2-5A.
The OSI model for a Novell network. The Data Link Layer has a path to Ethernet and Token Ring protocols.

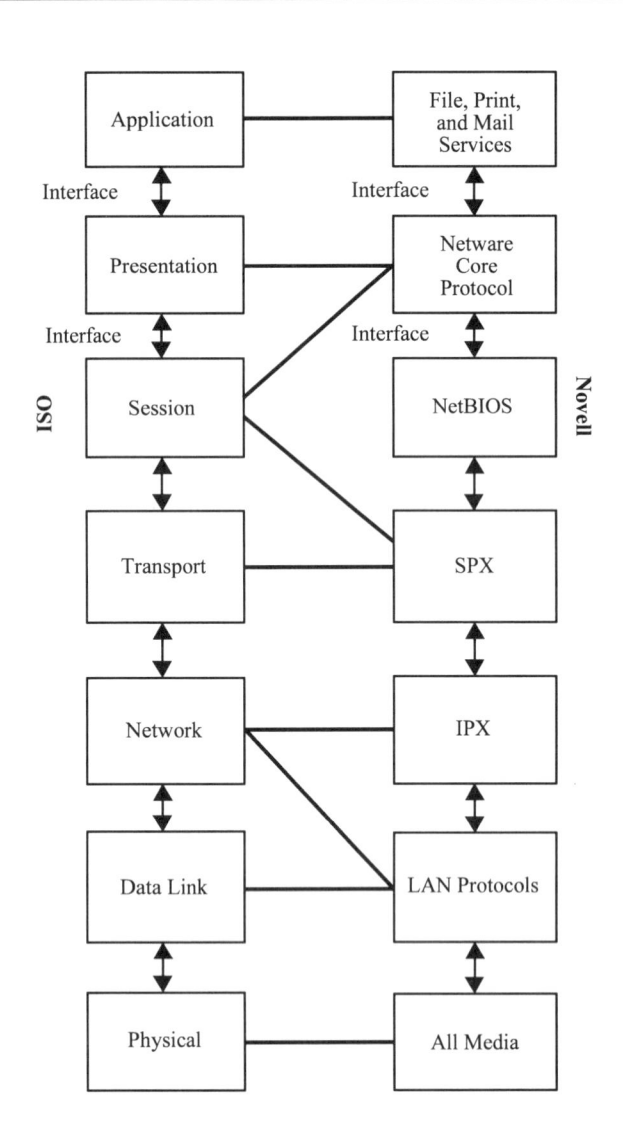

Bridges, Switches, and Routers. As we will find in the later chapters of this book, internetworking involves managing the growth of networks and connecting networks separated by some distance. The growth of a business may mean that a local-area network will also grow with the addition of nodes and servers. With this growth, traffic congestion in terms of data flowing from a variety of sources may become a problem. In addition, the growth of a business may require the connection of networks both separated by distance and using different architectures. In all cases, the need to share and manipulate information continues to exist.

Chapter 2: Layers, Protocols, and Reference Models

```
        ISO                          AppleTalk
   ┌──────────────┐            ┌──────────────────┐
   │ Application  │────────────│ File, Print,     │
   │              │            │ and Mail         │
   │              │            │ Services         │
   └──────┬───────┘            └────────┬─────────┘
      Interface                     Interface
   ┌──────┴───────┐            ┌────────┴─────────┐
   │ Presentation │────────────│ AppleTalk        │
   │              │            │ Filing           │
   │              │            │ Protocol         │
   └──────┬───────┘            └────────┬─────────┘
      Interface                     Interface
   ┌──────┴───────┐            ┌────────┴─────────┐
   │ Session      │────────────│ AppleTalk        │
   │              │            │ Session          │
   │              │            │ Protocol         │
   └──────┬───────┘            └────────┬─────────┘
   ┌──────┴───────┐            ┌────────┴─────────┐
   │ Transport    │            │ AppleTalk        │
   │              │            │ Transaction      │
   │              │            │ Protocol         │
   └──────┬───────┘            └────────┬─────────┘
   ┌──────┴───────┐            ┌────────┴─────────┐
   │ Network      │            │ Datagram         │
   │              │            │ Delivery         │
   │              │            │ Protocol         │
   └──────┬───────┘            └────────┬─────────┘
   ┌──────┴───────┐            ┌────────┴─────────┐
   │ Data Link    │            │ LocalTalk,       │
   │              │            │ EtherTalk,       │
   │              │            │ TokenTalk        │
   └──────┬───────┘            └────────┬─────────┘
   ┌──────┴───────┐            ┌────────┴─────────┐
   │ Physical     │            │ Cables and       │
   │              │            │ Connectors       │
   └──────────────┘            └──────────────────┘
```

Figures 2-5B.
The OSI model for an AppleTalk network. The Data Link Layer has a path to LocalTalk, EtherTalk, and TokenTalk protocols.

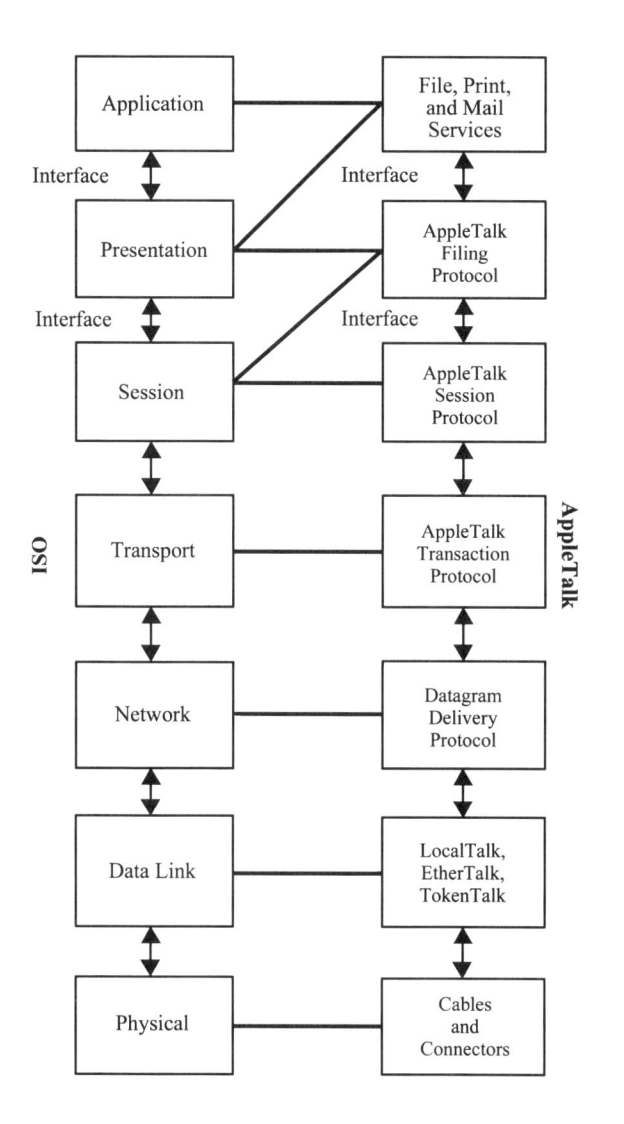

Internetworking relies on equipment such as bridges, switches, and routers. The first two devices function within the Data Link Layer, while routers function at the Network Layer of the OSI model. While bridges determine whether a frame has the proper destination and also provide filtering, a switch consists of a large number of high-speed ports that connect either network segments or individual devices on a port-by-port basis. Switches evaluate the destination address and forward individual frames to the correct port.

Routers interconnect networks that have the same communications architecture. But the networks may have different lower level architectures and depend on the presence of the correct protocols for proper operation. Moreover, router operation also depends on the presence of a network architecture that includes a Network Layer.

The Network Layer

Rather than sending frames across the network, the Network Layer sends packets of information. The Network Layer handles any problems associated with delivering a packet of information form one node of the network to another. Although this process may seem simple, the network may include any number of nodes. The packet must travel through intermediate nodes, because no direct connection exists between the source and destination nodes.

Going back to the previous section, we found that the Data Link Layer only communicates with the peer process found at the end of the communications link. A process found within the Network Layer communicates with processes found at the other ends of all communication links connected to the transmitting node. To accomplish this, the Network Layer places information taken from the Transport Layer into the header portion of the data packet. With this, the header contains protocol information used by the peer Network Layer process.

From there, the Network Layer passes the packet to the Data Link Layer for placement into a frame. If the data travels to an intermediate node, the Network Layer in that node has the responsibility for forwarding the packet to the proper destination node. In some instances, the Network Layer must have the capability to handle packets to and from different types of nodes. In turn, this capability relies on the capability of the Network Layer to work with different communications protocols and types of addressing.

Theory to Function: The Network Layer

When we move to the Network Layer in both models, however, the differences become even more apparent. In Figure 2-5A, the Network Layer has a path to the LAN protocols and to the Internetwork Packet Exchange (IPX) Layer. The IPX establishes the connectivity between networks and works as part of the network operating system. In comparison, the Network Layer shown for the AppleTalk model only has a path to the Datagram Delivery Protocol. With datagrams, the network does not establish a session before sending the data packet from one node to another.

With this, AppleTalk supports communication between two sockets. In the most basic terms, a socket functions as an interface for communication between a system and a network. A socket consists of a port and an address. During network operation, application programs use sockets to communicate across a network to a peer process.

The Transport Layer

As Figure 2-3 shows, the Transport Layer resides as the fourth layer in the OSI model. In brief, the Transport Layer sees the entire network and uses the Physical, Data Link, and Network Layers to establish end-to-end communications for the higher levels of the model. End-to-end communications involve peer processes at either end of the connection communicating through a common protocol. As a result, the Transport Layer has the primary task of moving messages from one end of a network to the other. Processes within the Transport Layer recognize all intermediate nodes as adjacent nodes and rely on the lower levels of the OSI model to pass the data through the intermediate nodes.

The Transport Layer has the primary task of moving messages from one end of a network to the other.

Network Layer to Transport Layer Services

As we learned in Chapter 1, we can classify networks according to the area that the network covers. For example, a wide-area network may cover a campus, a city, or an entire region. Networks may consist of subnetworks that link together to form the larger networks. Since the Network Layer exists at the junction between subnetworks and full networks, it provides a number of services for the Transport Layer.

During the interaction between the Network and Transport Layers, the Network Layer establishes a unified addressing scheme. As a result, each node on the network has a unique address, which becomes part of a total,

During operation, several processes may occur simultaneously within a node. With addressing, the Transport Layer ensures that a connection occurs to each of the specific processes.

consistent addressing scheme for the network. In addition, the integration of Network and Transport functions also facilitates the use of the circuit-switched and packet-switched networks that we will discuss in later chapters. While a circuit-switched network establishes virtual circuits between pairs of nodes and transfers data in both directions across the circuits, a packet-switched network relies on the sending of packets that contain the full network address of the destination node.

Transport Layer Services

Outside of the interaction with the Network Layer, the Transport Layer also provides a set of network services that includes addressing, connection management, data-flow control, and buffering. During operation, several processes may occur simultaneously within a node. With addressing, the Transport Layer ensures that a connection occurs to each of the specific processes. Connection management establishes and releases the connections.

Data flowing on the network must travel at a rate accepted by the network hardware. With flow control, the Transport Layer considers the capacity of the hardware to handle data transmission and adjusts accordingly. In relation to flow control, the Transport Layer also ensures that the correct amount of memory remains available for buffers in the receiving node.

The Transport Layer also matches the resources found in the lower layers to the upper layers of the OSI model. With some applications, the upper layers may require slower services than those provided by the channel. To accommodate this requirement, the Transport Layer uses multiplexing, or the interleaving of different message packets. At the receiving end, the Transport Layer sorts the packets and re-creates the original message through the use of user IDs stored in the message headers.

Other applications may establish a condition in which the upper layers require faster service than that available with a single channel. Paralleling the flow of data across the channel multiplies the effective flow of the data for the higher layers. Generally, an operating system such as UNIX will provide the multiprogramming capability that allows the lower layers to operate in parallel.

The Session Layer

After the Transport Layer has established a connection between two nodes, the Session Layer sets a number of parameters that allow actual communication between the nodes to occur. The parameters apply not only to the nodes, but also to the higher layers of the OSI model. Before the transfer of information takes place, the Session Layer ensures that communication can occur and that the nodes do not attempt to communicate simultaneously. Then the Session Layer manages the communication by breaking it into usable parts. When the communication ends, the Session Layer provides an orderly method for the break to occur.

During operation, the Application Layer may need to access the services given by the Session Layer. When the Session Layer provides services for the Application Layer, the services pass through the Presentation Layer as part of a suite. Regardless of the need for one or all services, the suite passes through as a whole.

Sessions

We define a session as simply the communication that occurs between two nodes after a connection has taken place. Several sessions may occur during one connection, or one session may require the establishment of several connections. The Transport Layer allows the multiplexing of several sessions onto a single connection. Sessions occur through dialogs that begin the communication and provide an orderly ending to the communication.

Simultaneous communication occurs in both directions. The communication remains orderly, however, through the use of tokens. As the communication takes place, the nodes pass the token back and forth. The node with the token can transmit data, while the node waiting for the token receives the data. With this, the passing of the token signals the nodes that a change in the direction of data flow will occur.

The Session Layer also manages the starting and stopping of the data transmission. During the transfer of information from one node to another, the receiving node must receive notification that the transmission has reached the end of the file. Control information embedded in the file header informs the node that the transmission has reached the end of the file. Without the activity management given by the Session Layer, several different files transferred during one session could appear as one large file. In this sense, each file becomes an activity.

The Presentation Layer

Computer systems interpret data according to different character codes. For example, an IBM mainframe communicates with a different set of codes than an Apple Macintosh computer. As the data flows from one type of system to another, the Presentation Layer translates the different character codes so that the systems may communicate.

Character coding involves the placement of command sets within a structured code. In addition, the format of the document, the format of floating point numbers, or the type of program used to compile the language data transmitted from computer to computer may vary. When considering the format of the document, a special language format may display a document in a certain manner. The format of floating point numbers consists of a given number of bits in a specific order within a word or double word. When moving from computer to computer, the number of bits varies. During the compiling of data, the representation of data may have different forms. In addition, the Presentation Layer encrypts, decrypts, and authenticates the data to prevent unauthorized access to information and to confirm the source of the information. The Presentation Layer also compresses data passed by the Application Layer to save space within the channel during transmission. At the receiving end, the Presentation Layer decompresses the data.

The Application Layer

At the highest level of the OSI model, the Application Layer contains network applications. As we have worked through the remainder of the OSI model, we found that each layer offered a variety of services. Network applications that include electronic mail, file-transfer, and file-sharing

As the data flows from one type of system to another, the Presentation Layer translates the different character codes so that the systems may communicate.

software use the services to accomplish specific tasks. Since each application relies on protocols for communication, the Application Layer must have a method for accessing services for the protocols. To do this, the Application Layer passes unmodified messages from the application to the Presentation Layer.

Theory to Function: The Transport, Session, Presentation, and Application Layers

The same comparison becomes effective when viewing the Transport, Session, and Presentation Layers for both models. In Figure 2-5A, the Transport Layer has a path to the Sequential Packet Exchange Layer, while the Transport Layer for the AppleTalk model has a path to the AppleTalk Transaction Protocol Layer. Within the Novell terminology, the SPX Layer provides the Transport Layer functions for a Novell Netware System.

The AppleTalk Transaction Protocol (ATP) conducts end-to-end interaction within an AppleTalk network and provides request-response transactions. ATP interactions involve independent transactions on the network. In addition, the transactions found within the ATP layer establish the basis for services seen in the upper layers of the AppleTalk model.

Again moving back to Figure 2-5A, we see that the Session Layer for the Novell model has a path to both the SPX and the Netware Core Protocol (NCP). The Session Layer shown for the AppleTalk model has paths to the AppleTalk Session Protocol layer and the AppleTalk Filing Protocol layer. Novell's Netware Core Protocol transmits information between clients and servers. Going back to the IPX Layer of the Novell OSI model, the IPX Layer provides Network Layer functions by transporting the NCP messages. The AppleTalk Filing Protocol Layer allows workstation users on the AppleTalk network to share files.

While the Presentation Layer for the Novell model connects only to the Netware Core Protocol, the Presentation Layer for the AppleTalk model has paths to the AppleTalk Filing Protocol and the File, Print, and Mail Services. Both models have the Application Layer connected to the File, Print, and Mail Services. Application programs request and manipulate

Low-level software applications called device drivers link devices used at the Physical Layer to the protocols found in the Network Layer. Microsoft Windows NT uses the Network Device Interface Specification to provide the linkage.

files by using native file system commands found within the operating system of the workstation. Printers connected through a network allow users to share print resources, while print sharing applications ensure that no bottlenecks occur within the process.

In contrast to the OSI models shown for Novell Netware and AppleTalk networks, the relationship between Microsoft Windows NT and the model remains relatively basic. The network operating system provides transparent support for client applications in a distributed environment. Moving to *Figure 2-6*, we see that Windows NT supports network-interface cards and connectors through the Physical Layer and network protocols such as Ethernet and Token Ring through the Data Link Layer.

Low-level software applications called device drivers link devices used at the Physical Layer to the protocols found in the Network Layer. Microsoft Windows NT uses the Network Device Interface Specification to provide the linkage. As seen with the Novell OSI model, the Transport Layer supports SPX. In addition, the Transport Layer also supports the Transmission Control Protocol and the Transport Device Interface (TDI). The TDI allows the software to establish an interface with multiple transport protocols.

At the top of the Microsoft OSI model, the Session Layer allows the loading of redirectors for Windows NT, Novell Netware, and Banyan VINES. Redirectors work in a client/server environment and decide whether a request for some type of computer service should involve an individual workstation or the network server. While the Windows NT application layer supports client/server applications as well as peer-to-peer processing, the Network File System (NFS) protocol allows servers to share disk space and files.

The TCP/IP Reference Model

The Transmission Control Protocol/Internet Protocol (TCP/IP) reference model contains a suite of protocols that allow the interconnection of networks manufactured by different vendors. As an example of the cross-vendor support available through TCP/IP, Macintosh computers rely on MacTCP

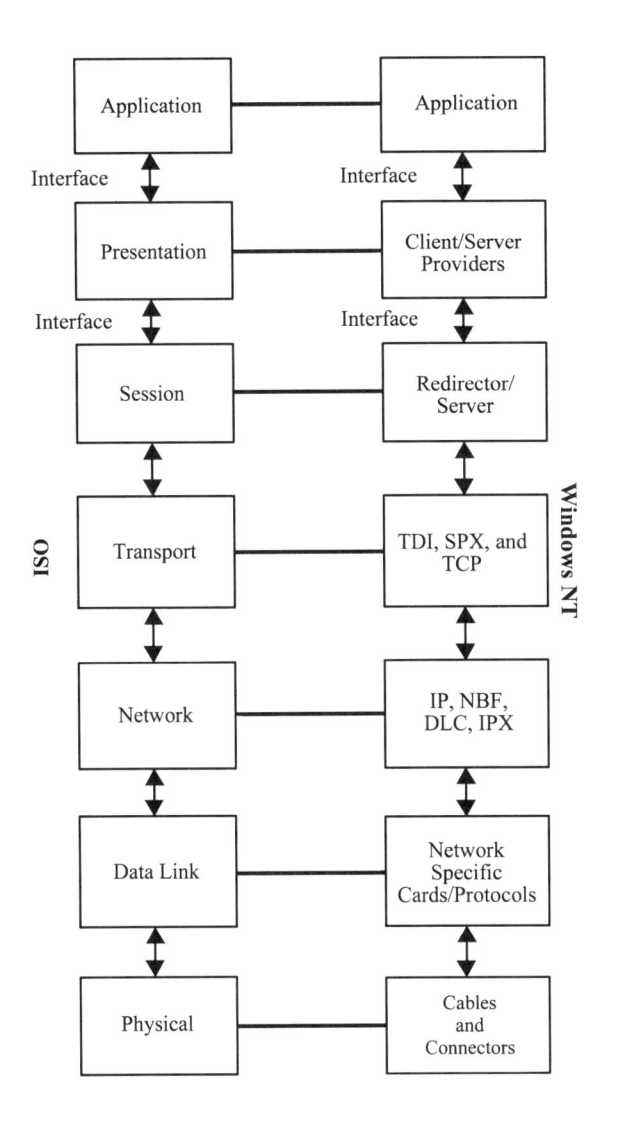

Chapter 2: Layers, Protocols, and Reference Models

Figure 2-6.
The OSI model for Windows NT, which supports network-interface cards and connectors through the Physical Layer and network protocols through the Data Link Layer.

for connection to the Internet and for access to IBM-compatible computers or networks using TCP/IP.

Figure 2-7 shows the relationship between TCP/IP and the OSI model. As the diagram shows, both models share the concept of utilizing a stack of independent protocols. In addition, the layers extending through the Transport Layer of each model establish end-to-end communications that transport service to processes. In both the OSI and TCP/IP reference models, layers

Telecommunication Technologies

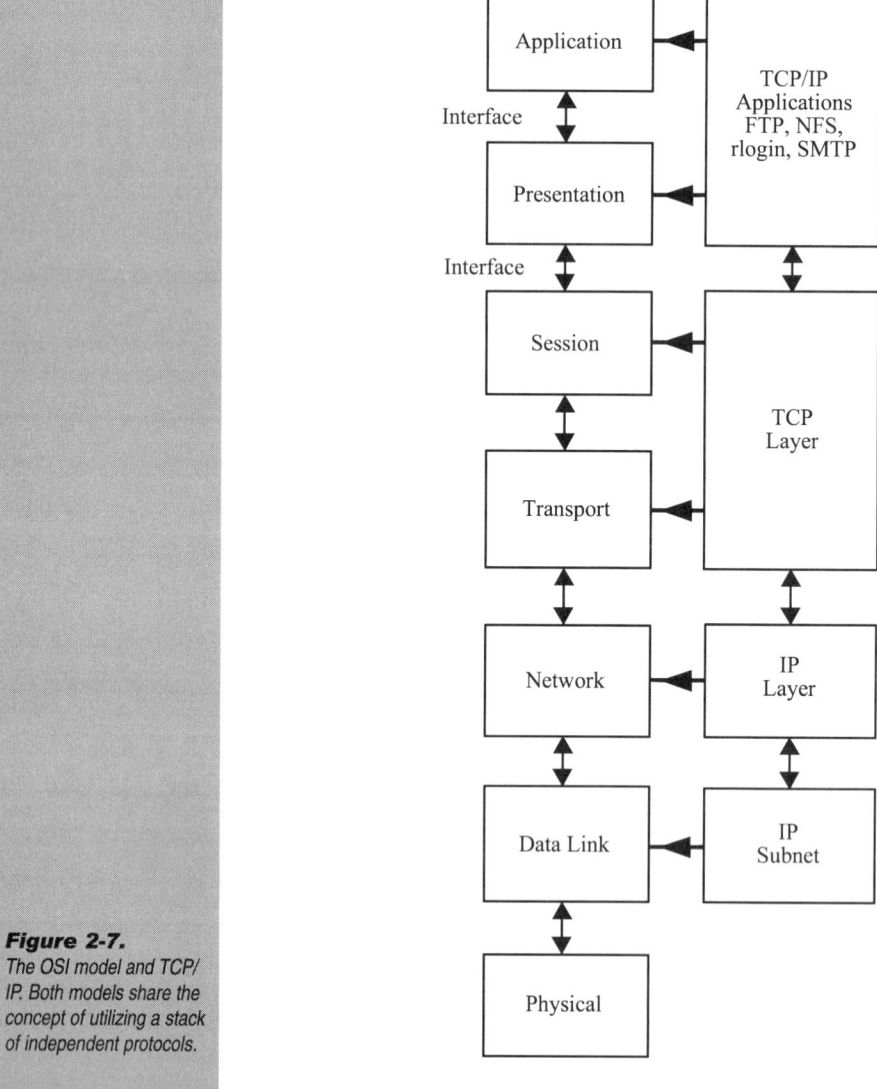

Figure 2-7.
The OSI model and TCP/IP. Both models share the concept of utilizing a stack of independent protocols.

extending above the Transport Layer provide support for applications. Both models function independently of any network specifications.

The Host-to-Network Layer

Rather than using a Physical Layer, as seen with the OSI reference model, the TCP/IP model relies on the Host-to-Network Layer at the lowest level of operation. The Host-to-Network Layer contains the protocol that allows a host such as a web server or mainframe computer located in a data center to connect to the network. In addition, the Host-to-Network Layer protocol allows the host to send packets across the network.

The Internet Layer

When considering the TCP/IP reference model, the Internet Layer provides the basis for the model. While the Host-to-Network Layer allows a host to send packets, the Internet layer permits the packets to travel into any network and remain independent of the network destination. The Internet Layer defines the packet format and protocol seen within the Internet Protocol. As with the OSI reference model Network Layer, the Internet Layer also allows the network to avoid data congestion and ensures that the network delivers the packets to the proper destination.

Theory to Function: The TCP/IP Internet Layer

The Internet Protocol has rapidly become a standard for transporting data, voice, and video across networks. Every device connected through TCP/IP is assigned a four-byte identification code, or IP address, that identifies a network on the Internet and a host segment attached to a specific network. The numeric IP addresses are classified as:

- Class A—an address structure supporting networks with more than 16 million nodes.
- Class B—an address structure supporting networks with up to 65,536 nodes.
- Class C—an address structure supporting networks with up to 256 nodes.

The Transport Layer

The Transport Layer allows peer entities located on the source and destination hosts to communicate. Using the connection-oriented Transmission Control Protocol (TCP), the Transport Layer delivers the data stream originating on one computer to any other machine connected to the Internet without error. The sending TCP process breaks the data stream into discrete messages and passes each message onto the Internet Layer while the receiving TCP process reassembles the received messages into an output stream. Along with TCP, the Transport Layer also uses the connectionless User Datagram Protocol (UDP) for applications that do not rely on the Transmission Control Protocol.

The Application Layer

Again comparing the OSI and TCP/IP reference models, both use an Application Layer that contains all high-level protocols. The TCP/IP model does not include Session or Presentation Layers. Along with the SMTP and FTP functions, the Application Layer also covers the Domain Name Service (DNS) used to map host names to network addresses and the HyperText Transport Protocol (HTTP) used for fetching pages on the World Wide Web.

Theory to Function: The TCP/IP Application Layer

Internet-based electronic mail connections occur through a source system establishing a TCP connection at a destination system. The Simple Mail Transport Protocol (SMTP) accepts the incoming transmission and copies the electronic mail messages into the appropriate destination mailboxes. An SMTP server transfers text-only electronic mail between host servers along with a line of text that provides an identity and a ready-to-receive confirmation.

The Internet relies on the File Transfer Protocol (FTP) to access and transfer files from a host server to another host or to a client. FTP servers located on global networks allow users to log in and download files. Since the FTP interface remains rather difficult to use, many web-based applications have begun to shift to HTTP servers for file transfers.

Summary

As Chapter 2 illustrated, layers, protocols, and reference models function as an organizing framework for the functionality and operation of a network. With respect to actual network operation, two key points have yet to be emphasized: (1) Data communications and telecommunications networks use all or portions of the models. (2) In most—but not all—of those uses, the interaction occurs between individual layers. As a result, the tasks associated with individual layers—as outlined in the previous sections—may not remain as distinct as depicted in those sections.

With the definitions of layers, protocols, and reference models, Chapter 2 established a framework for discussing local-area network, wide-area network, and internetworking technologies. Throughout this book, the description of a technology begins with a comparison of the specific networking model with the general OSI reference model. Each of the following chapters also relies on the opening discussion about protocols and protocol stacking when describing specific network characteristics.

Throughout this book, the description of a technology begins with a comparison of the specific networking model with the general OSI reference model.

Chapter 3
Cabling, Connectors, and Cabling Systems

Introduction

In Chapter 2, we defined the different layers and protocols of the OSI and TCP/IP reference models. With Chapter 3, we will concentrate on the lower layers of the models as we consider different types of wired transmission media, cable characteristics, and connecting points. The chapter brings the discussion full circle as it considers structured cabling systems and methods used to test network cabling.

The many different types of cables that have become available give silent testimony to the evolution of networking. Since networking applications may require more bandwidth or extended distances between nodes, the matching of the correct cable to the application becomes more than essential. As the chapter shows, network cabling may take the form of twisted-pair copper lines, coaxial cable, fiber optics, or a combination of fiber optics and copper called hybrid fiber/coax.

Both the Electronic Industries Association/Telecommunications Industry Association (EIA/TIA) and the International Standards Organization/International Electrotechnical Commission (ISO/IEC) create industry standards for cabling along with voice and data systems. The standards not only cover cabling but also cable-delivery methods, such as the pathways and spaces found at the installation. With a standardized approach to cabling and cable-delivery methods, data networks have gained increased manageability and flexibility. Before standardization, a network system might contain different types of connectors and cables and have little appreciable planning for the cable routing.

The many different types of cables that have become available give silent testimony to the evolution of networking.

Factors that Affect Cabling

All high-speed networking standards require compliance with generic cabling specifications and with additional parameters defined within the specifications and standards for the network-interface products. The additional parameters define the actual electrical and digital signaling characteristics that should occur within a well-designed, consistent networking system. In addition, all high-speed standards conform to signal-to-noise ratios (SNR) and maximum noise thresholds. The standards also consider pair skew and propagation delay characteristics that occur at high frequencies.

Bandwidth

Every amplifier has bandwidth, or a frequency range over which the amplifier has relatively constant gain. Amplifiers have constant gain over a specific range of frequencies called a band. Since this range of frequencies fits within upper and lower limits, the frequencies have a bandwidth. The upper and lower limits of the bandwidth are set by the cutoff frequencies of the amplifier. For example, an amplifier that has constant gain for a range of frequencies extending from 40 MHz to 44 MHz has a bandwidth of 4 MHz. To take this a step further, a signal covering a wide band of frequencies can carry more information than a signal covering a narrow band of frequencies.

Data Rate

When viewing network specifications, some may make the assumption that bandwidth and data rate measure the same quantities. In fact, bandwidth measures the capability of a network to carry information within a specified band of frequencies, while data rate measures the capability of the network to carry data at specific speeds. While we measure bandwidth in megahertz (MHz), we measure the data rate in megabits per second (Mbps). Older networking technologies such as 10BaseT Ethernet and 16-Mbps Token Ring networks offer a 1:1 relationship between bandwidth and data rate. In comparison, newer networking technologies use sophisticated encoding schemes that deliver higher data rates over lower frequencies. For example, Fast Ethernet delivers data at a rate of 100 Mbps over a bandwidth of 80 MHz.

Amplifiers have constant gain over a specific range of frequencies called a band. Since this range of frequencies fits within upper and lower limits, the frequencies have a bandwidth.

Impedance

Impedance measures the ratio of voltage to current at any given point along a cabling pair. For optimum performance, the impedance of the transmitter, receiver, and cable should match throughout the operating frequency range to ensure that the maximum energy transfer occurs between transmitter and receiver. Any impedance changes along the length of the cable or any impedance mismatches between the cable and termination equipment result in a loss to transmitted signals. In addition, the generation of echo signals reflects back at the point of impedance change or mismatch. The reflected or echo signals become noise to the receiver equipment.

Structural Return Loss

Structural return loss (SRL) measures the uniformity of impedance along the length of a cable. Low SRL values indicate that changes in pair impedance have occurred, thus affecting the uniformity of the impedance. Factors that affect cable-impedance uniformity cover any change to the construction of the pair along the cable pair length. Those changes include:

- Conductor diameter
- Insulation thickness
- Dielectric constant of the insulation
- Evenness of the pair twist

Attenuation

Attenuation refers to the loss of power as a signal travels down a cable. In brief, attenuation measures the amount of signal loss between a transmitter and a receiver in terms of decibels. At the receiver end of a cable pair, the transmitted signal at a particular frequency must always have a greater value than the total crosstalk appearing at the receiver end. Attenuation becomes more of a factor during the transmission of high frequencies throughout a system and may occur because of kink or bend in the cable, poor termination, or due to the use of a cable grade that cannot support the data transmission rate.

Attenuation measures the amount of signal loss between a transmitter and a receiver in terms of decibels.

Crosstalk

With twisted-pair cabling, crosstalk exists as the major noise contributor and occurs when one pair induces undesired noise on another pair. Since both attenuation and crosstalk increase as frequency increases, the crosstalk signal will become stronger than the transmitter signal at a given frequency. Consequently, this frequency defines both the upper limit of the usable bandwidth of the system and the highest operating frequency for the system.

Typically, we measure the crosstalk for the worst cabling pair in a cable, because most applications use only one pair for transmitting a signal and another pair for receiving the signal. When working with a four-pair cable, we measure crosstalk for each of the six-pair combinations. The "worst pair crosstalk" equals the worst of these six individual measurements.

Attenuation to Crosstalk Ratio. The attenuation-to-crosstalk ratio (ACR) measures the possible performance of a network on a cable installation by examining the difference—or signal headroom—between crosstalk and attenuation at different frequencies. For example, a cable that has 44 decibels of crosstalk and 20 dB of attenuation at 100 MHz will have an ACR value of 24 dB at 100 MHz. When we compare this measurement to the EIA/TIA specification for an attenuation-to-crosstalk ratio at 100 MHz of 10 dB, we find that the cabling doesn't provide the characteristics needed for good network performance. Improving either or both the crosstalk or attenuation values also improves ACR values. Generally, cable manufacturers have improved crosstalk by twisting the pairs more tightly.

Unfortunately, tightly twisting the pairs increases attenuation as the signal travels down the cable. The tighter twists lengthen the conductors and require the signal to travel farther. In turn, greater loss of power or attenuation occurs. Since attenuation directly affects signal strength, any attempt to improve ACR must also improve attenuation. When considering ACR values, it becomes important to balance attenuation and crosstalk, rather than only looking at the total ratio.

Near-end Crosstalk. Near-end Crosstalk (NEXT) stands as the most important indicator of a proper installation. Measured in decibels, NEXT represents a ratio of signal strength to the noise generated by

---

Near-end Crosstalk (NEXT) stands as the most important indicator of a proper installation. Measured in decibels, NEXT represents a ratio of signal strength to the noise generated by crosstalk.

crosstalk. A strong desired signal and a weak noise signal produce a high NEXT value. As a result, high NEXT measurements indicate that the network system operates properly. Low NEXT measurements indicate the presence of high crosstalk interference, the lack of proper cable termination, and poor network performance.

NEXT occurs because current flowing through the data communications cabling creates an electromagnetic field. The presence of the field interferes with the signal carried on adjacent cables. Twists in the cabling should cancel opposing fields. A loose twist, however, will allow excessive crosstalk to exist.

Power Sum Crosstalk. With networking technologies increasingly utilizing four cabling pairs, power sum crosstalk measurements gain importance. Power sum crosstalk determines the total crosstalk a pair receives from all the other pairs in a cable. Power sum crosstalk for the near end (PS NEXT) consists of the combined noise caused by the three near-end transmitters on the fourth-pair near-end receiver. Power sum crosstalk from the far-end (PS EL-FEXT) consists of the combined noise caused by the far-end transmitters on the fourth-pair near-end receiver.

Cable Balance

Proper cable balance decreases the opportunity for the cable to pick up environmental noise and ensures good crosstalk performance. During operation, a cable pair carries a differential mode signal, while a common mode signal exists between the pair and ground. With a high ratio of differential mode signal to common mode signal, the cabling has a better pair balance. While a cable with good balance will absorb less noise from the surrounding environment, it also radiates less noise.

An unbalanced cabling pair produces a low ratio between the differential mode signal and the common mode signal. Without the proper balance, the cable may function as an antenna and allow outside noise to interfere with the desired signals. Interference that enters the cabling system may generate a signal between ground and each conductor of a pair. As a result, the two common mode signals appear as noise on the pair in the differential mode. In addition, an unbalanced cabling pair generates more

Longitudinal conversion transfer loss (LCTL) measures the amount of cable balance by comparing the signal that appears across the pair to the signal between the ground and the pair.

crosstalk coupling between the pairs. Within a cable, a pair with good balance will have less crosstalk coupling with other pairs.

LCL and LCTL. Longitudinal conversion loss (LCL) evaluates cable balance at high frequencies. Also referred to as near-end unbalance attenuation, LCL measures cable balance by comparing the signal appearing across the pair to the signal applied between ground and the pair, where the applied signal and the across-pair signal are at the same end of the cable. Another measurement tool, called longitudinal conversion transfer loss (LCTL), measures the amount of cable balance by comparing the signal that appears across the pair to the signal between the ground and the pair. Also defined as far-end unbalance attenuation, LCTL places the applied signal at the opposite end of the cable from the measured-pair signal.

Pair Skew

Pair skew applies to cabling technologies that rely on multiple pairs for transmitting data signals. Essentially, the signals divide between pairs and reunite at the receiving end. If the signals don't arrive at the receiving end simultaneously, signal skewing occurs and transmission errors exist. When considering pair skew, propagation delay—or the measure of the time needed for the signal to travel to the receiver—becomes a factor. Propagation delay compares the capability of the cable to carry the signal against the theoretical speed of electricity.

Twisted-Pair Cabling

Almost every building in America equipped for telephone services uses twisted-pair cabling to carry the telephone and other communications signals. Since the signals have become more complex, and because more sources of interference have surfaced, the twisted-pair cabling industry has experienced change. Even with the application of fiber-optic cabling, twisted-pair cabling must now have the capability to carry high data-rate signals and to reject noise interference from industrial and telecommunications sources such as electric motors, power lines, high-power radio, and radar signals.

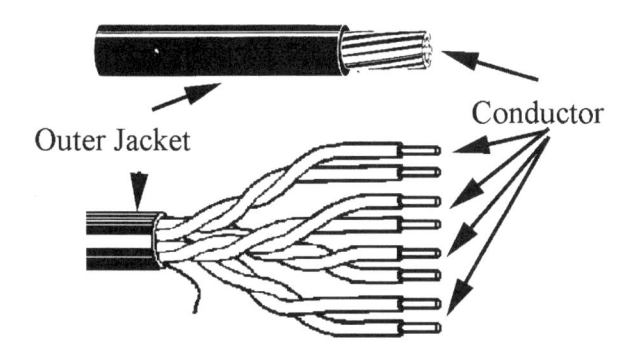

Figures 3-1.
Unshielded twisted-pair (UTP) cable. UTP has become one of the most popular types of transmission media.

Twisted-pair cabling begins with two copper wires encased in color-coded insulation and twisted together to form one twisted pair. Then the manufacturer packages multiple twisted pairs in an outer jacket to make the twisted-pair cable. Varying the length of the twists minimizes the possibility of interference between pairs packaged in the jacket.

Unshielded Twisted-Pair Cabling

Twisted-pair cabling used for telephone and local-area network applications provides limited bandwidth and works with the baseband transmission of signals. Shown in *Figures 3-1*, unshielded twisted-pair (UTP) cabling has become one of the most popular types of transmission media and originated with the cabling used for telephone connections. Since UTP combines low cost, ease of installation, flexibility, and the capability for carrying relatively high data rates, data communications and telecommunications installations continue to use the unshielded twisted-pair standard for millions of network connections.

To obtain optimal performance, UTP cable should work as part of a well-engineered structured cabling system. For example, the installation of UTP cabling requires the use of balance transformers called baluns, or the use of media filters. In either case, the equal induction of noise into the two conductors cancels the noise out at the receiver.

Shielded Twisted-Pair Cabling

Shielded twisted-pair (STP) cable provides additional protection against noise interference and includes screened twisted-pair cable and foil twisted-

Figure 3-2.
Shielded twisted-pair (STP) cable, which encases the signal-carrying wires within two shields.

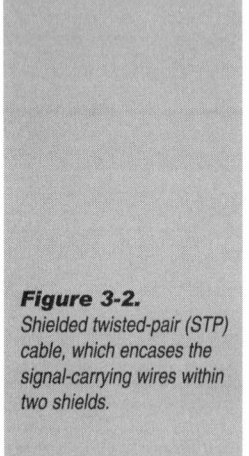

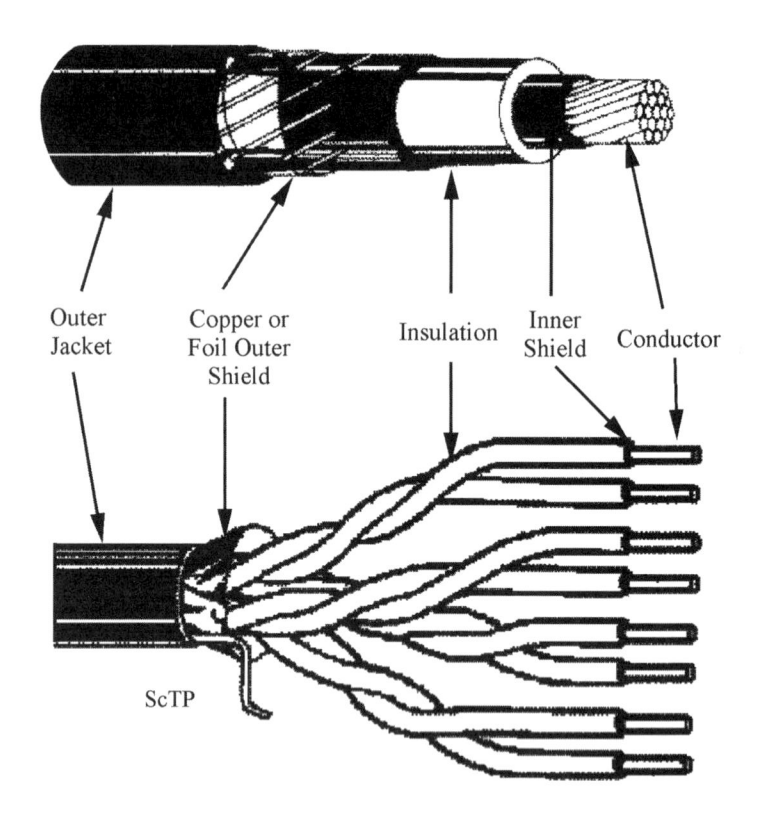

pair cable. While some similarities occur, STP cable has a slightly different design and manufacturing process than UTP cable. As shown in *Figure 3-2*, STP cable encases the signal-carrying wires within two shields.

The shields act as an antenna and convert received noise into current flowing along the shield. In turn, the current induces an equal and opposite current flowing in the twisted pairs. As long as the two currents remain symmetrical, the currents cancel one another and deliver no net noise to the receiver. Any break in the shield or difference between the quantity of current flowing along the shield and the quantity of the current flowing in the twisted pairs establishes noise.

As a result, the effectiveness of STP at preventing radiation or blocking interference depends on the proper shielding and grounding of the entire end-to-end link. The effectiveness of the shielding becomes more important due to the characteristics of STP cable. For example, the attenuation of STP cabling may increase at high frequencies. In addition, crosstalk and signal noise may increase without compensation for the shield.

Shield effectiveness depends on the shield material and thickness, the type and frequency of the electromagnetic interference, the distance from the noise source to the shield, the continuity of the shield, and the grounding structure. For some applications, STP cables use a thick braided shield that increases the weight, thickness, and difficulty of installation for the cable.

Referring to the bottom portion of Figure 3-2, we see that other STP cables—called screened twisted pair (ScTP) and foiled twisted pair (FTP)—rely on a relatively thin overall outer foil shield and have a decreased thickness and cost. The difficulty of installation remains, however, because the cabling has a minimum-bending radius. In addition, the ScTP and FTP cabling has a maximum pulling tension force intended to prevent the tearing of the shield.

Table 3-1 describes the color codes used for shielded twisted-pair cabling. STP cabling consists of four 24-gauge cables that, in turn, consist of 100 twisted pairs. As mentioned, ScTP and FTP cabling wrap a foil shield around the cable package. Fully shielded twisted-pair cabling encloses each of the four cables with an individual foil shield. An overall shield complements the individual shield.

The dependence on proper shielding also carries over to the installation of connectors and other hardware for STP cable. Improperly shielded connectors, connecting hardware, or outlets can cause the degradation of the overall signal quality and noise immunity. A well-installed shielded cabling system requires the full and seamless shielding of every component within the system, along with good grounding practices.

Pair Designation	Pair Color Code
Pair 1	White/Blue = Blue
Pair 2	White/Orange = Orange
Pair 3	White/Green = Green
Pair 4	White/Brown = Brown

Table 3-1.
Shielded twisted-pair cabling designations and color codes.

An improperly grounded system can become a primary source of emissions and interference. The signal frequency of the application dictates whether the system should have grounds at one or both ends. High-frequency signal applications require continuous grounding at both ends of the cable run. Moreover, the length of the ground conductor also affects performance, with excessive length degrading the ground connection. In all cases, the application establishes the grounding characteristics for STP cabling.

UTP and STP Categories

Due to changing application requirements, UTP and STP cabling has become available in different specifications called categories. Although basic telephone cable—or direct-inside wire—continues to provide value for voice connections, improvements in the manufacture of UTP cabling include variations in the twists, individual wire sheaths, or overall cable jackets. The application requirements and the capability to manufacture improved cabling have led to the development of the three EIA/TIA-568 standard-compliant categories shown in *Table 3-2*.

Table 3-2.
Twisted-pair cabling categories and signal bandwidth specifications.

Category	Maximum Signal Bandwidth
Cat 1	Up to 1 MHz (Used only for analog voice transmission)
Cat 2	Up to 4 MHz (Used for low-speed IBM Token Ring networks)
Cat 3	Up to 16 MHz (Widely used for digital voice transmission and for some 10BaseT Ethernet networks)
Cat 4	Up to 20 MHz (Has few existing uses)
Cat 5 Cat 5e (Category 5 Extended)	Up to 100 MHz (Used for a wide range of 10BaseT, 100BaseT, and 1000BaseT networking applications)
Cat 6	Up to 250 MHz (Intended for high-speed network applications)
Cat 7	Up to 600 MHz (Intended for high-speed network applications)

Note: *TIA standards use the term "category" to specify both components and cabling performance. ISO/IEC standards use the term "category" to describe component performance—that is, cable and connecting hardware. The term "class" describes cabling characteristics such as link and channel performance.*

Cat 3 and Cat 4 Twisted-Pair Cabling. As Table 3-2 shows, the six categories of transmission performance specified for cables, connecting hardware, and links are Cat 3 through Cat 7. Cat 3 cabling transmission requirements specify the capability to carry an upper frequency limit up of 16 MHz. For Cat 4 cabling, the upper frequency limit moves up to 20 MHz. Due to the similarities between Cat 3 and Cat 4 cabling and the introduction of other cable standards, Cat 4 cabling has remained virtually unused.

Cat 5 Twisted-Pair Cabling. Cat 5 cabling has a larger diameter and—as a result—higher bandwidth capabilities than smaller gauge cables. As shown in Table 3-2, Cat 5 cabling has an upper frequency limit of 100 MHz. The newer superset of Cat 5, called extended Cat 5 or Cat 5E, also has an upper frequency limit of 100 MHz.

As Cat 5 cabling has become the standard for networking applications, some variations of the standard have exhibited varying degrees of performance. Jacketing material, conductor insulation, and physical shape can cause performance inconsistencies among Cat 5 cable products from different manufacturers. As a result, a wide range of inconsistencies may exist within the Cat 5 designation.

To counter these problems, manufacturers have used the "meets or exceeds Cat 5 standard" designation to show that cabling surpasses the standards set for Cat 5 cabling by a factor of two. Attenuation-to-crosstalk ratio has become the benchmark for evaluating any product called Cat 5 cabling.

Under actual operating conditions, Cat 5 cable becomes severely attenuated at frequencies above 100 MHz. As the signal frequency increases, attenuation and NEXT increase, because the electromagnetic field that interferes with signals on adjacent wires builds as the frequency increases. In comparison, the attenuation-to-crosstalk ratio decreases with frequencies of 115 MHz and above. NEXT levels exceed the signal level and cause the ACR to change from a positive to a negative value.

Cat 6 and Cat 7 Twisted-Pair Cabling. Another new standard, Cat 6, establishes the best performance range delivered by unshielded and screened twisted-pair cabling. Cat 6/class E cabling has an upper frequency range of 250 MHz and requires an eight-position modular jack interface at the work area. In addition, Cat 6/class E components and cabling will offer backward compatibility for applications running on lower categories and classes.

The new Cat 7/class F describes a new performance range for fully shielded twisted-pair cabling and has a bandwidth of 600 MHz. When introduced, Cat 7/class F cabling and components require a new modular plug and socket design and offer the same backward compatibility seen with Cat 6/class E cabling and components.

Modular Jacks and Plugs

Four basic modular jack styles exist in data communications. Rather than refer to the jacks as an RJ-45 or RJ-11 jack, the more correct method refers to the number of cable positions found in the jack. For example, we refer to the jacks commonly used for UTP cabling as eight-position modular outlets. The letters "RJ" designate the jack as a registered jack and actually refer to specific wiring configurations called Universal Service Wiring Codes. A technician could wire each of the basic jacks shown in this section for several different types of RJ designations. *Figure 3-3* shows the four modular plug and jack styles, while *Table 3-3* lists the wiring configurations that can match the plugs and jacks.

Figure 3-3.
The four modular plug and jack styles of data communications.

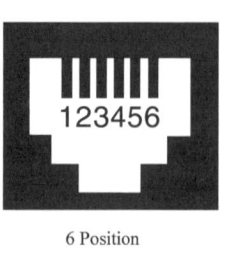

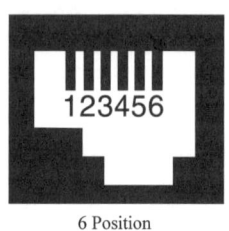

6 Position

8 Position

6 Position Modified

8 Position Keyed

Chapter 3: Cabling, Connectors, and Cabling Systems

Type of Modular Jack	Associated RJ Wiring Configurations
6-position	RJ11C (1-pair), RJ14C (2-pair), RJ25C (3-pair)
Modified 6-position jack or MMJ	DEC data equipment
8-position	RJ61C (4-pair), RJ48C
Keyed 8-position	RJ45S, RJ46S, RJ47S

Table 3-3.
Modular jack designations and wiring configurations.

Regardless of the type of jack or connector used, we need to follow a simple set of rules when connecting cables and terminating the connectors. When working with UTP cabling, maintain the pair twists as close as possible to the termination point. For Cat 3 and Cat 4, don't untwist the cabling beyond one inch from the termination point. For Cat 5 and above, don't untwist the cabling beyond one-half inch from the termination point. When removing the insulation from the cabling, remove only the amount of jacket needed for the termination of the individual pairs.

Always match the pairing of the wires in the modular plug with pairs in the modular jack or other backbone connections. If the pairs don't match, the transmitted data may pair with incompatible signals. Depending on the application, modular cables may have straight-through, reversed, or crossover wiring configurations. *Figure 3-4* sketches different pinout configurations for modular jacks.

One application for modular cables involves using the cables as "patches" between modular patch panels. *Figure 3-5* shows some patch cords. For this application, the modular cables should always have a straight-through configuration. Another application for modular cabling involves the connection of a personal computer, telephone, FAX machine, or other electronic equipment to a modular jack similar to the one shown in *Figure 3-6*.

Depending on the manufacturer's specifications, the modular cables may have either a reversed or a straight-through configuration. Most voice systems use the reverse configuration. The third application for modular cabling involves the use the cables and jacks for Ethernet, Token Ring, ATM, FDDI, and T1 connections and uses the crossover configuration. *Table 3-4* lists the pin connections for the straight-through, reverse, and different crossover configurations, while *Figure 3-7* illustrates each of the crossover configurations.

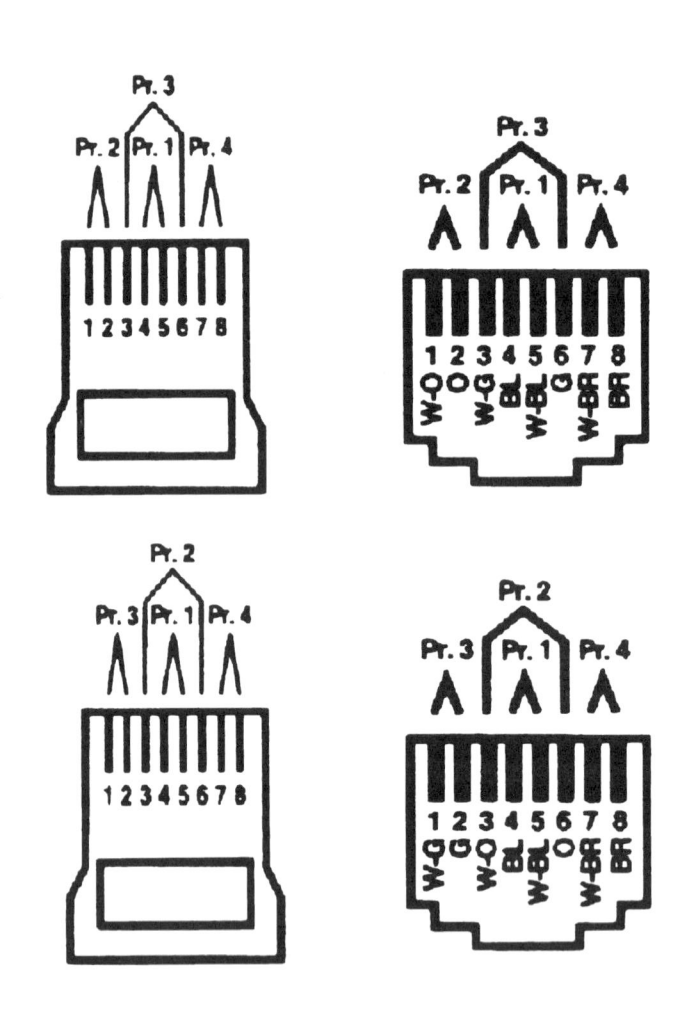

Telecommunication Technologies

Figure 3-4.
Different pinout configurations for modular jacks.

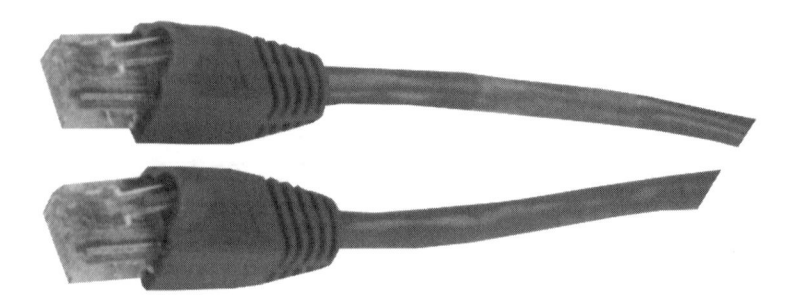

Figure 3-5.
Patch cords.

62

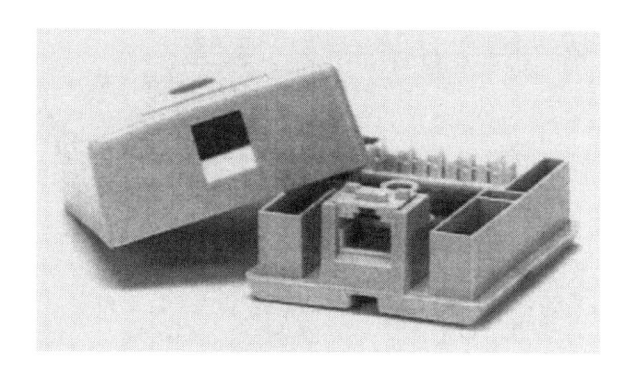

Chapter 3: Cabling, Connectors, and Cabling Systems

Figure 3-6.
Jack and plug cross-connections.

10BaseT, 100BaseT

ATM, FDDI

Token Ring

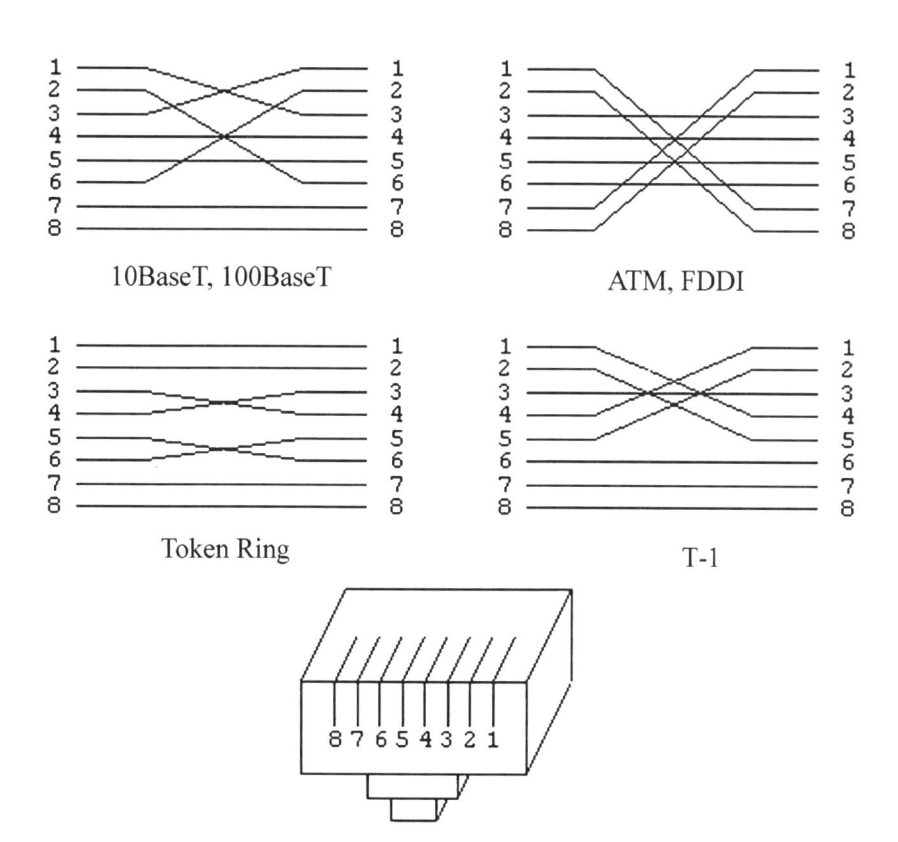

T-1

Figure 3-7. Crossover configurations for twisted-pair cable.

63

Telecommunication Technologies

Configuration	Pin Connections
Straight-through	Pin 1 to Pin 1
	Pin 2 to Pin 2
	Pin 3 to Pin 3
	Pin 4 to Pin 4
	Pin 5 to Pin 5
	Pin 6 to Pin 6
	Pin 7 to Pin 7
	Pin 8 to Pin 8
Reverse	Pin 1 to Pin 8
	Pin 2 to Pin 7
	Pin 3 to Pin 6
	Pin 4 to Pin 5
	Pin 5 to Pin 4
	Pin 6 to Pin 3
	Pin 7 to Pin 2
	Pin 8 to Pin 1
Crossover (10BaseT, 100BaseT)	Pin 1 to Pin 3
	Pin 2 to Pin 6
	Pin 3 to Pin 1
	Pin 4 to Pin 4
	Pin 5 to Pin 5
	Pin 6 to Pin 2
	Pin 7 to Pin 7
	Pin 8 to Pin 8
Crossover (ATM, FDDI)	Pin 1 to Pin 7
	Pin 2 to Pin 8
	Pin 3 to Pin 3
	Pin 4 to Pin 4
	Pin 5 to Pin 5
	Pin 6 to Pin 6
	Pin 7 to Pin 1
	Pin 8 to Pin 2
Crossover (Token Ring)	Pin 1 to Pin 1
	Pin 2 to Pin 2
	Pin 3 to Pin 4
	Pin 4 to Pin 3
	Pin 5 to Pin 6
	Pin 6 to Pin 5
	Pin 7 to Pin 7
	Pin 8 to Pin 8
Crossover (T1)	Pin 1 to Pin 4
	Pin 2 to Pin 5
	Pin 3 to Pin 3
	Pin 4 to Pin 1
	Pin 5 to Pin 2
	Pin 6 to Pin 6
	Pin 7 to Pin 7
	Pin 8 to Pin 8

Table 3-4. Pin connections for twisted-pair cable.

Looking at *Figure 3-8*, we can check for a straight-through or reverse configuration by aligning the plugs side by side and comparing the wire colors from left to right. Plugs wired for the straight-through configuration have the colors for both plugs appearing in the same order. Plugs wired

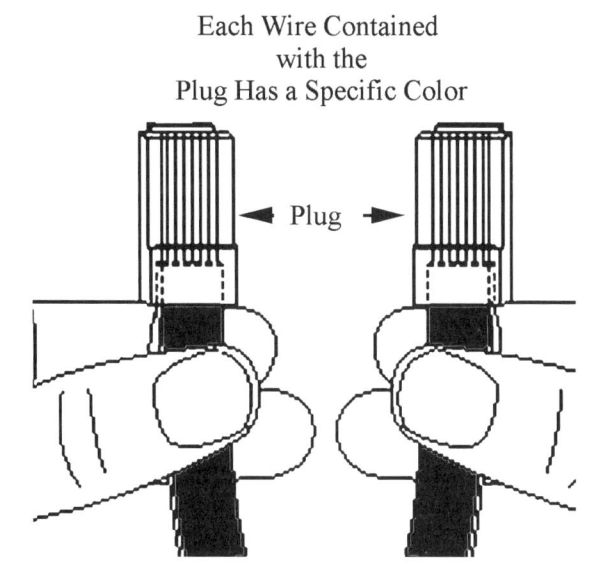

Figure 3-8.
Checking for a straight-through or reverse configuration by aligning the plugs side by side and comparing the wire colors from left to right.

for the reverse configuration will have the colors for the second plug appearing in the opposite order when compared to the first plug. In comparison to the straight-through and reverse configurations, plugs wired for the crossover configuration have the far left colored wire found at one end of the cable appearing as the third colored wire at the other end of the cable.

Terminating UTP Cables

With traditional UTP cables, contractors must remove the outer jacket, untwist and separate the pairs, lay them out in proper sequence onto a connecting device, and terminate. This process can take anywhere from 30 seconds to one minute, depending on the proficiency of the individual installer. *Figure 3-9* shows a punchdown tool and punchdown block used in this process.

The punchdown tool uses a spring-loaded mechanism to push a shielded wire between the jaws of a retaining clip found on the block. Forcing the wire between the jaws of the retaining clip slices away the outer covering of the wire and establishes electrical contact. Since punchdown blocks vary with age, always specify a Cat 5 punchdown block for new installations or upgrades.

Telecommunication Technologies

Figure 3-9.
A punchdown tool and punchdown block used in terminating UTP cables.

Punchdown Tool

Punchdown Block

Coaxial Cable

Pictured in *Figure 3-10*, coaxial cable has a solid center conductor and is surrounded by an insulating spacer. In turn, a tubular outer conductor consisting of either braided or foiled material surrounds the inner conductor and insulator. Another outer layer covers the entire cable assembly and provides both insulation and protection. Coaxial cable works well for data transmission because of its wide bandwidth and ability to carry multiple data, voice, and video streams simultaneously.

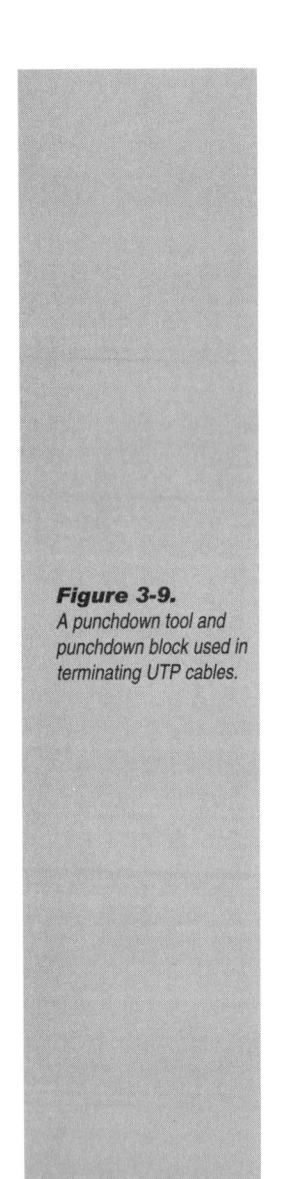

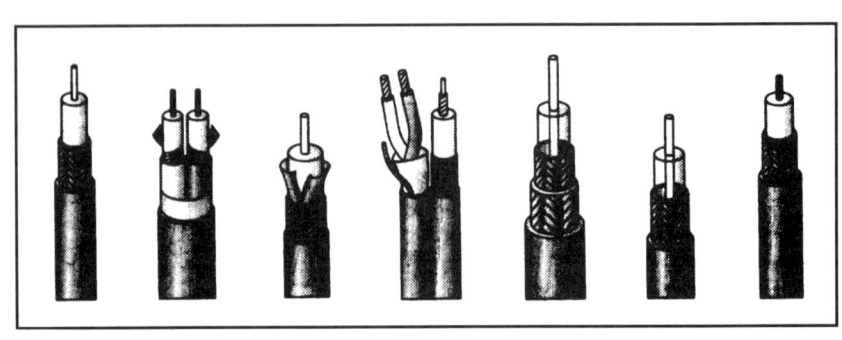

Figure 3-10.
Coaxial cables, which have a solid center conductor and are surrounded by an insulating spacer.

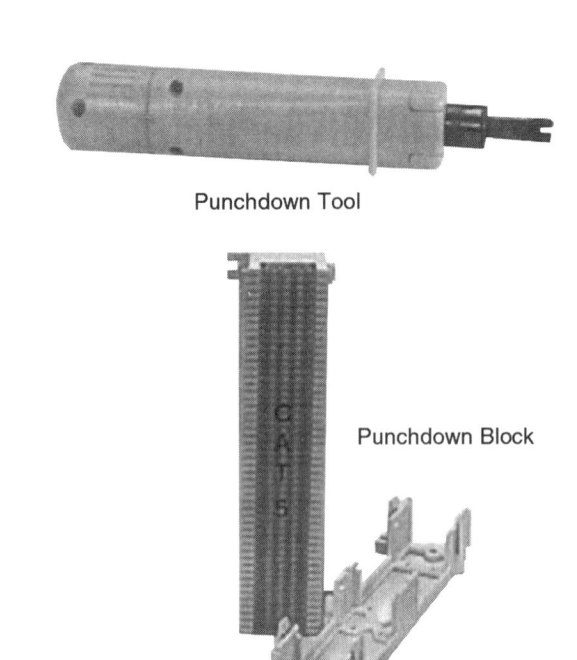

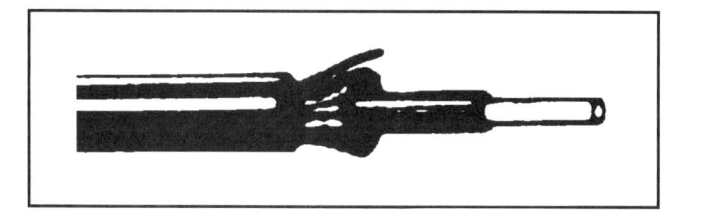

Chapter 3: Cabling, Connectors, and Cabling Systems

Figure 3-11.
Thinnet cable, which supports a data transmission rate of 10 Mbps at a maximum segment length of 185 meters.

Thinnet

Some network applications require a type of coaxial cable called Thinnet, which provides a 50-ohm impedance and a five-millimeter diameter. Shown in *Figure 3-11*, Thinnet supports a data transmission rate of 10 Mbps at a maximum segment length of 185 meters. Due to its smaller diameter, Thinnet offers lower cost, lighter weight, flexibility, and easy installation. But Thinnet has the drawback of transmission characteristics that support a shorter maximum segment length and fewer connected nodes per segment than the thicker coaxial cable types.

Thicknet

Compared to Thinnet, Thicknet offers longer segment lengths and the capability to support more nodes per segment. When used for 10Base5 applications, Thicknet supports maximum segment lengths of 500 feet and 50 nodes per second. As with Thinnet, a Thicknet coaxial cable has a 50-ohm impedance. As the name implies, however, the cable has a larger diameter at 10 millimeters. Thicknet Ethernet cables have bright outer jack colors punctuated with black bands placed at 2.5-meter intervals to mark the installation points for transceivers. *Figure 3-12* shows a Thicknet cable.

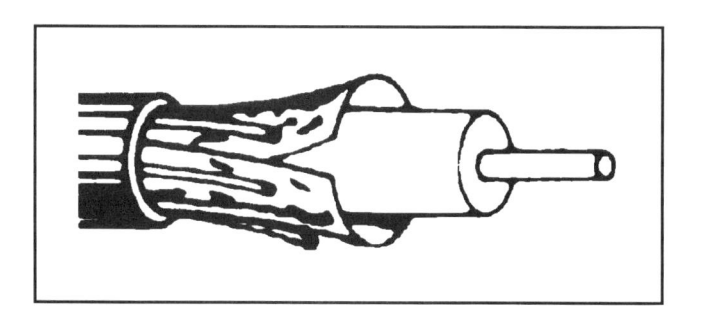

Figure 3-12.
Thicknet cable, which offers longer segment lengths and the capability to support more nodes per segment than does Thinnet cable.

67

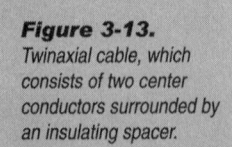

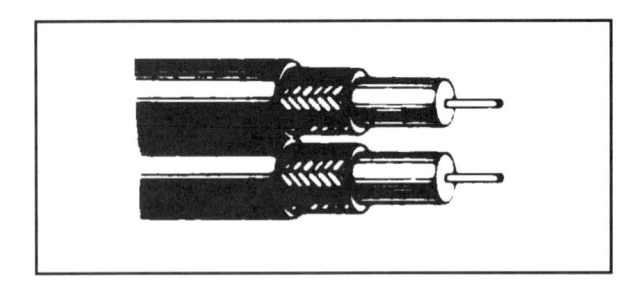

Figure 3-13.
Twinaxial cable, which consists of two center conductors surrounded by an insulating spacer.

Twinax

Figure 3-13 shows another type of coaxial cable called Twinax, or twinaxial. Twinax consists of two center conductors surrounded by an insulating spacer. As with the other coaxial cable types, an outer conductor surrounds the inner conductors and insulation. Then another layer of insulation covers the entire assembly. Given its construction, Twinax offers the durability of coaxial cable and the balanced transmission characteristics of twisted-pair cabling. The 150-ohm Twinax works as a "short haul" cable with the 1000Base-CX media system. Although Twinax has better transmission characteristics than twisted-pair media, it supports segment lengths of only 25 meters for 1000Base-CX, because of the very high 1.25-Gbps data transmission rate seen with that standard.

Coaxial Cable Connectors

BNC Connectors

Looking at *Figure 3-14*, we see that the BNC Tee connector resembles the letter "T", with the horizontal part of the "T" including male connectors that mate with the female BNC coaxial connectors on each end of the attaching cable sections. *Figure 3-15* shows a set of patch cables with the female BNC connectors. The vertical part of the "T" includes a female BNC connector that either plugs directly into the Ethernet network-interface card in the computer station or to an external thin Ethernet transceiver. Disconnecting a node from the network allows the replacement of the "T" connector with the barrel connector shown in *Figure 3-16*. The barrel connector establishes a straight-through connection.

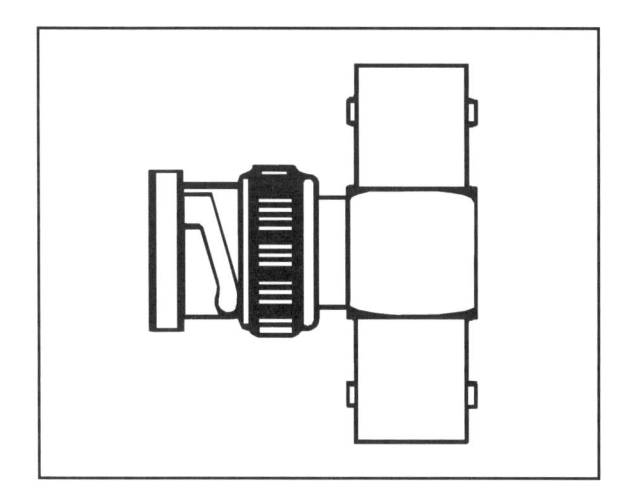

Chapter 3: Cabling, Connectors, and Cabling Systems

Figure 3-14.
A BNC T-connector.

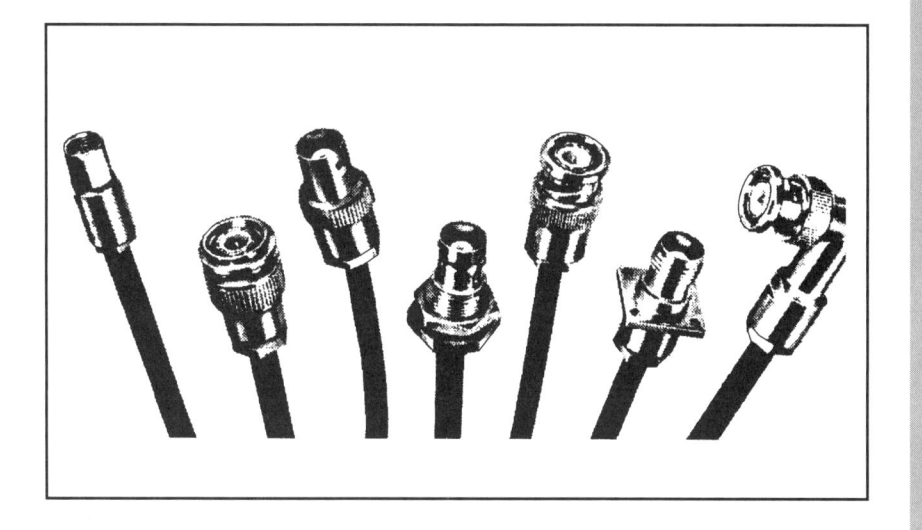

Figure 3-15.
A set of BNC patch cables.

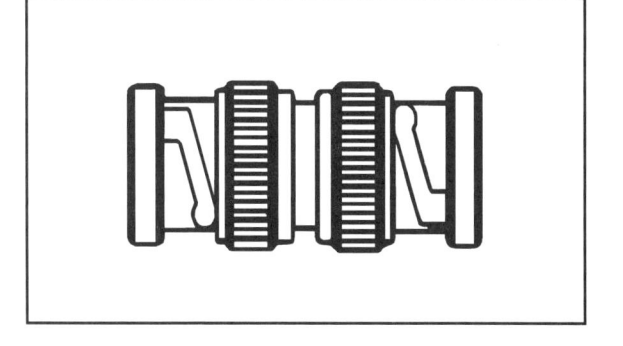

Figure 3-16.
A BNC barrel connector.

69

Telecommunication Technologies

Figure 3-17.
N-connectors, which are used in other coaxial cable networks.

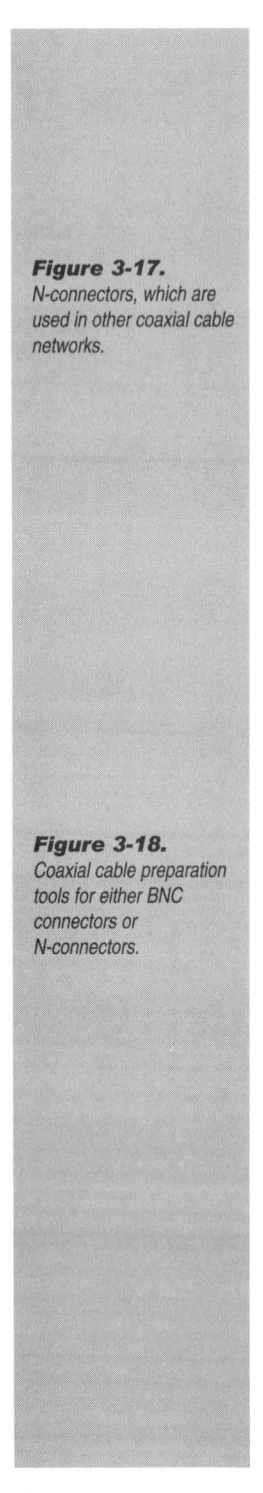

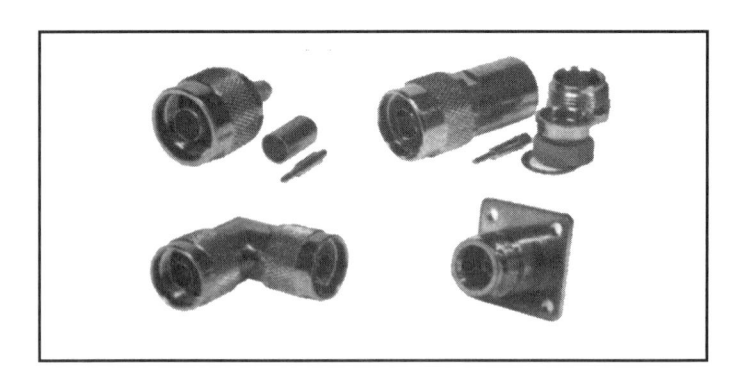

N-Connectors

Other coaxial cable networks use "N-type" connectors similar to the ones shown in *Figure 3-17*. Each end of a segment in a network must have an N-type coaxial connector and N-type 75-ohm terminators installed. Two sections of a segment join together through the use of connectors that mate together through an N-type barrel connector. *Figure 3-18* shows tools used to prepare coaxial cables for either BNC connectors or N-connectors.

Figure 3-18.
Coaxial cable preparation tools for either BNC connectors or N-connectors.

Fiber-Optic Cables

Many building-to-building and WAN installations involve the use of fiber optics. With this, an insulator encloses a bundle of glass threads. Each thread can transmit information at almost the speed of light. Along with the advantage of speed, fiber optics also offers benefits such as:

- Greater bandwidth
- Less susceptibility to interference
- The ability to carry digital signals
- Reduced size and weight

Along with those basic advantages, optical fiber also offers several specific benefits not seen with copper media. In particular, the use of fiber-optic cabling enhances network security, reliability, and performance, because fiber does not emit electrical signals. In comparison, copper cabling emits electrical signals and remains susceptible to tapping and the unauthorized access to carried data. In addition, fiber-optic cabling offers immunity to radio-frequency interference (RFI) and electromagnetic interference (EMI).

Until recently, the use of fiber optics hadn't become widespread, due to the installation cost and the fragile characteristics of the medium. Despite those factors, fiber optics have become more popular for LANs and telephone networks. Nearly every telephone company in the nation has replaced or plans to replace existing copper lines with fiber optics. *Table 3-5* displays the differences between the media types.

Even though costs have decreased, the per-mile cost for laying fiber-optic cables continues to surpass the cost of laying copper lines. The cost savings seen with fiber-optic cables occur at the per-circuit level when used for long-

Transmission Media	Data Rate	Cable Length
Unshielded twisted-pair	1-100 Mbps (dependent on category)	0.1 km
Shielded twisted-pair	16 Mbps	0.3 km
Coaxial cable	70 Mbps	Greater than 1 km
Fiber optics	100 Mbps	Greater than 1 km

Table 3-5.
Transmission types, data rates, and maximum cable lengths.

Telecommunication Technologies

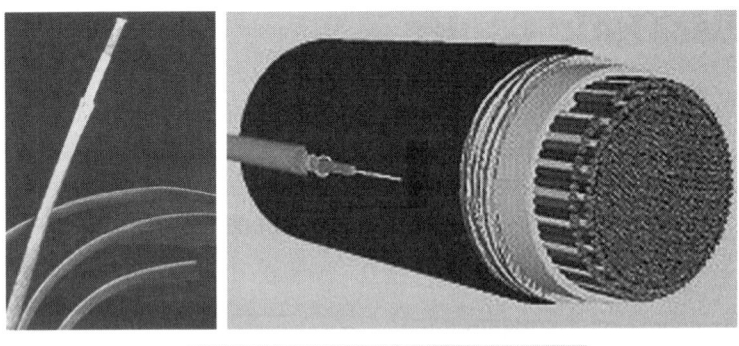

Figure 3-19.
Several types of fiber-optic cables.

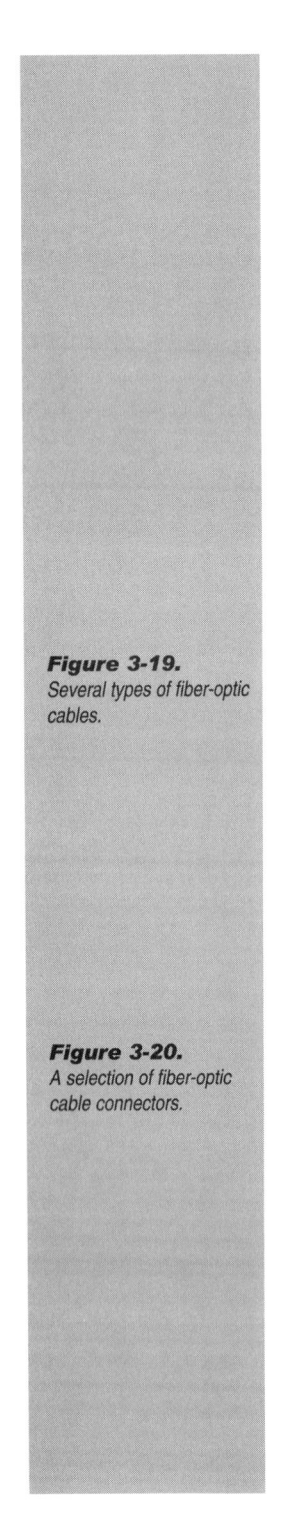

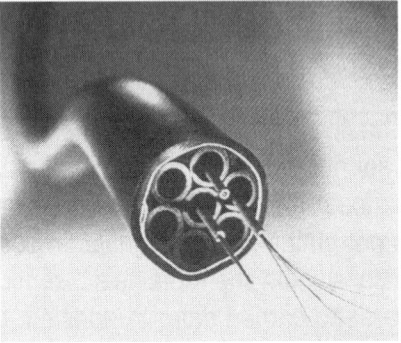

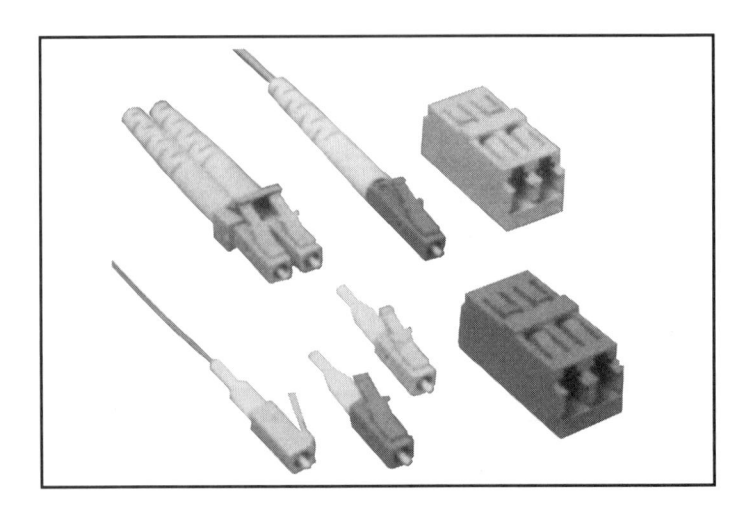

Figure 3-20.
A selection of fiber-optic cable connectors.

distance applications. At shorter distances, such as a local loop, the use of fiber-optic cables becomes impractical because of the low number of circuits. *Figure 3-19* shows several types of fiber-optic cables, while *Figure 3-20* shows a selection of fiber-optic cable connectors.

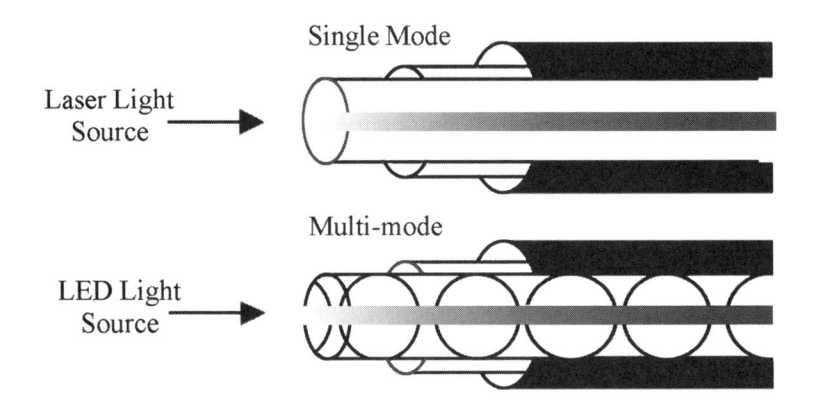

Figure 3-21.
Fiber-optic cable modes. Both the single mode and the multimode have a total strand diameter of about 125 microns.

Single-Mode and Multimode Fiber-Optic Cables.

Manufacturers produce fiber-optic cables according to mode specifications. We can define a mode as a ray of light that enters the fiber at a specific angle. In terms of communications, two types of fiber-optic cabling—multimode and single-mode—receive the most attention. Moving to *Figure 3-21*, we see that both types have a total strand diameter of about 125 microns. Multimode fiber-optic cable, however, has a larger core transmission medium than single-mode fiber-optic cable.

Applications for the multimode type allow multiple modes of light—produced by light-emitting diodes (LEDs)—to propagate throughout the fiber. With each mode of light entering the fiber at a different angle, each arrives at the destination at different times and becomes part of a characteristic called modal dispersion. Due to modal dispersion, the bandwidth capability and transmission distances with multimode fiber-optic cabling remain limited. As a result, applications for multimode fiber provide connectivity within a building or a small geographic area.

As shown in Figure 3-21, a laser beam provides the light source for single-mode fiber. The thinner core transmission medium found in the single-mode fiber-optic cable allows only one mode to travel down the strand. As a result, modal dispersion does not occur. In addition, regenerating the signal at points along the span becomes easier and the cable can carry data at a faster rate than seen with the multimode type. Single-mode fiber-optic cables have become the standard for long-distance networks.

Structured Cabling Systems

Many networks rely on point-to-point connections from servers to hubs and then to clients. These connections form a network topology or general layout for the network. For example, the point-to-point cable connections for a network may result in a configuration that resembles a star. The star topology places a hub in a centralized location and has point-to-point links connecting from the center. With the network established as a star, maintenance becomes simpler and faster because of the ease in isolating cable problems to a single link.

Within network designs, a cable plant covers the selection of the transmission media, the physical placement and installation of the media, and the termination of the media at the node and telecommunications closet. Structured cabling systems locate hubs in central wiring closets that remain dedicated to telecommunications functions. If used in an office environment, cables link from the hub to wall outlets in each office. In the office, a "patch" cable connects the computer station to the wall outlet.

A structured cabling system (SCS) consists of a set of cabling and connectivity products that allow the integration of the voice, data, and video systems with other management systems contained in a building. The management systems may include safety alarms, security access subsystems, and energy management systems. Conduit, cable trays, raceways, and other installation tools support and protect the cabling investment. All structured cabling systems cover the:

- Building entrance
- Equipment room
- Wiring closet
- Horizontal and vertical cabling
- Backbone cabling
- Work areas

In addition, the SCS emphasizes an open architecture and conforms to national and international standards through the use of standardized media and standardized layout, as well as the use of connection interfaces. The process of constructing a structured cabling system begins with early

Structured cabling systems locate hubs in central wiring closets that remain dedicated to telecommunications functions.

planning and a decision to select the cabling as the first part of the system. The EIA/TIA and the ISO/IEC have created industry standards for cabling voice and data systems that address the installation of cabling along with the establishment of cable pathways and spaces.

The Building Entrance

Every building equipped with data communications and telecommunications cabling has an entrance way where outside cabling links with the backbone cabling contained within the building. The equipment room typically houses the building entrance in the form of a punchdown panel or router connection. Cables found at the building entrance include fiber-optic cables that provide connections between buildings and copper cabling that connects to other wide-area networks.

The Equipment Room

Equipment rooms offer security for servers, uninterruptible power supplies, and other telecommunications-related equipment not located in the wiring closet. Only the network manager and key personnel from the Information Systems department should have access to the equipment room. IA/TIA standards establish guidelines for the placement and mounting of equipment, racks, overhead cable trays, and power supplies. In addition, the guidelines specify types of cabling used in the equipment room.

The Wiring Closet

Given the standards set by those bodies, technicians can design a wiring closet that matches cabling and cable-delivery methods and serves as a terminating point for vertical and horizontal cables. With this, the integration of cabling and delivery methods becomes much more effective. Following this practice creates a safe and secure area for all cabling and reduces the need for separate installations that may consume additional space.

Since a telecommunications closet functions as the common access point for cabling pathways, the construction of a telecommunications closet and the placement of all systems within the closet simplifies maintenance. The

Telecommunication Technologies

use of standardized cabling architecture allows the design of a single delivery method that supports various horizontal cables in the workspace. More specifically, the method of delivery involves the use of one cable-tray system for telecommunications cabling and the use of conduit to protect critical voice, data, and video services.

Every wiring closet features an equipment rack similar to the one shown in *Figure 3-22*. The rack provides easy access to any hubs, switches, or routers installed in the closet. Along with equipment, the rack may also frame a patch panel that allows for interconnections between the equipment and the remainder of the building. *Figure 3-23* shows a patch panel. Referring to *Figures 3-24* and *3-25*, we see that the cable tray accommodates the installation of cabling from the equipment rack through the wire-management system enclosed by the wiring closet. As *Figure 3-26* shows, the wiring closet and wire-management system become part of the entire building plan.

Table 3-6 lists the design criteria for a wiring closet. As the table shows, the criteria emphasize the location and construction of the closet. In addition, the criteria also focus on serviceability issues. *Table 3-7* lists EIA/TIA cabling and cabling-delivery requirements for a telecommunications closet.

Figure 3-22.
An equipment rack, featured in every wiring closet.

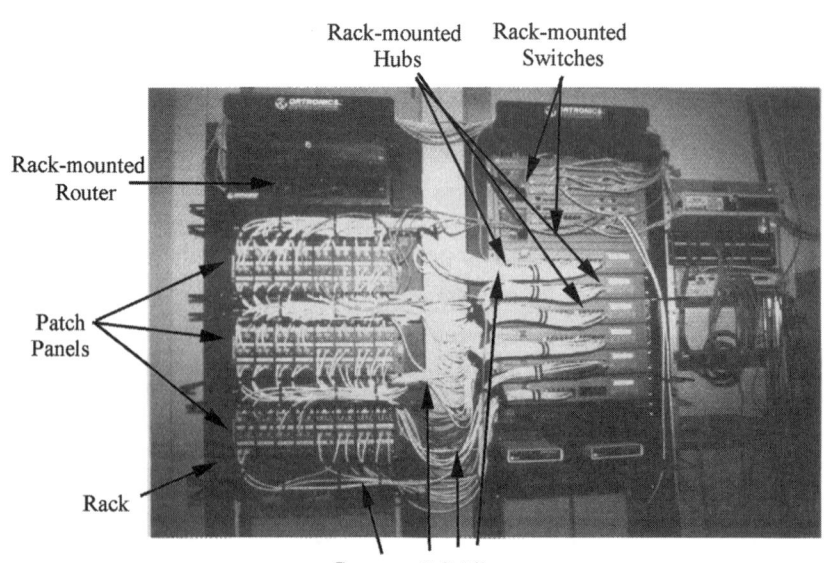

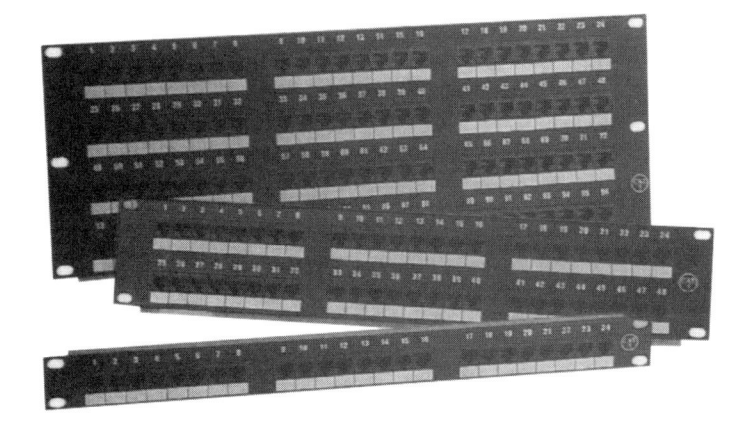

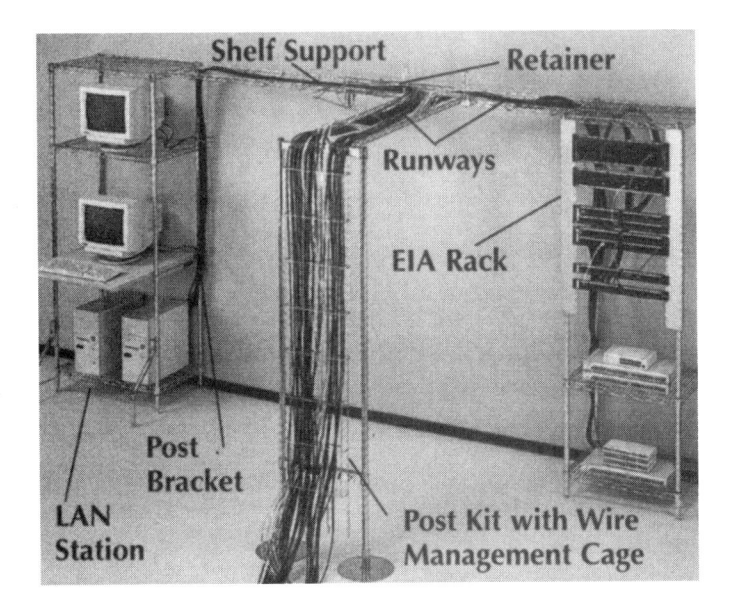

Chapter 3: Cabling, Connectors, and Cabling Systems

Figure 3-23.
A patch panel, which allows for interconnections between the equipment and the rest of the building.

Figures 3-24.
Cable trays, which accommodate the installation of cabling from the equipment rack through the wire-management system.

Figures 3-25.
A wire-management system.

77

Telecommunication Technologies

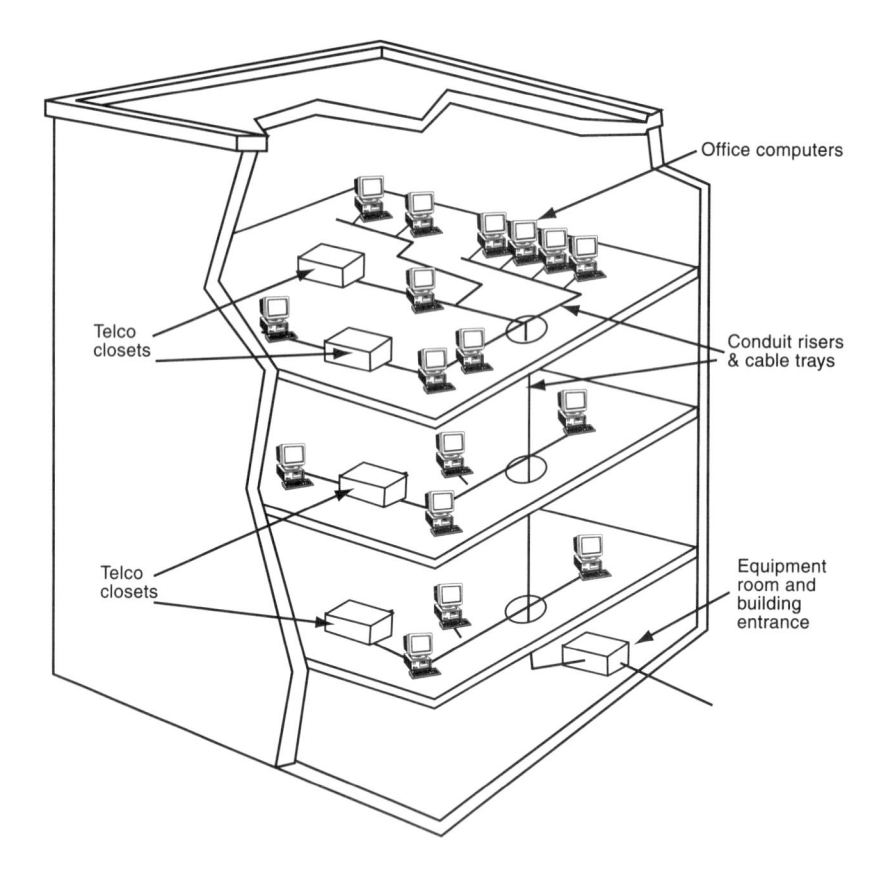

Figure 3-26.
A building plan, of which the wiring closet and wire-management system are part.

Table 3-6.
Wiring closet design criteria.

- Minimum one closet per floor to house telecommunications equipment/cable terminations and associated cross-connect cable and wire.
- Located near the center of the area being served.
- Horizontal pathways shall terminate in the telecommunications closet on the same floor as the area served.
- Accommodate seismic zone requirements.
- Two walls should have 20 mm (0.75 in.) AC plywood 2.4m (8 ft.) high.
- Lighting shall be a minimum of 500 lx (50 foot candles) and mounted 2.6m (8.5 ft.) above floor.
- Minimum of two dedicated 120V 20A nominal, nonswitched, AC duplex electrical outlet receptacles, each on separate branch circuits.
- Additional convenience duplex outlets placed at 1.8m (6 ft.) intervals around perimeter, 150 mm (6 in.) above floor.
- Access to the telecommunications grounding system as specified by ANSI/TIA/EIA-607.
- Temperature maintained at the same level as adjacent the office area.

Table 3-7.
TIA/EIA specifications for telecommunications closets.

- The cabling system shall be configured in a physical STAR layout.
- The telecommunications closet shall be on the same floor as the work area.
- Any adapters required (baluns, Ys, Ts, pair changers, hybrid couplers, active or passive adapters, terminating resisters) shall be external to the work-area telecommunications outlet/connector.
- No more than one transition point (i.e., UTP cable to under-carpet cable) shall be used per horizontal run.
- No bridge taps or splices shall be used. All work-area cables shall be home run to the telecommunications closet.
- Each work area shall have a minimum of two telecommunications outlets.
- One outlet shall terminate a Category 3 or higher four-pair, 100-ohm, 24-AWG, UTP cable.
- The second outlet shall terminate one of the following:
  1) A Category 3 or higher four-pair, 100-ohm, 24-AWG, UTP cable or a two-pair, 150-ohm STP cable.
  2) STP-A cable (which shall be grounded per the TIA/EIA-607 requirements).
  3) A two-strand 62.5/125 µm, graded-index, multimode, optical fiber cable.
- The maximum horizontal run shall not exceed 90 meters (295 feet) regardless of media chosen.
- The total amount of jumpers and/or patch cords on a horizontal run shall not exceed six meters (20 feet).
- The number of patch cords per horizontal run shall not exceed two.
- Work-area and telecommunications closet terminations shall be configured as T-568A or T-568B.
- All eight conductors of the cable shall be terminated on both ends.
- Pair untwisting at the termination points shall not exceed 0.5" for Category 5 installations and shall not exceed 1.0" for Category 4 installations.
- The maximum pull tension shall not exceed 25 lbs. for a four-pair 24-AWG cable.
- All four-pair 24-AWG cable bend radii shall be no less than four times the outer diameter of the cable.
- All multipair cable bend radii shall be no less than 10 times the outer diameter of the cable.
- Installers shall avoid excessive twisting and kinking of the cable.
- Installers shall avoid skinning, scraping, cutting, and other physical damage to the cable.
- Installers shall strip off only as much jacket as necessary to terminate wires.
- TSB-67 as published by TIA/EIA shall be used as a guideline in field testing.

Vertical and Horizontal Cabling

Going back to Figure 3-26, we see that the cable-distribution system for any building includes both vertical and horizontal cabling, with the vertical cabling connecting between floors and the horizontal cabling connecting equipment on the floor. The 100-length limit for horizontal cables includes 90 meters between the punchdown block in the wiring closet and the wall plate, with the remainder designated for jumper cables and patch cables. Patch cable length limits include six meters within the wiring closet and three meters within the work area consisting of the wall plate to the node.

Equipment such as suspended cable trays and conduit not only supports the cabling, but also protects the cabling from abrasion and chemicals. Cable trays suspend from the ceiling or locate in an area above the ceiling and allow the easy addition or removal of cables from the system. Conduits usually exist between or underneath floors. Since conduits offer little or no access to the enclosed cabling, usage of conduits remains limited to applications such as an entranceway or a riser between floors. When installing cabling, always clearly identify each cable with labels placed at the jacks, closet, and panel. Labels should associate the connector with the nearest termination point and panel designation in the wiring closet. *Table 3-8* defines the coding used for a wire carrying the label 1613P:05:017.

Backbone Cabling

Cable installed from the wiring closet to another wiring closet, building, or series of wiring closets forms the backbone for the network. Due to the distances between wiring closets and potential bandwidth requirements, installers rely on fiber-optic cable for a backbone. Fiber-optic cabling also provides immunity against noise that may occur within the connections between closets.

Table 3-8.
Cable labeling procedures.

Portion of ID	Numbering Sequence	Purpose
Closet ID	1613P	Designates the closet identification
Panel ID	05	Designates the particular panel or block within the closet
Jack	017	Designates the unique jack on the particular panel

Testing the Cable Installation

Most—if not all—cable testers offer ease-of-use. Through the use of the tone generator often found in cable testers, a technician can plug the tester into the cable at the PC end, go to the wiring closet, and wave the tone probe over the cables. The ringing sound identifies the terminating end of the cable.

When conducting tests, one member of a two-person team can work from the telecommunications closet, unplug a cable from the hub or patch panel, and attach the cable to the tester. The other member of the team connects the remote unit of the tester to the terminating end of the cable located at the user's work area. From there, the technicians can run an autotest, or a series of predetermined tests, and grade the system. Test results store within the testing unit. When moving to the next cable, the person at the terminating end moves to the next wall jack to repeat the process.

As we learned in the opening sections of this chapter, the components and cabling that make up a telecommunications network must have uniform impedance values. Faults within the cable can change impedance. Consequently, a load with one impedance value will reflect or echo part of a signal being carried by a cable with a different impedance level and cause failures. Because of this, both vendors and installers test to ensure that the impedance, resistance, and capacitance values found within the networking cabling comply with standard cable specifications.

Cable testers check for impedance mismatches during tests for cable faults. For example, a break in a wire creates an open circuit and infinitely high impedance at the break. When a high-frequency signal emitted from a cable tester encounters this high impedance, the signal reflects back to the tester. In contrast, a short circuit within a cable displays as a zero impedance value. With this, the impedance mismatch also reflects a high-frequency signal. But the signal reflected by the short circuit will have an inverted polarity.

Most cable-testing devices can provide an approximate measurement of the distance to the cable fault. To do this, the tester relies on a cable value called the nominal velocity of propagation (NVP). The NVP measures the

Most cable-testing devices can provide an approximate measurement of the distance to the cable fault. To do this, the tester relies on a cable value called the nominal velocity of propagation (NVP).

rate at which a current can flow through the cable and expresses the measure as a percentage of light speed. Then the cable tester multiplies the speed of light by the NVP and by the total time taken for the pulse to reach the fault and reflect back to the tester. To show the one-way distance, the tester divides the measurement by two.

When checking the electrical length of a cable installation, a tester applies the same concept by using time domain reflectometry to measure length. Yet this test requires that one end of the cable not have any termination. During the test, the open end will register as infinite impedance and reflect a pulse back to the tester. Then the tester factors the response time into the formula used to estimate the overall electrical length of the wire. Cable testers don't have the capability to check the first 20 feet of a cable, because a pulse transmitted by the tester reflects back to the device before the signal transmission concludes. As a result, the tester cannot provide an accurate measurement.

Noise stands as one of the biggest problems for a cabling system. A cable tester transmits different frequencies to check the ability of the system to dampen the effects of noise. Due to attenuation, the transmitted signals vary from highest strength at the transmission point to lowest strength at the destination point. As a result, the magnetic field of a signal transmitted from a device through one wire may overwhelm a signal arriving at the same device on the wire pair.

Purchasing a Cable Tester

As network technologies have progressed to the high-speed transmission of data, network managers have faced the challenge of knowing whether projected or installed cabling systems will handle the increased data rates. Because of this, test-equipment vendors have added a wider range of tests and tools for taking and interpreting measurements. New cable testers provide LCD display screens, cable grading and troubleshooting capabilities, and also work as fault locators, Ethernet monitors, cable toners, and voice sets.

Given the new functionality, selecting the right cable tester for a given application presents a challenge. Much of the challenge arises from the different test and evaluation philosophies held by test-equipment vendors. While some vendors point to the need to certify that a cabling system can handle specific high-speed applications, others emphasize testing to ensure that cabling exceeds current standards. Within this second set, vendors may define specific levels at a specific frequency or recommend the taking of measurements over a range of frequencies.

In addition to capabilities, network managers and technicians must also consider the physical size, portability, and cost of cable test equipment. For example, handheld cable testers used for testing Cat 5 cabling range in weight from one pound to more than three pounds. *Figure 3-27* shows two popular handheld cable testers. With and without various options, purchase costs for full-function cable testers range from $2,500 to $5,500. Smaller handheld units that offer continuity testing rather than total test functionality may sell for as little as $110.

Cable Tester Specifications and Test Sets

With the major points listed in *Table 3-9*, Telecommunications Systems Bulletin 67 authored by the EIA/TIA establishes standards for cable testing. Every cable tester must have the capability to run a suite of four tests: NEXT, wiremap, length, and attenuation. Minimum performance requirements depend on the cable type undergoing testing. In addition, TSB-67 also defines two degrees of accuracy defined as Levels I and II.

Figure 3-27.
Cable testers. With and without various options, purchase costs for full-function cable testers range from $2,500 to $5,500.

Function	Description
NEXT	Measured on all six-pair combinations from both ends of the link.
Wiremap	Verifies that the pin-and-connector pairs on either end of the link match, as specified by the configuration.
Length	Verifies that the distance of the run does not exceed TSB-67 limits.
Attenuation	Verifies that the maximum attenuation value, as defined in the cable specification, is not exceeded.

Table 3-9.
TIA/EIA TSB-67 required tester functions.

TSB-67 defines two test configurations called Basic Link and Channel. Basic Link testing covers the permanent portion of the cabling from the wall outlet to the first point of termination in the telecommunications closet. As a result, the test occurs before the installation of any network hardware. Channel testing covers the entire cable run, including the cable, patch cords, and all connections.

Despite the different philosophies, a cable tester has basic functions and may work with various cable types. When purchasing a cable tester, verify whether the cable will accurately test all or only a portion of the cable types discussed in this chapter. In addition, check for the availability of a fiber-optic probe attachment and automatic bidirectional NEXT measurement. The tester should have the capability to process and store tests quickly. While some products may run an entire suite of tests within eight seconds, other testers require up to 45 seconds for the same task.

Since networks feature multiple links, the ability to store and manipulate test results has become critical. Cable testers may store only the last test performed or may store as many as 2,000 tests. In addition, some cable testers allow the downloading of results to a personal computer so that installers can provide customers with a printed summary of each test link.

Regardless of the speed, the true measure of tester usefulness is accuracy. Vendors may express accuracy in terms of maximum and typical accuracy levels. If a tester conforms to TSB-67 requirements, the device must provide an indication of a marginal pass or fail. If a cable passes any of the tests by a margin that is lower than the accuracy of the tester, a "marginal" rating occurs. Although most cable testers offer TSB-67 Level II accuracy for Basic Link tests, few offer Level II accuracy for Basic and Channel tests.

Manufacturers ship cables that can carry data at frequencies far higher than the 100 MHz defined in the Commercial Building Telecommunications Cabling Standard 568A specification. Cables carrying high-frequency signals become more susceptible to noise caused by signal reflections. Given the functions shown in *Table 3-10*, many cable testers check the actual bandwidth of the cable and give an indication of how the cable will handle high-speed technologies.

Summary

Chapter 3 described the factors that affect various cabling systems used in local-area and wide-area networks. In addition, the chapter defined and compared the characteristics of twisted-pair cabling, coaxial cabling, and fiber-optic cabling. The descriptions of each cabling type also included an overview of the jacks, plugs, and connectors that join the cables to network devices.

Chapter 3 also outlined the requirements for structured cabling systems and showed major components such as racks, conduit, and patch panels. With this approach, the chapter demonstrated that the telecommunications closet and wire-management system must integrate into the entire building plan. The chapter concluded with a discussion of the cable testers and test methods used to ensure the proper operation of connected networks.

Since networks feature multiple links, the ability to store and manipulate test results has become critical.

Function	Explanation
Power sum NEXT	PS NEXT measures the signal coupled from all adjacent pairs. New technologies such as gigabit Ethernet use all four pairs
Propagation delay	The time required for electrical signals to travel from one end of the cabling link to the other
Delay skew	The difference in time required for a signal to travel the length of the link over each of the wire pairs
Attenuation-to-crosstalk ratio	ACR measures the strength of the signal in relation to the noise on the same pair. A high ACR yields a lower error rate and a higher quality connection
PS-ACR	The ratio of attenuation to PS NEXT
Impulse noise	Noise induced by nearby equipment such as light fixtures
Capacitance	Measures the signal distortion caused by the interaction between electrons on two nearby wires
Impedance	Measures opposition to the flow of electrical current. Impedance is measured for each pair in the cable
Return loss	The power of the signal reflections measured at the cable relative to the power of the transmission
Loop resistance	Resistance measured in a loop through each pair in the cable
Test to frequencies as high as 155 MHz	Accurately predicts if a cable system can support 155-Mbit/s ATM traffic
Cable grading	Quantifies link performance against minimum Cat 5 requirements
Measure across multiple frequencies	Identifies cable that minimally performs beyond required specifications
Measure Ethernet traffic	Locates unused ports and measures utilization, collisions, and length of frames
Tone generation	Traces a particular cable from node to telecommunications closet
Topology autotest	Conducts tests according to cabling or network type
Auto-troubleshooting mode	Gives a detailed analysis when any test in the autotest series fails
Failure location	Uses crosstalk analysis to pinpoint the location of a failure
Upgradeability	Upgrades through either software or EPROM

Table 3-10.
Desirable cable tester functions.

Chapter 4

Working with LANs

Introduction

A local-area network (LAN) is a high-speed data network that spans a relatively small area such as a single building or group of buildings. Most local-area networks connect workstations, personal computers, and peripherals. Each node, or individual computer, in a local-area network works independently, but also has the capability to access data and devices anywhere on the network.

LANs offer computer users many advantages, such as shared access to devices and applications, file exchange among connected users, and communication among users via electronic mail and other applications.

LAN Protocols

LAN protocols function at the lowest two layers of the OSI reference model and use CSMA/CD or token passing to access the physical network medium. In the carrier sense multiple access collision-detect scheme, network devices contend for use of the physical network. Ethernet networks use the CSMA/CD media access protocol.

In the token-passing media-access scheme, network devices access the physical medium based on possession of a token. Token Ring and FDDI networks depend on the passing of a token.

LAN Transmission Methods

Local-area networks may use unicast, multicast, or broadcast data transmission methods. In each type of transmission, the transmitting node

Local-area networks offer computer users many advantages, such as shared access to devices and applications, file exchange among connected users, and communication among users via electronic mail and other applications.

sends a single packet to one or more nodes. In a unicast transmission, the transmitting node sends a single packet to a destination on a network. Prior to transmission, the source node addresses the packet by using the address of the destination node. From there, the transmitting node sends the packet onto the network, where the packet passes to its destination.

With multicast transmission, the network copies a single data packet and then sends the packet to a specific subset of network nodes. As with unicast transmission, multicast transmission addresses the packet prior to sending. But the source node addresses the packet by using a multicast address. Once addressing has occurred, the transmitting node sends the packet onto the network. Then the network makes copies of the packet and sends a copy to each node that's part of the multicast address.

A broadcast transmission works much like the multicast transmission. A key difference exists, however, in that the network copies the single data packet to all nodes on the network. The source node addresses the packet by using a broadcast network.

Network Topologies

The topology of a network describes the design of the network. Network topologies consider interoperability (or the compatibility of resources on the network) and integration (or how hardware, software, and other resources work together). With all this, the architecture establishes a standardized approach to accessing information and increases the value of the technologies.

Bus Topology

The bus topology uses a common cable—and no additional hardware—to connect all computers and peripheral devices on the network. Referring to *Figure 4-1*, we see that the bus consists of a single pair of wires or a wire and shield that make up a single circuit. At either end of the bus, a terminator provides the resistance needed to establish the load for attached devices.

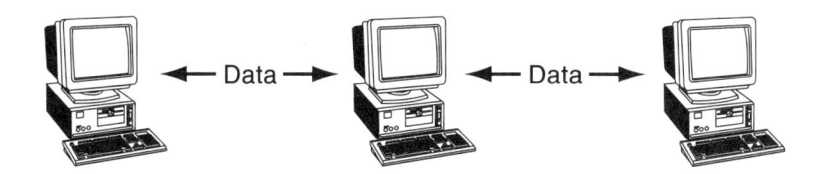

Figure 4-1.
The bus topology, which uses a common cable to connect all computers and peripherals to the network.

Ring Topology

As shown in *Figure 4-2*, the ring topology takes its name from the shape of the separate point-to-point links that make up the network. Each node attached to the ring network has one input and one output connection. As a result, each node connects to two links. Repeater circuitry in each node immediately passes signals received on the input connection to the output connection. As a result, data travels in only one direction on the ring.

With each node in an active state, all nodes have the ability to place new messages on the ring. As the messages circle the ring, each has an address for a specific node that allows the node to copy the message. If a node fails,

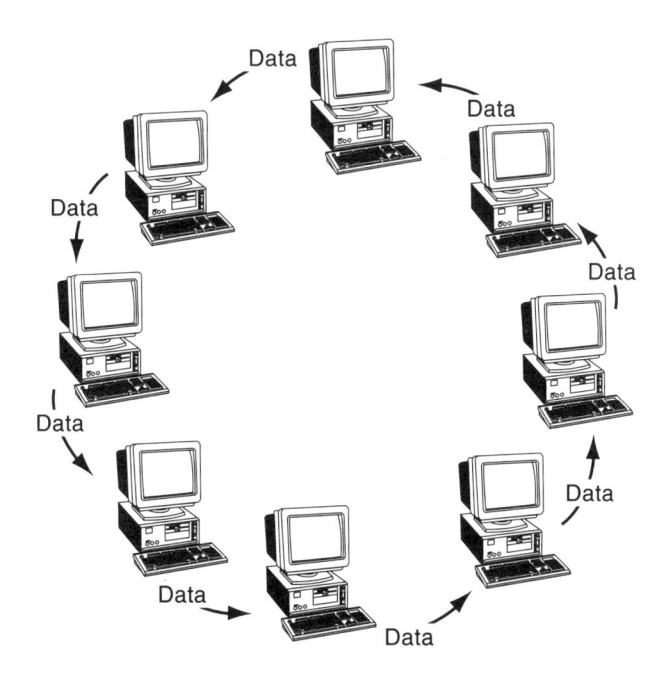

Figure 4-2.
The ring topology, which takes its name from the shape of the separate point-to-point links that make up the network.

89

then the failed node doesn't repeat signals from the input. The failure of any node breaks the ring and causes the transfer of data to stop.

Star Topology

In a star topology, all stations connect to a central wiring concentrator called a hub. As seen with the bus topology, the star topology relies on the sending of packets from one station to another with the packets repeated to all ports on the hub. The sending and repeating of the data packets allows all stations to see each packet sent on the network. Yet only the station with the appropriate packet address receives the data. In a star topology, each station is connected to hub by an individual length of twisted-pair cable. The cable is connected to the station's network-interface card at one end and to a port on the hub at the other end. *Figure 4-3* represents the star topology.

Network Cloud

In Chapters 1 and 3, several figures used a cloud to depict the connection of one network to a larger network or set of networks. The network cloud

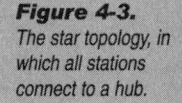

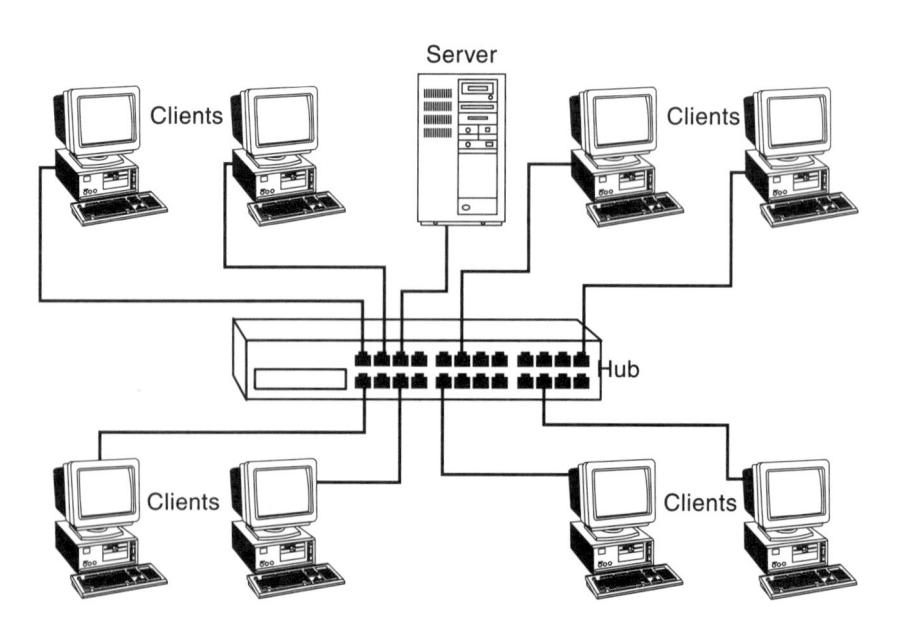

Figure 4-3.
The star topology, in which all stations connect to a hub.

represents a network in which individual users don't have access to the individual components of the network.

Peer-to-Peer Processes

In the layered architecture, a process may only communicate with a peer process. As a result, each layer of the OSI model establishes a different protocol for common services and a different set of protocols for the processes. Cooperation at the process level occurs within the different networking methods called peer-to-peer, client/server, and master/slave.

Peer-to-Peer Networks

A peer-to-peer process provides a communication format in which neither process controls the other. In addition, the same protocol operates with data flowing in either direction. Peer layers communicate through a common protocol that matches the services provided by the layers. The bus network shown in Figure 4-1 functions as a peer-to-peer network.

Client/Server Processes

A client/server process establishes a format in which one process serves as a client, while the other operates as a server. Each client process uses the shared protocol to request that the server perform some task. Since the client/server process allows the sharing of networked resources, the client has the capability to take advantage of a resource—such as processing power, storage, or software—that resides on the server.

The roles of client and server never reverse within the processes. While the processes share a common protocol, the protocol defines different methods for communications sent from the client and communications sent from the server. But peer-to-peer processes may occur within a client-server environment when two clients communicate with one another.

Client/Server Networks

With the client/server network shown in *Figure 4-4*, a group of personal computers attaches to a server for the purpose of sharing resources. During

Figure 4-4.
A client/server network, in which a group of personal computers attaches to a server for the purpose of sharing resources.

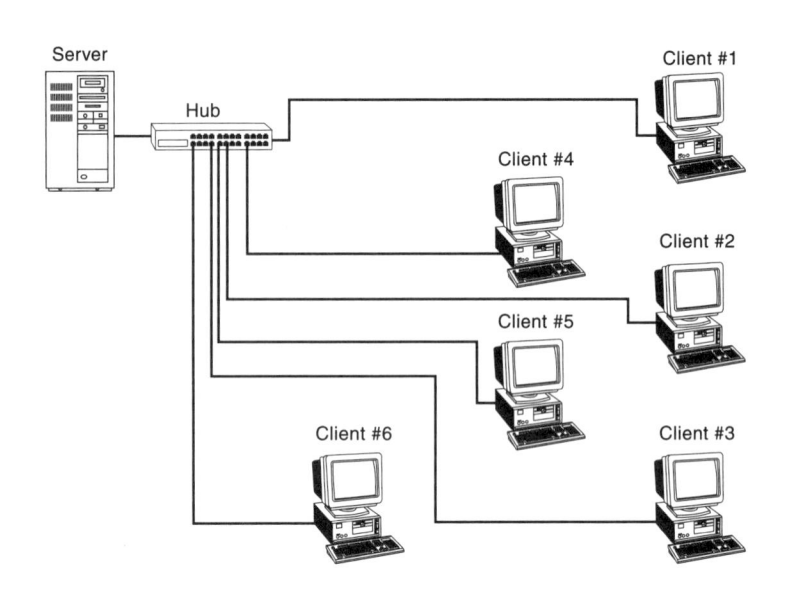

operation, any of the personal computers—or clients—can request that the server perform a processing task. With this, the task divides into two separate parts. For example, a client process may create a report, while the server process prints the report.

The clients and server work through a shared protocol. Each transaction between the client and server takes the form of request/reply pairs in which the client initiates a request and the server responds with an acknowledgment of the request. If the server cannot perform the request, it responds with an error message.

Operation of a Client/Server Network

With client/server networks, a client computer has lesser capabilities, while the server offers greater storage capacity, more random-access memory, greater processing power, and may use a different operating system such as Windows NT server or UNIX. Each client in the network includes a graphical user interface—such as Windows or the Macintosh operating system—that provides easy access to word processing, spreadsheet, e-mail, and presentation graphics applications while attached to the network. Each client on the network has simultaneous access to the applications found on the server. Client/server networks may include different types of personal computers and often attach IBM-compatible and Macintosh computers to the same server. The software found at the server and client levels, the transmission media, and the network protocols establish cross-platform connectivity.

Master/Slave Processes

Master/slave processes indicate the use of controls rather than services. With the master/slave process, one node has greater computing intelligence than the other nodes. While the protocol remains the same for either direction of communication, the master node retains complete control. *Figure 4-5* shows a master/slave network.

Interconnections for Local-Area Networks

Devices commonly used in LANs include network-interface cards, repeaters, hubs, extenders, bridges, switches, and routers. While network-interface cards, repeaters, extenders, and hubs operate at the Physical Layer, devices such as bridges, switches, and routers function at the Data

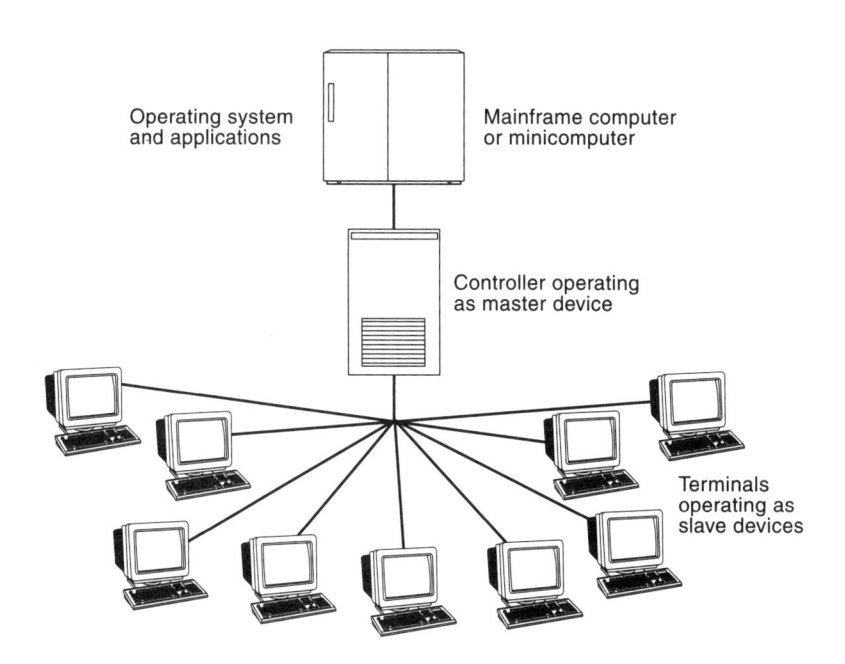

Figure 4-5.
A master/slave network, in which one node has greater computing intelligence than the other nodes.

Link Layer. In brief review, the Data Link Layer controls data flow, handles transmission errors, provides physical addressing, and manages access to the physical medium.

The operation of bridges, switches, and routers also takes us to the separation of the Data Link Layer as shown by the Institute of Electrical and Electronic Engineers. On one hand, the Media Access Control (MAC) sublayer allows media access to occur through either contention or the passing of tokens. The Logical Link Control (LLC) sublayer covers framing, flow control, error control, and addressing for the MAC sublayer.

Network-Interface Cards

The physical connection to the network occurs through the placement of a network-interface card (NIC) inside the computer and connecting the card to the transmission cable. *Figure 4-6* shows a network-interface card. Since the expansion slot connects the NIC to the processor and memory of the computer, it allows the transfer of data to and from the computer to the network.

Figure 4-6.
A network-interface card (NIC), which allows the transfer of data to and from the computer to the network.

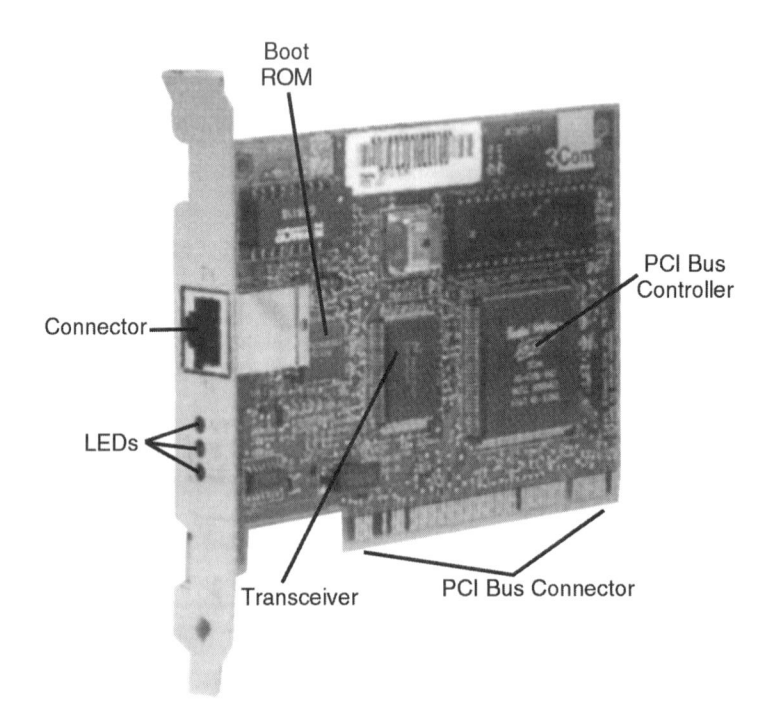

Network Type	Network Speed	Computer Bus Type
Ethernet	10 Mbps	8-bit, 16-bit, 32-bit, 64-bit ISA, EISA, PCI, MCA, USB
Fast Ethernet	100 Mbps	16-bit, 32-bit, 64-bit ISA, EISA, PCI, MCA, USB
Token Ring	4 Mbps, 16 Mbps	8-bit, 16-bit, 32-bit ISA, EISA, PCI, MCA
FDDI	16 Mbps	16-bit, 32-bit ISA, EISA, PCI, MCA

Table 4-1.
Network-Interface card configuration issues.

Data-transfer methods for a network-interface card include programmed input/output transfers in which the NIC stores data at the CPU for transfer to RAM. In addition, NICs transfer data by sharing the computer memory and interrupting the processor. Two other methods for data transfer involve bus mastering (or the transfer of data from the NIC to the system memory) and direct memory access (in which the NIC interrupts the system microprocessor from routine tasks). The microprocessor initiates the transfer to the system memory. As shown in *Table 4-1*, network-interface cards arrive configured for the type of network, the speed of the network, and the type of computer bus.

Once the physical connection is in place, the network software manages communications between stations on the network and the server. Network-interface cards support network operating systems such as Windows NT, Novell Netware, LAN Manager, VINEs, and AppleTalk. During the installation of the network operating system, the configuration of the software involves the selection and setup for the installed card.

While many computer systems rely on a separate network-interface card, some manufacturers have begun to integrate the Ethernet controller functions onto the PC motherboard through the use of a single integrated circuit. *Figure 4-7* shows the schematic for a single-chip, 32-bit Ethernet controller that addresses high-performance system application requirements. The device operates as a flexible bus-mastering device that functions in any networking application. Bus-master architecture provides high data throughput in the system and low CPU and bus utilization. The integrated Ethernet controller offers low operating and standby current for power-sensitive applications such as network-ready PCs, printers, FAX modems, and bridge/routers.

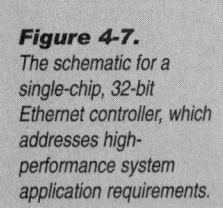

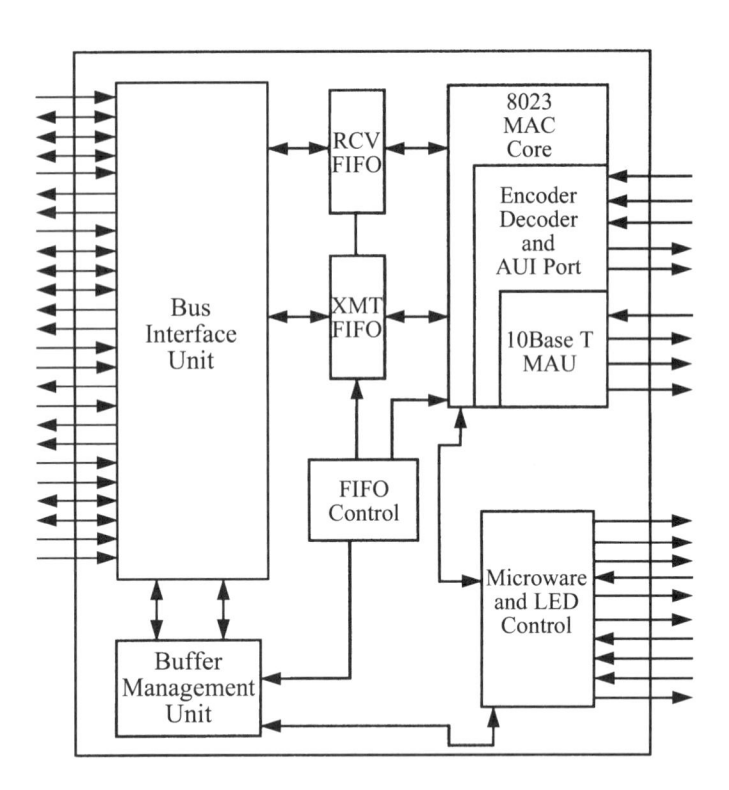

Figure 4-7.
The schematic for a single-chip, 32-bit Ethernet controller, which addresses high-performance system application requirements.

Repeaters

A repeater functions at the Physical Layer and interconnects the media segments of an extended network. Repeaters receive signals from one network segment and amplify, retime, and retransmit those signals to another network segment. With this, the repeater allows a series of cable segments to act as a single cable. The amplification gained through the use of repeaters prevents signal deterioration caused by long cable lengths and large numbers of connected devices. *Figure 4-8* is a photograph of a repeater.

Repeaters cannot perform complex filtering and other traffic processing. Moreover, the amplification given by a repeater not only applies to desired signals, but also to noise and network errors. Due to these factors, as well as the possibility of timing errors, most network designs limit the usage of repeaters. *Figure 4-9* depicts the connection of a repeater between two network segments, while *Figure 4-10* shows a schematic diagram of an integrated circuit that provides all repeater functions.

Chapter 4: Working with LANs

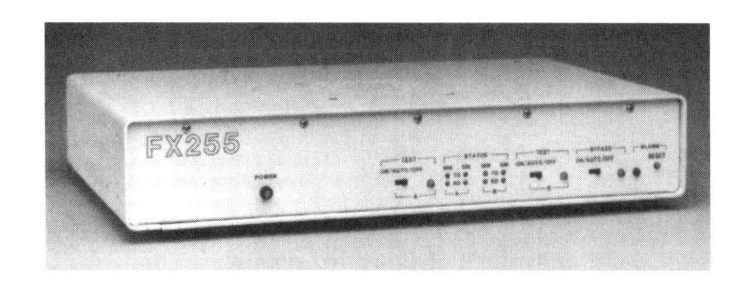

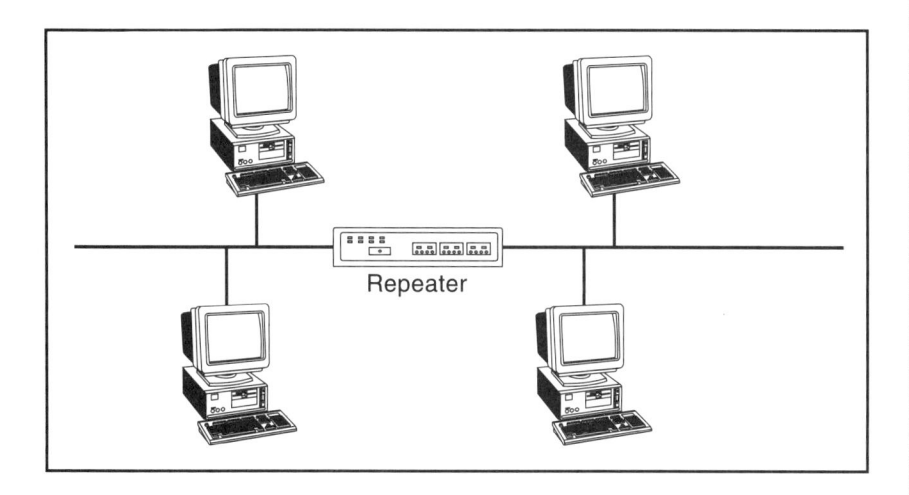

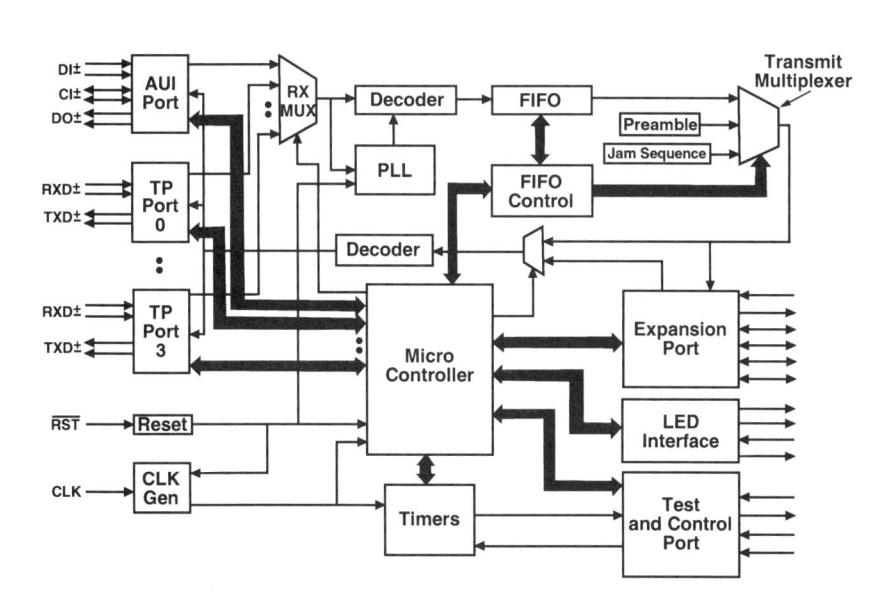

Figure 4-8.
A photograph of a repeater.

Figure 4-9.
Connecting a repeater between two network segments.

Figure 4-10.
A schematic diagram of an integrated circuit (IC) that provides all repeater functions.

97

The VLSI integrated circuit provides a system-level solution to designing nonmanaged multiport repeaters. In addition, the circuit integrates repeater functions specified in Section 9 of the IEEE 802.3 standard and Twisted Pair Transceiver functions complying with the 10BaseT standard.

Moving back to Figure 4-10, we see that the device provides four twisted-pair ports, uses one AUI port for direct connection to a Media Access Controller, and requires a single +-5V supply. AMD fabricated the device using CMOS technology.

Extenders

An extender operates as a remote-access multilayer switch that connects to a host router. Extenders forward traffic from all the standard network-layer protocols and filter traffic based on the MAC address or network-layer protocol type. But extenders don't segment LAN traffic or create security firewalls. *Figure 4-11* illustrates the connection of several extenders to a host router.

Figure 4-11.
Connecting several extenders to a host router.

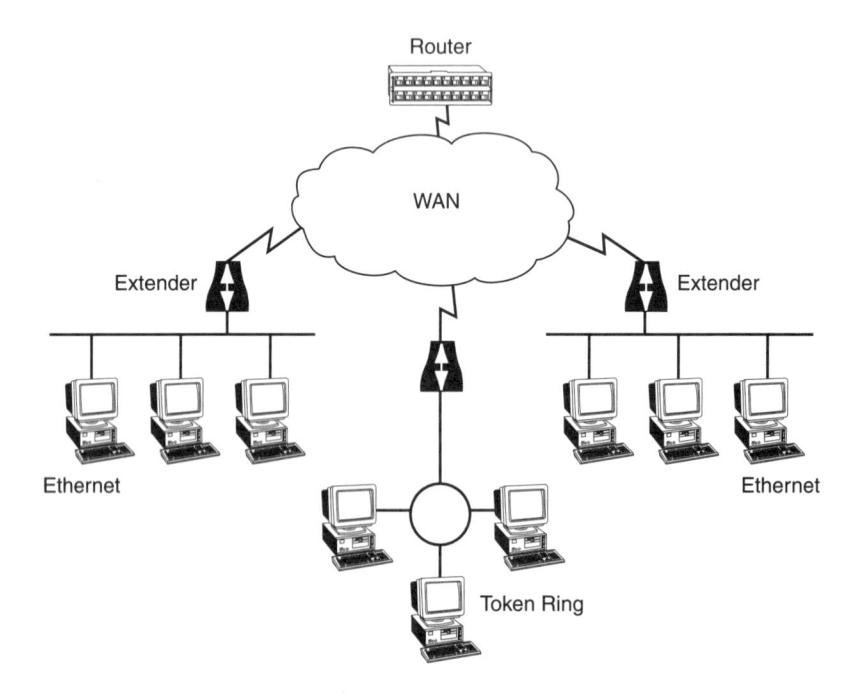

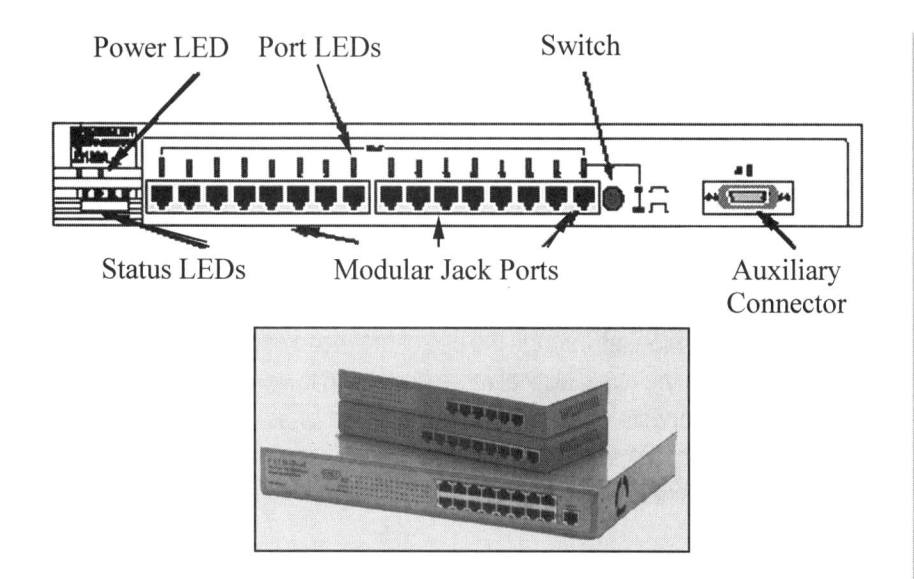

Figure 4-12.
A network hub connects multiple user stations through dedicated cables and establishes electrical interconnections.

Hubs

Referring to *Figure 4-12*, we see that a hub connects multiple user stations through dedicated cables and establishes electrical interconnections. With this, the hub operates much like a multiport repeater. As shown in Figure 4-3, hubs create a physical star network while maintaining the logical bus or ring configuration of the LAN.

Passive hubs concentrate multiple connections into a single device and use electrical relays to connect all users. In some network installations, a number of passive hubs may connect together in a daisy chain arrangement. Intelligent hubs add management capabilities to the connections provided by the passive hubs. With this, the hub provides status reports about the connections, compiles statistics about connection usage, and supports the connection of Ethernet and Token Ring cards.

Hubs install in wiring closets centrally located in a building. Hubs also:

- detect when a node isn't responding and "locks it out" so that the ring can continue to operate when a node fails. This happens automatically when the hub senses a node isn't responding.

- provide a "bridge" to other rings. With this, the hub sends messages addressed to nodes on other rings across the bridge circuits of those rings and accepts messages from other rings for its nodes.

Stackable Hubs. As the name implies, a stackable hub consists of hubs that stack on top of one another and interconnect through short cables. Even though separate hubs make up the stackable hub, the network "sees" the hub as a single logical device. Stackable hubs offer increased network management capabilities through the use of managed and manageable hubs, flexibility through the availability of more ports, and lower prices per port. The economy and benefits obtained through the use of stackable hubs occurs because the packets traveling between the devices don't depend on the electronics contained within the repeater.

Looking at *Figure 4-13*, we see that the repeater units in the hubs connect and form a single device. Each manufacturer of stackable hubs implements the intrahub connections in different ways. In all instances, the patch cables connect repeaters found within the hubs and carry the network data and control information. But hubs working within the stack must come from the same manufacturer. Packets moving between the hubs do not depend on the type of device or manufacturer.

Figure 4-13.
Stackable hubs. The repeater units in the hubs connect and form a single device.

Bridges

First available in the early 1980s, bridges connected and enabled packet forwarding only between the same type of networks. As technologies have improved, bridges have performed the same tasks between different types of networks. In addition, bridges provide filtering for frames based on any Layer 2 field or can reject unnecessary broadcast and multicast packets. For example, a network administrator can program a bridge to reject all frames delivered from a particular network. *Figure 4-14* shows the building of a packet containing application information by the host. The host also encapsulates the packet in a frame for transit over the medium to the bridge. At

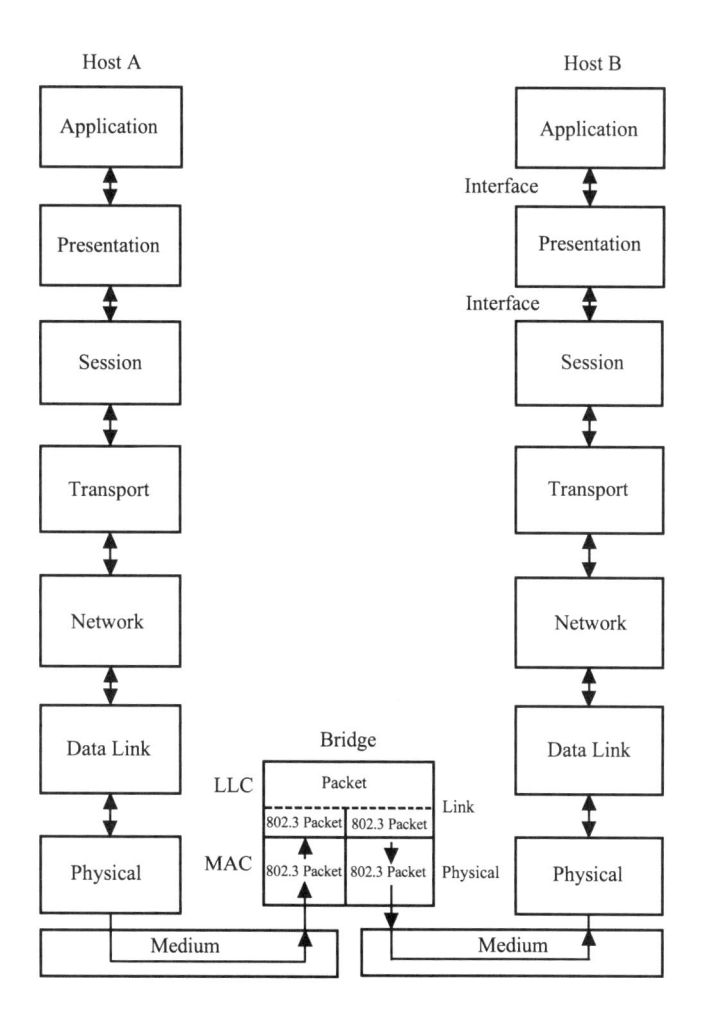

Figure 4-14.
Building a packet containing application information from a host.

this point, the bridge strips the frame of the header at the MAC sublayer of the link layer. Then the bridge passes the frame up to the LLC sublayer for further processing. After this processing, the packet passes back down for encapsulation into a header for transmission on the network to the next host. Due to different types of network support for certain frame fields and the use of different protocol functions within different networks, a bridge may not produce perfect data translation.

Three types of bridging exist: transparent, translational, and source-route. Connections between Ethernet-based LANs and WANs have prompted the use of transparent bridging. As the name implies, translational bridging translates between the formats and transit principles of Ethernet and Token Ring networks. Source-route bridging occurs with Token Ring networks and combines the algorithms of transparent bridging and source-route bridging to enable communication in mixed Ethernet/Token Ring environments.

Bridges provide these functions through the use of various Data Link Layer protocols that dictate specific flow control, error handling, addressing, and media-access algorithms. During operation, a bridge analyzes incoming frames, makes forwarding decisions based on information contained in the frames, and forwards the frames toward the destination. With source-route bridging, the frame contains the entire path to the destination. In comparison, transparent bridging forwards frames one transaction at a time toward the destination.

Switches

Like bridges, switches enable the interconnection of multiple physical LAN segments into a single larger network and forward traffic based on MAC addresses. Switches provide a significant improvement in transaction speed, however, because the operation occurs within hardware rather than through software. Many types of switches exist for different applications such as local-area networking, ATM networks, and internetworking between LANs and WANs. *Figure 4-15* shows a network switch.

A LAN switch provides much higher port density at a lower cost than traditional bridges. For this reason, LAN switches can accommodate network designs featuring fewer users per segment. As a result, the use of

Chapter 4: Working with LANs

Figure 4-15.
A network switch, which enables the interconnection of multiple physical LAN segments into a single larger network.

a switch in a LAN increases the average available bandwidth per user. *Figure 4-16* illustrates a simple network in which a LAN switch interconnects 10-Mbps and 100-Mbps Ethernet LANs.

Bridges and switches provide several advantages through the division of networks into segments. The forwarding of only a given percentage of traffic diminishes the traffic experienced by devices on all connected segments. In addition, a bridge or switch will act as a firewall for some potentially damaging network errors. Bridges and switches also

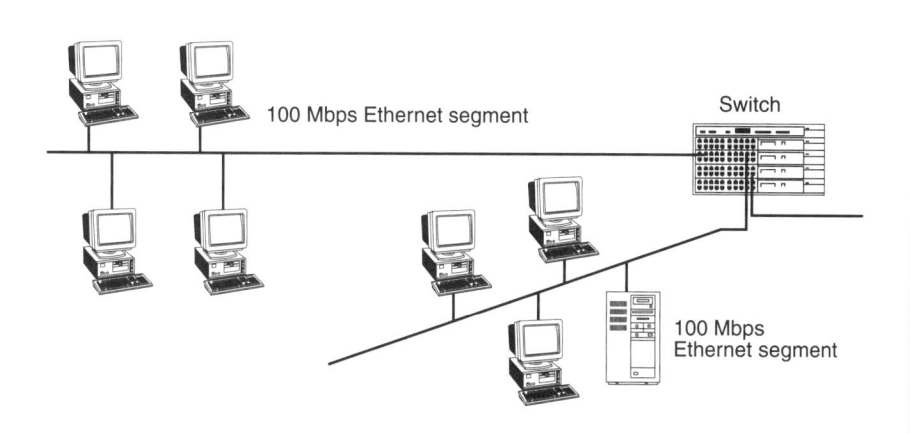

Figure 4-16.
A switch interconnects 10-Mbps and 100-Mbps Ethernet LANs.

accommodate communication between a larger number of devices than would be supported on any single LAN connected to the bridge. With this, a bridge or switch extends the effective length of a LAN and allows the attachment of distant nodes.

LAN switches provide transparent bridge-like functions in terms of learning a network topology, forwarding, and filtering. In addition, the switches also support dedicated communication between devices, multiple simultaneous conversation, full-duplex communication, and media-rate adaption. Dedicated collision-free communication between network devices increases the speed of a file transfer. Multiple simultaneous conversations can occur by either forwarding or switching several packets at the same time. As a result, network capacity increases by the number of supported conversations.

Full-duplex communication effectively doubles the data throughput. Media-rate adaption refers to the ability of the switch to translate between 10-Mbps and 100-Mbps data rates. In addition, media-rate adaption allocates bandwidth on an as-needed basis. With this, the installation of a switch requires no change to existing hubs, network-interface cards, or cabling.

As we have seen, LAN switches interconnect multiple network segments and provide dedicated, collision-free communication between network devices, while supporting multiple simultaneous conversations. In addition, LAN switches provide high-speed switching of data frames. To accomplish this, switches use either store-and-forward switching or cut-through switching when forwarding traffic.

The cut-through switching method reduces latency by eliminating error checking. From there, the switch looks up the destination address in its switching table, determines the outgoing interface, and forwards the frame toward its destination. The switch copies only the destination address—consisting of the first six bytes following the preamble—into its on-board buffers.

With the store-and-forward switching method, the switch copies the entire frame into its onboard buffers and computes the cyclic redundancy check (CRC). If the frame contains a CRC error—such as having less than 64 bytes or more than 1,518 bytes—then the switch discards the frame. If the frame doesn't contain any errors, the switch looks up the destination address in its forwarding, or switching, table and determines the outgoing interface. Then the switch forwards the frame toward its destination.

Some switches accept the configuration to perform cut-through switching on a per-port basis until the reaching of a user-defined error threshold. Then the switch automatically changes to store-and-forward mode. When the error rate falls below the threshold, the port automatically changes back to store-and-forward mode.

Switched Networks and Microsegmentation

Most network administrators strive for microsegmentation—that is, the fewest possible users per network segment. Microsegmentation creates dedicated segments with one user per segment through the use of a LAN switch. With microsegmentation in place, each user receives instant access to the full bandwidth and doesn't have to contend for available bandwidth with other users. As a result, the collisions that normally occur in a shared Ethernet environment that depends on hubs do not occur.

During operation, the switch forwards frames based on either the Layer 2 address or Layer 3 address of the frame. Because of this, we can refer to a LAN switch as a frame switch. Ethernet, Token Ring, and FDDI networks take advantage of LAN switching. *Figure 4-17* shows the configuration for a LAN switch that provides dedicated bandwidth to devices and illustrates the relationship of Layer 2 switching to the OSI Data Link Layer.

Switch Categories

Efficient network management requires the evaluation of the needed amount of bandwidth for connections between devices. With this, a network manager can ensure that the network accommodates the data flow of network-based applications. Much of the decision-making in terms of switch selection depends on bandwidth requirements.

We can categorize LAN switches according to the proportion of bandwidth allocated to each port. Symmetric switching provides evenly distributed bandwidth to each port. During operation, a symmetric switch

Microsegmentation creates dedicated segments with one user per segment through the use of a LAN switch.

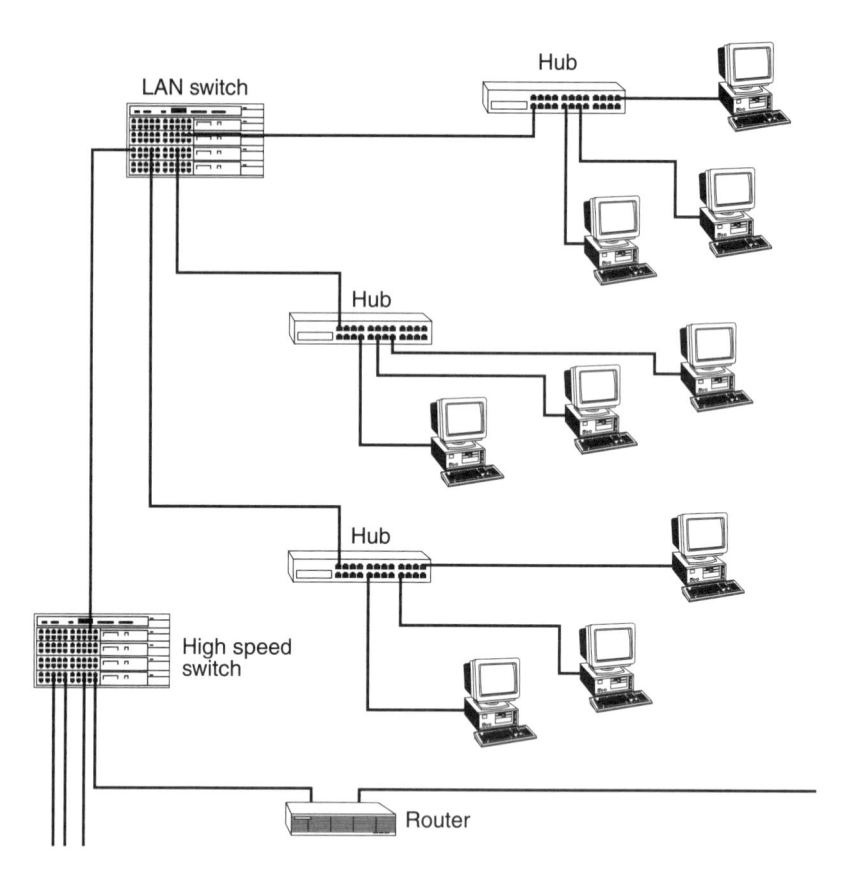

Figure 4-17.
A LAN switch providing dedicated bandwidth to devices, illustrating the relationship of Layer 2 switching to the OSI Data Link Layer.

establishes switched connections between ports with the same bandwidth. As a result, symmetric switching remains optimized for a reasonably distributed traffic load.

Asymmetric switching provides unlike, or unequal, bandwidth between some ports. As opposed to the symmetric switch, an asymmetric LAN switch establishes switched connections between ports of unlike bandwidths. We can also refer to asymmetric switching as 10/100 switching. Asymmetric switching works well with client-server traffic where multiple clients simultaneously communicate with a server. This type of traffic flow requires the dedication of more bandwidth to the server port so that a bottleneck won't occur at any particular port.

In addition, we can also categorize switches according to the OSI layer at which the device filters and forwards—or switches—frames. The switch categories include Layer 2, Layer 2 with Layer 3 features, and multilayer. A Layer 2 switch performs switching and filtering based on the OSI Data Link Layer MAC address. Along with those characteristics, the Layer 2 switch also functions similar to a multiport bridge but has a much higher capacity and supports full-duplex operation. As with bridges, a Layer 2 switch remains completely transparent to network protocols and user applications.

A Layer 2 LAN switch with Layer 3 features makes switching decisions based on more information than just the Layer 2 MAC address. While the switch includes Layer 2 switching functions, it also may incorporate Layer 3 traffic-control features. Some of those features include broadcast and multicast traffic management, security through access lists, and IP fragmentation.

A multilayer switch makes switching and filtering decisions on the basis of Data Link Layer and network-layer addresses. During operation, this type of switch dynamically decides whether to switch or route incoming traffic. A multilayer LAN switch switches within a workgroup and routes between different workgroups.

Ethernet Networks

In the mid-1970s, the Xerox Corporation, the Intel Corporation, and the Digital Equipment Corporation combined their resources to develop Ethernet technologies. The term "Ethernet" refers to the family of baseband local-area network implementations that includes Ethernet networks operating at 10 Mbps, 100-Mbps Fast Ethernet networks, and Gigabit Ethernet networks operating at 1,000 Mbps. The Ethernet standard offers a combination of tremendous flexibility and its relative simplicity in terms of implementation and understanding. The underlying transmission scheme used with the Ethernet standard continues to stand as one of the principal means of transporting data. The Ethernet system consists of:

- The physical medium used to carry Ethernet signals between computers
- A set of medium access control rules embedded in each Ethernet interface that allows multiple computers to fairly arbitrate access to the shared Ethernet channel
- An Ethernet frame that consists of a standardized set of bits used to carry data over the system

CSMA/CD Protocol

Ethernet networks use the Carrier Sense Multiple Access/Collision Detect (CSMA/CD) protocol for carrier transmission access. In an Ethernet network, multiple access means any device may attempt to send a frame at any time. Each device senses the condition of the line. If the line has an idle condition, the device begins to transmit its first frame. If another device attempts to transmit data at the same time, a collision occurs. Collision detect occurs and the network discards the frames. In addition, each device waits a random amount of time and retries until successfully transmitting the data.

With the broadcast-based environment found with Ethernet networks, all stations see all frames placed on the network. Following any transmission, each station must examine every frame to determine whether the destination of the frame corresponds with the station. Frames intended for a given station pass to a higher-layer protocol. The Ethernet frame fields include the following information.

- Preamble—The alternating pattern of 1s and 0s tells receiving stations that a frame is in transit. The Ethernet frame includes an additional byte that is the equivalent of the Start-of-Frame field.

- Start-of-Frame (SOF)—The delimiter byte ends with two consecutive 1 bits that serve to synchronize the frame-reception portions of all stations on the LAN. SOF is explicitly specified in Ethernet.

- Destination and Source Addresses—The first three bytes of the addresses are specified by the IEEE on a vendor-dependent basis. The last three bytes are specified by the Ethernet vendor. The source address is always a unicast (single-node) address. The destination address can be unicast, multicast (group), or broadcast (all nodes).

- Type—The type specifies the upper-layer protocol to receive the data after Ethernet processing is completed.

- Data—After Physical-Layer and link-layer processing is complete, the data contained in the frame is sent to an upper-layer protocol, which is identified in the Type field. Although Ethernet Version 2 doesn't specify any padding, Ethernet expects at least 46 bytes of data.

- Frame Check Sequence (FCS)—This sequence contains a four-byte cyclic redundancy check (CRC) value, which is created by the sending device and is recalculated by the receiving device to check for damaged frames.

10-Mbps Ethernet Networks

10-Mbps Ethernet networks transmit data over twisted pair, coaxial cable, or wireless transmission media. While the network-interface card supports the Ethernet standard through the use of a transceiver that performs Physical Layer functions, cables or wireless transmissions connect the nodes to the networks. We can refer to the connecting cable as an attachment unit interface (AUI) and the transceiver as a media attachment unit (MAU).

Figure 4-18 shows the most basic Ethernet network configuration. In the figure, several workstations connect together on a single Ethernet segment. Based on the 10Base2 standard, the network includes four network nodes and a single server. This particular network relies on Thinnet coaxial cabling and a network interface card that contains the transceiver.

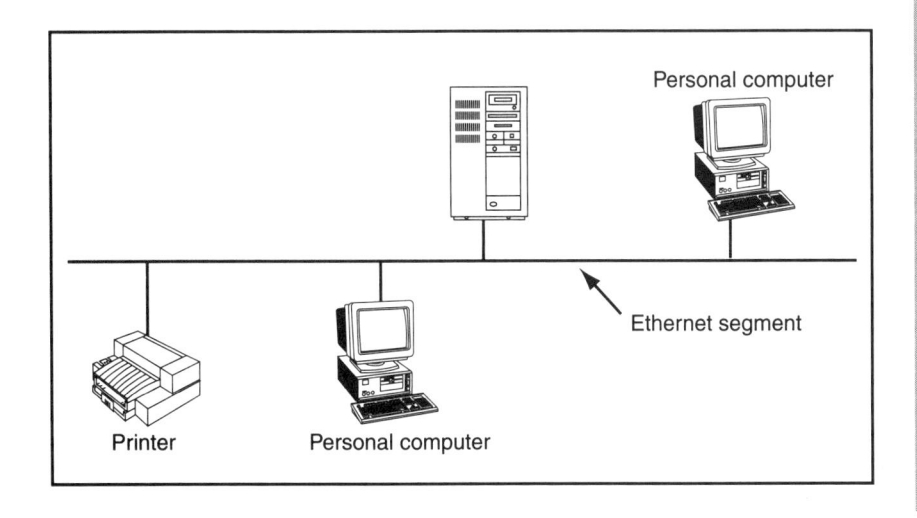

Figure 4-18.
The most basic Ethernet network configuration, in which several workstations connect together on a single Ethernet segment.

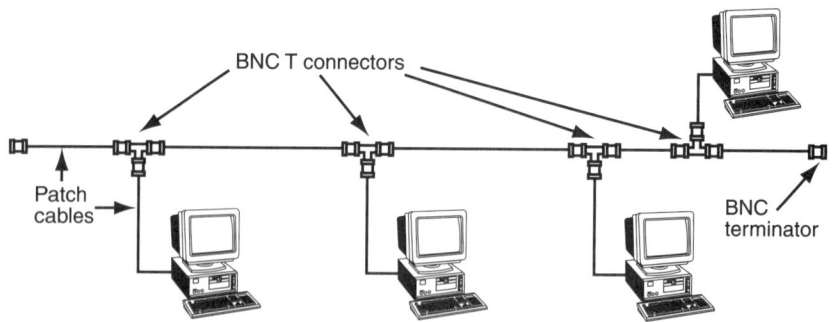

Telecommunication Technologies

Figure 4-19.
A 10Base5 Ethernet network, which relies on thicknet coaxial cabling and external transceivers.

The 10Base5 Ethernet network shown in *Figure 4-19* relies on Thicknet coaxial cabling and external transceivers. A transceiver takes the digital signal found at the workstation and converts the signal into the format needed for physical cabling. Each segment of the network terminates at both ends.

Category 3 and Category 5 twisted-pair cabling supports the 10BaseT data transmission standard. If we break 10BaseT into parts, the "10" represents the 10-megabit-per-second transmission rate, while the "Base" signifies baseband communication. The "T" shows that the standard works through twisted-pair cabling. Since the 10BaseT standard uses baseband communication, one cabling pair carries transmitted data, while the other pair carries received data.

While 10BaseT over Category 3 cable has a maximum length of 100 meters, the use of Category 5 cabling for the 10BaseT standard increases the cable length to 150 meters. One cable contains the two pair of wires and may bundle multiple pairs together. Each end of the 10BaseT pair set terminates with an eight-position jack.

Figure 4-20 shows the configuration of a 10BaseT Ethernet network. In the figure, an Ethernet hub connects eight workstations together into a star topology. The use of the hub—rather than a switch—establishes a shared domain where all nodes attached to the hub share the 10-Mbps bandwidth. UTP cabling functions as the transmission media and uses RJ-45 connectors. 10BaseT networks offer the following advantages.

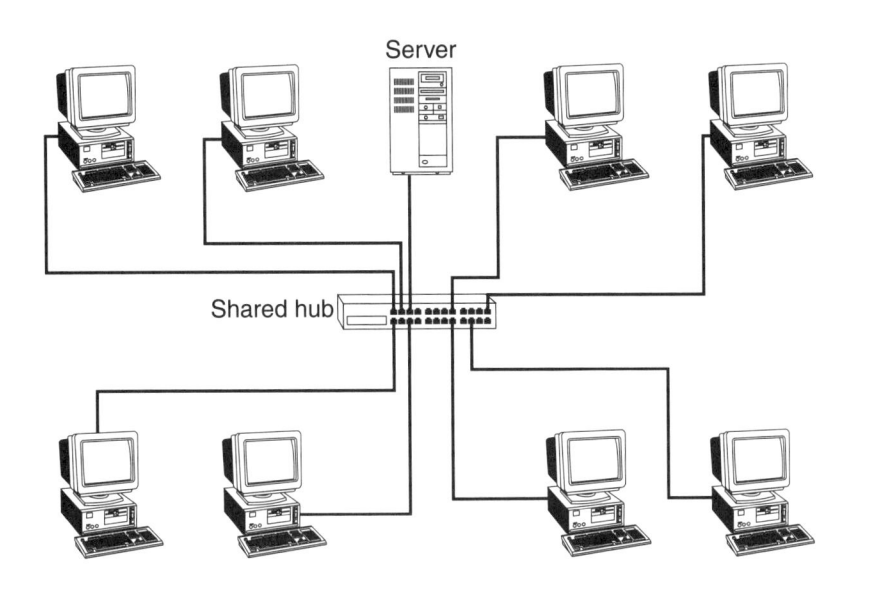

Figure 4-20.
The configuration of a 10BaseT Ethernet network. An Ethernet hub connects eight workstations together into a star topology.

- Intelligent hubs
- Easily added or removed nodes
- Easier troubleshooting

Figure 4-21 shows that hub-to-hub connections may exist due to the need to attach more computers to the network. Once the original hub reaches capacity, the network cannot accept additional computers. The

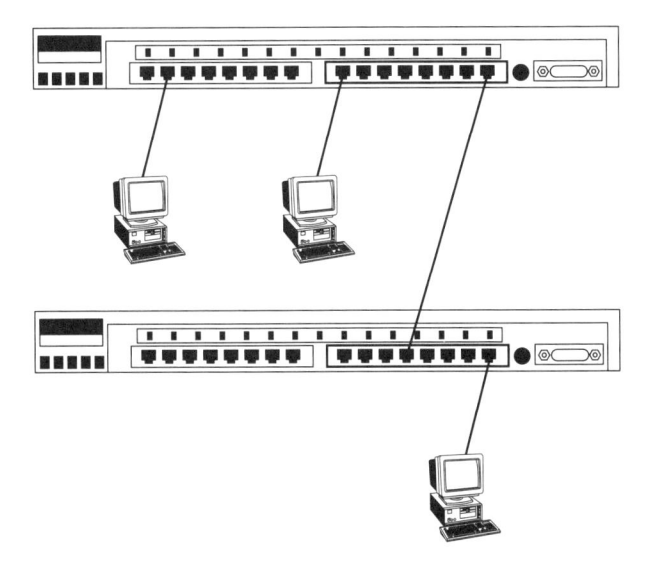

Figure 4-21.
Hub-to-hub connections, which may exist due to the need to attach more computers to the network.

addition of another hub provides more physical ports for network connectivity. But the additional hub also becomes part of the collision domain. Each additional port provided through the new hub contends for the shared bandwidth. As network traffic increases, the interconnected hubs will provide inadequate performance.

Figure 4-22 shows the solution for this problem with the replacement of the standard hub with a switching hub. With Ethernet switching, the MAC-layer address determines the switch port as the destination for the frame. In addition, a virtual connection occurs between the sending and receiving ports. As this determination and connection occurs, no other ports become available as destinations for the frame. Consequently, the switched network avoids any chance for data collisions. The dedicated connection remains in place only for the time needed to pass the frame between the sending and receiving stations.

If the destination port for a frame appears busy, the station holds the data in a buffer. When the destination port becomes free, the station releases the frame and sends the frame to the correct port. The use of buffers works well unless the buffer fills to capacity. When this occurs, the network may lose frames. *Figure 4-23* shows the expansion of network capacity through the connection of hubs to a central switch. In the figure, each hub connects to a switch port and the network logically arranges users into workgroups on a hub-by-hub basis. The switch does not forward frames between hubs.

Figure 4-22
Replacing a standard hub with a switching hub in order to curtail the effects of increased network traffic.

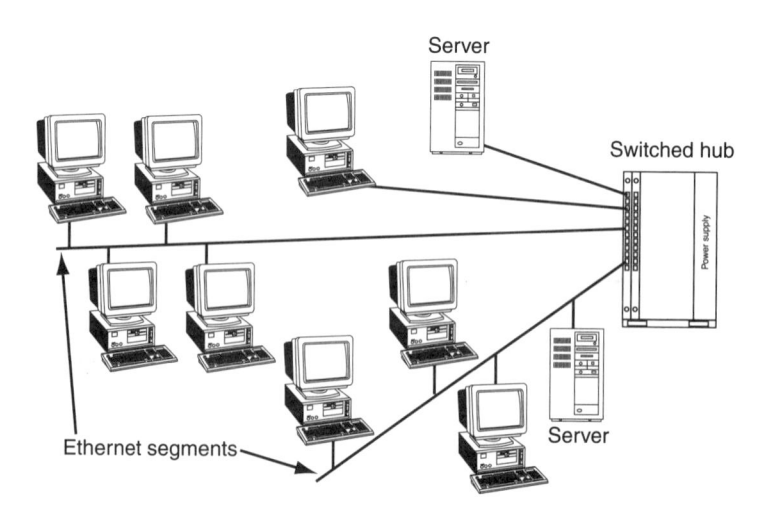

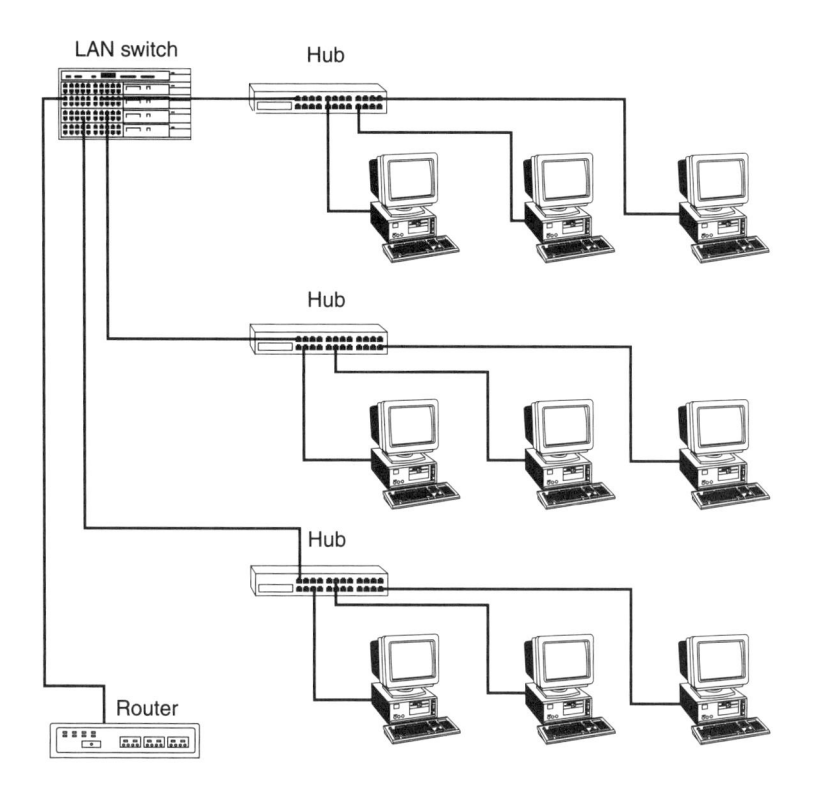

Figure 4-23.
Connecting hubs to a central switch in order to expand network capacity.

The figure also shows that the server has a dedicated port on the switch. With this, entire segments become switched and multiple users attach to the same switch port through the same hub. Since the server requires access to multiple segments, the device provides better functionality through a dedicated port. Without the use of a dedicated port for the server, network performance would decrease. Moreover, the figure illustrates the use of a switch that offers different port speeds. As the figure shows, the server attaches to a 100-Mbps port, while the workgroup hubs attach to and share 10-Mbps bandwidth.

Fast Ethernet Networks

Fast Ethernet provides a 100-Mbps high-speed LAN technology that offers increased bandwidth. The 100BaseT and 10BaseT standards rely on the same IEEE 802.3 MAC access and collision detection methods and have the same frame format and length requirements. Like 10BaseT, the

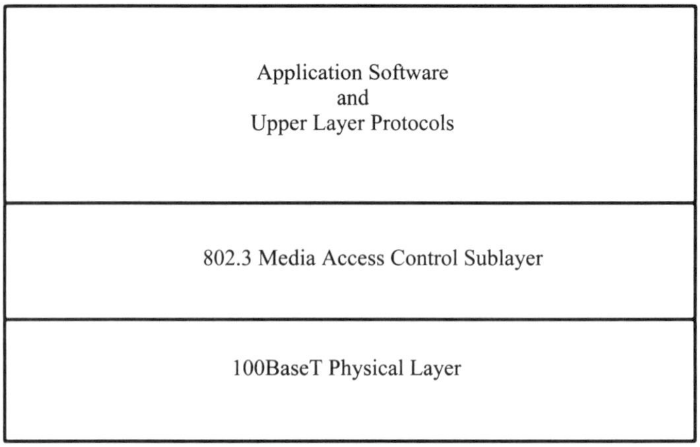

Figure 4-24.
100BaseT and the MAC sublayer.

Application Software and Upper Layer Protocols
802.3 Media Access Control Sublayer
100BaseT Physical Layer

100BaseT standard operates over unshielded twisted-pair and shielded twisted-pair cabling.

The 100BaseT Ethernet Standard relies on Category 5 twisted-pair cabling and carries a signal at a data rate of 100 Mbps. Applications for 100BaseT over Cat 5 cabling have a maximum cable length of 100 meters. In comparison to the 100BaseT standard, 100BaseT2 supports a data-transmission rate of 100 Mbps over Category 3 twisted-pair cabling, rather than requiring the use of Cat 5 cabling.

In addition, 100BaseT supports all applications and networking software currently running on 802.3 networks and supports dual speeds of 10 and 100 Mbps using 100BaseT fast-link pulses. As a result, Fast Ethernet hubs and adapter cards detect and support 10 and 100 Mbps data-transfer rates. *Figure 4-24* illustrates how the 802.3 MAC sublayer and higher layers run unchanged on 100BaseT.

The two standards differ in terms of network diameter because of the adherence to the same methods and requirements. While a 10-Mbps Ethernet network may have a maximum diameter of 2,100 meters, a 100-Mbps Ethernet network has a maximum diameter of 205 meters. This occurs because the valid minimum-length frame transmission defines the collision domain. In turn, the frame transmission governs the maximum distance between two end stations on a shared segment. As the speed of network operation increases, both the minimum frame-transmission time and the maximum diameter of a collision domain decrease. The bit budget of a collision domain consists of the maximum signal-delay time of the various networking components, the MAC layer of the station, and the physical medium.

The increased data-transfer rate seen with 100 Mbps combines with the physical media propagation speed to limit the diameter by 10 times. A station transmitting on a 100-Mbps network at 10 times the speed of a station transmitting on a 10-Mbps network must have a maximum distance that measures 10 times less.

The 100BaseT2 standard uses two pair of twisted-pair cabling and a dual-duplex baseband transmission scheme. With dual-duplex baseband transmission, data simultaneously transmits over each wire pair in each direction. Dual duplex baseband transmission uses the complex Five-level Pulse Amplitude Modulation (PAM5x5) signal-encoding scheme that transmits data through a quinary (five-level) signal. With this, four bits of information transmit per signal transition on each wire pair. As a result, the combination of a 25-megabaud transmission rate and two pairs of cabling supports the full duplex transmission of data at a rate of 100 Mbps.

As with 100BaseT2, 100BaseT4 supports the transmission of data at a rate of 100 Mbps over Category 3 or better cabling. With 100BaseT4, however, four pairs of cabling carry the signal. One pair carries transmitted data, while another pair remains dedicated to carrying received data. Two bidirectional pairs carry or receive data. When using four cabling pairs, one dedicated pair allows collision detection on the network, while the three remaining pairs remain available for carrying the data transfer.

Data transmission over a 100BaseT4 network occurs through an "8B6T" signal-encoding scheme that converts eight bits of binary data into six "ternary" signals. During the transmission of data over twisted-pair wires, the ternary signal may have one of three binary values. As a result, the encoding scheme splits the 100 Mbps data-transmission rate over three twisted pairs with each pair carrying a 33.3-Mbps transmission rate. Ternary signaling requires only six bauds to transfer eight bits of information and provides a maximum signal-transmission rate of 25 megabaud on each of the twisted pairs. Since 25 megabauds translates into a maximum frequency of 12.5 MHz, the data-transmission rate falls within the 16-MHz limit supported by Category 3 cabling.

100VG-AnyLAN

In the last section, we found that the adherence to the same CSMA/CD protocol within 10BaseT and 100BaseT networks reduced the diameter of Fast Ethernet networks. Hewlett-Packard developed the 100VG-AnyLAN standard as an alternative to CSMA/CD for newer time-sensitive, high bandwidth applications such as multimedia. 100VG-AnyLAN has an access method based on station demand and provides an upgrade path from Ethernet and 16-Mbps Token Ring. The standard operates with four-pair Category 3 UTP, two-pair Category 4 or 5 UTP, STP, and fiber-optic cables.

During operation, the 100VG-AnyLAN standard uses a demand-priority access method that eliminates collisions and accepts heavier loading than seen with the 100BaseT standard. As opposed to the CSMA/CD protocol, the demand-priority access method establishes controlled access to the network through the hub. The 100VG-AnyLAN standard calls for a level-one hub, or repeater, that acts as the root and controls the operation of the priority domain. A 100VG-AnyLAN network arranges hubs in a hierarchical fashion. Each hub has at least one uplink port, while every other port can function as a downlink port. Since hubs can cascade three-deep when uplinked to other hubs, the interconnected hubs act as a single large repeater—with the root repeater polling each port in port order.

100VG-AnyLAN networks use a demand-priority operation in which a node waiting to transmit signals the request to the hub or switch. With an idle network condition, the hub immediately acknowledges the request and the node begins transmitting a packet to the hub. If the hub simultaneously receives multiple requests, then the device uses a round-robin technique to acknowledge each request in turn.

But the hub acknowledges and transmits high-priority requests—such as time-sensitive videoconferencing applications—ahead of normal-priority requests. A 100VG-AnyLAN hub does not grant priority access to a port more than twice in a row.

Figure 4-25 depicts a 100VG-AnyLAN network and points out the link-distance limitations, hub-configuration limitations, and maximum network-

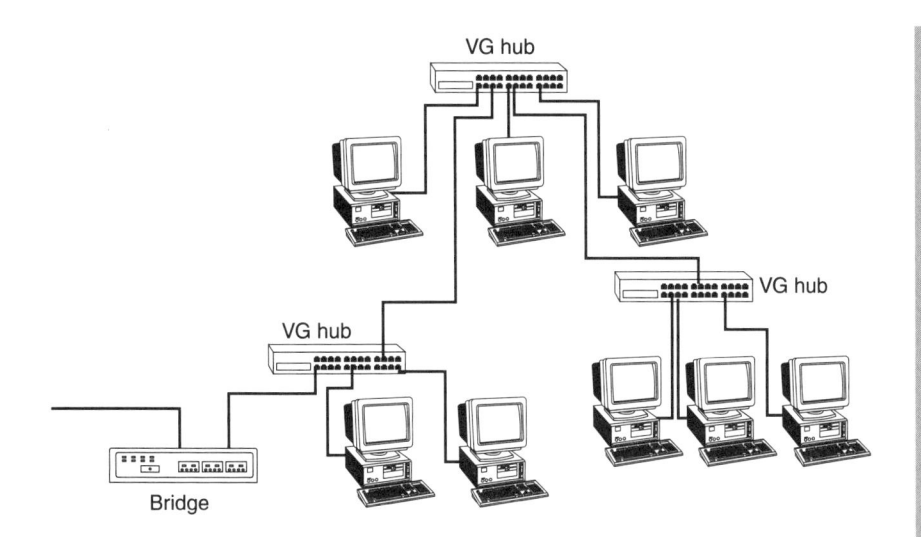

Figure 4-25.
A 100VG-AnyLAN network, which arranges hubs in a hierarchical fashion.

distance limitations. As shown in the figure, a 100VG-AnyLAN network has a 100-meter maximum distance from node to hub over Category 3 UTP cable and a 150-meter maximum distance when applied over Category 5 UTP cable. Again referring to Figure 4-25, we see that 100VG-AnyLAN networks have end-to-end network-distance limitations of 600 meters when operating over Category 3 UTP and 900 meters when operating over Category 5 UTP. Locating 100VG-AnyLAN hubs in the same wiring closet decreases end-to-end distances 200 meters over Category 3 UTP and 300 meters over Category 5 UTP.

Gigabit Ethernet Networks

Large networks feature a backbone or the portion of the network that carries the most significant traffic. In addition, the backbone uses a bridge to connect LANs or subnetworks together to form the larger network. As network applications have grown from utilizing 10BaseT standards to relying on the higher bandwidth and faster data-transfer speeds seen with 100-Mbps Ethernet networks, the need for high-speed backbone technologies has become more apparent.

The 1000BaseT standard—sometimes called Gigabit Ethernet—supports the transmission of data at a rate of 1,000 Mbps over Category

5 balanced copper cabling. The cabling has a maximum length of 100 meters and provides full-duplex baseband transmission over four pairs. Each cabling pair carries data at a transmission rate of 250 Mbps and supports baseband signaling at a modulation rate of 125 MHz. Gigabit Ethernet relies on PAM5 encoding for the transmission of the data over each cabling pair.

Several methods exist for using Gigabit Ethernet to increase bandwidth and capacity within a network. Gigabit Ethernet can improve Layer 2 performance by using throughput to eliminate bottlenecks and provides an answer as the building backbone for interconnection of wiring closets. *Figure 4-26* illustrates potential multilayer Gigabit switching designs. In this application, a gigabit multilayer switch aggregates the traffic for the building and provides connection to servers through Gigabit Ethernet or Fast Ethernet.

Gigabit Ethernet offers transparent compatibility with the lower speed Ethernet standards, while providing high-speed data transfer speeds of 1 Gbps. To accelerate speeds from 100 Mbps Fast Ethernet to 1 Gbps, several changes occur at the physical interface. Since Gigabit Ethernet remains identical to 10BaseT and 100BaseT Ethernet from the Data Link Layer upward, Gigabit Ethernet standard combines the best points of previous Ethernet standards and the Fibre Channel standard.

Figure 4-26.
Potential multilayer Gigabit switching designs, in which a gigabit multilayer switch aggregates the traffic for the building.

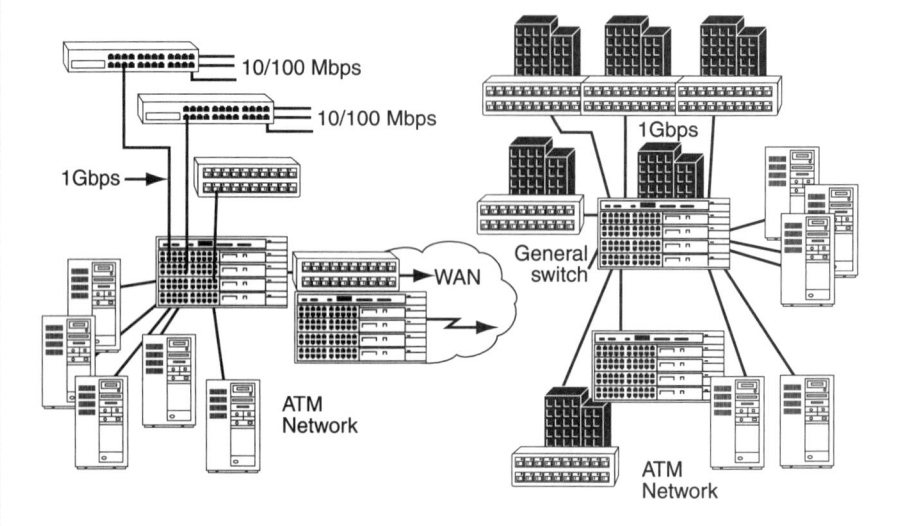

At the Physical Layer, the Gigabit Ethernet specification uses four forms of transmission media:

- 1000BaseLX—Long-wave (LW) laser over single-mode and multimode fiber
- 1000BaseSX—Shortwave (SW) laser over multimode fiber
- 1000BaseCX—Balanced shielded 150-ohm copper cable
- 1000BaseT—Category 5 UTP cable

The Gigabit Ethernet Interface Carrier (GBIC) allows network managers to configure each Gigabit Ethernet port on a port-by-port basis for short-wave and long-wave lasers as well as for copper physical interfaces. Given the GBIC configuration, switch vendors can build a single physical switch or switch module that the customer can configure for the required laser or fiber topology.

The MAC layer of Gigabit Ethernet has characteristics similar to those seen with earlier Ethernet standards. For example, the MAC layer of Gigabit Ethernet supports both full-duplex and half-duplex transmission. In addition, Gigabit Ethernet offers collision detection, maximum network diameter, and repeater rules. Gigabit Ethernet also provides support for half-duplex Ethernet while adding frame bursting and carrier extension.

Half-Duplex Gigabit Ethernet Operation

For half-duplex transmission, Gigabit Ethernet uses the same CSMA/CD protocol as seen with other Ethernet standards to ensure that stations can communicate over a single wire and that collision recovery can occur. Implementation of CSMA/CD for Gigabit Ethernet allows the creation of shared Gigabit Ethernet through hubs or half-duplex point-to-point connections. Since the CSMA/CD protocol remains delay sensitive, the use of CSMA/CD requires the creation of a bit-budget per-collision domain.

Acceleration of Ethernet standards to gigabit speeds has created some challenges in terms of the implementation of CSMA/CD. At speeds greater than 100 Mbps, smaller packet sizes become smaller than the length of the slot-time, or the time required for the Ethernet MAC layer to handle collisions. To remedy the slot-time problem, the Gigabit Ethernet standard uses carrier extension to add bits to the frame until the frame meets the required minimum slot-time. As a result, the smaller packet sizes can coincide with the minimum slot-time and allow seamless operation with the other Ethernet CSMA/CD standards.

The Gigabit Ethernet standard also uses frame bursting, or the transmission of a burst of frames without relinquishing control by an end station. As long as the network never appears idle, other stations defer to the frame burst. The end station using frame burst fills the interframe interval with extension bits so that the network never appears idle to other stations.

Full-Duplex Gigabit Ethernet Operation

Most of the issues that we have considered with half-duplex Gigabit Ethernet data transmissions negate the effectiveness of the high-speed Ethernet standard. Since the use of full-duplex Ethernet eliminates collisions, full-duplex Gigabit Ethernet doesn't require the use of the CSMA/CD protocol for flow control or as an access medium. As we have already discussed, full-duplex provides a method for transmitting and receiving data simultaneously on a single medium.

Typical full-duplex applications occur between two endpoints. For example, switch-to-switch connections, switch-to-server connections, and switch-to-router connections use full-duplex transmissions. The use of full-duplex data transmissions has easily and effectively allowed the bandwidth on Ethernet and Fast Ethernet networks to double from 10 Mbps to 20 Mbps and from 100 Mbps to 200 Mbps. Full-duplex transmission utilized in a Gigabit Ethernet environment promises to increase bandwidth from 1 Gbps to 2 Gbps for point-to-point links.

A Gigabit Ethernet Transceiver

Figure 4-27 is a block diagram for a low-cost, low-power Gigabit Ethernet transceiver. Used for data transmission over fiber or coaxial media in conformance with the IEEE 802.3z Gigabit Ethernet specification and Fibre Channel ANSI X3T11 at 1.0 Gbit/s and 1.25 Gbits/s, the transmitter section accepts parallel 10-bit encoded data. The transmitter uses this clock to synthesize the internal high-speed serial bit clock and places the serialized

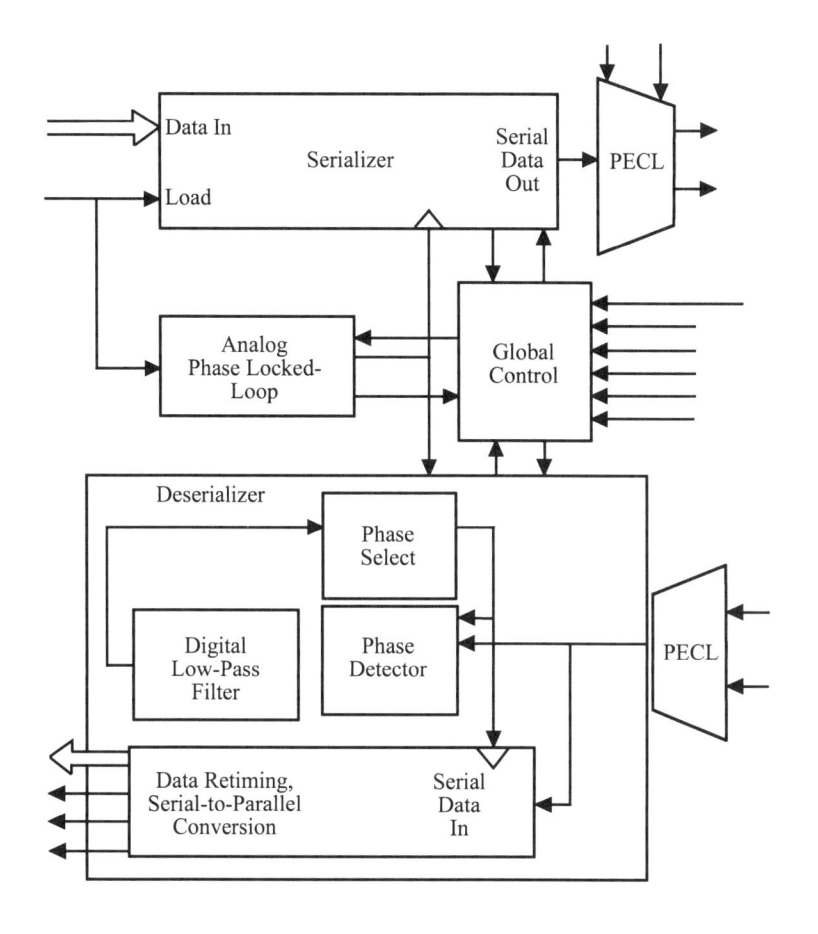

Figure 4-27.
A block diagram for a low-cost, low-power Gigabit Ethernet transceiver.

data at the differential PECL outputs. With the outputs terminated at 50 or 75 ohms, the transceiver can drive either an optical transmitter or coaxial media.

Moving to the second portion of the figure, the receive section accepts high-speed serial data at its differential PECL input port. The data feeds to the digital clock recovery section, which generates a recovered clock and retimes the data. After deserializing the retimed data, the receiver presents it as 10-bit parallel data on the output port.

Fast EtherChannel

The use of Gigabit EtherChannel can significantly increase the bandwidth available within the campus backbone to high-end wiring closets or to high-end routers. Fast EtherChannel bundles multiple Fast Ethernet ports together. Switches see the bundled ports as one high-bandwidth channel. Fast EtherChannel allows the bundling of up to four ports at a combined bandwidth of 800 Mbps. With support from major network adapter manufacturers, Fast EtherChannel works with high-end file servers.

Token Ring Networks

IBM developed the Token Ring network in the mid-1970s and continues to use the standard as its primary LAN technology. The IEEE 802.5 specification follows the IBM Token Ring development, but maintains a few key differences. While the IBM standard specifies a star topology with all stations attached to a multistation access unit (MSAU), the IEEE 802.5 standard does not specify a topology. In addition, IBM Token Ring networks rely on twisted-pair cabling, while the IEEE 802.5 standard doesn't specify a media type. *Figure 4-28* shows an IBM Token Ring network with workstations connected to MSAUs and forming one large ring.

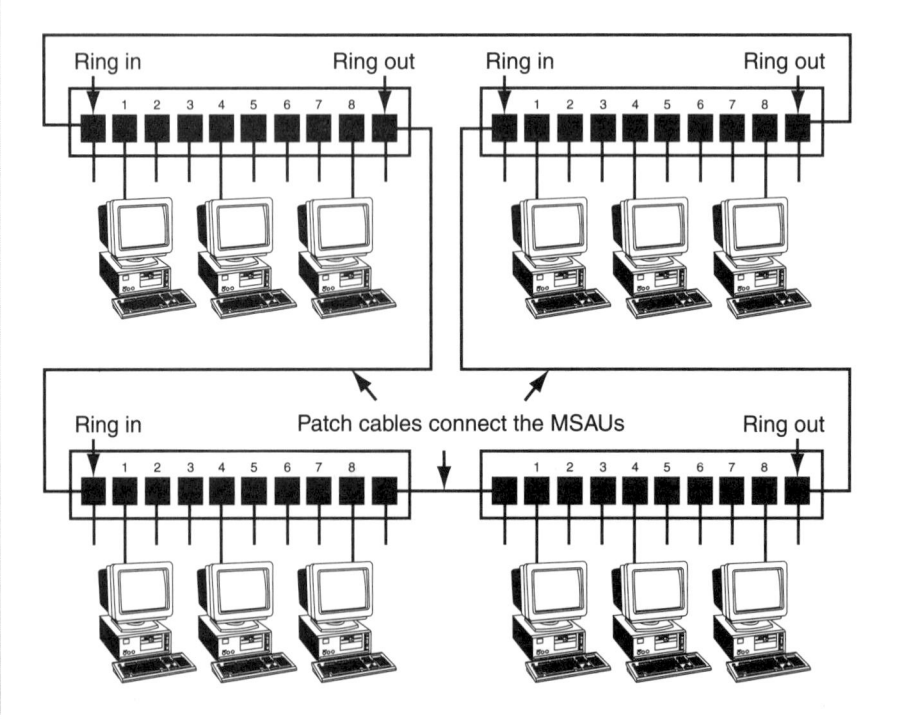

Figure 4-28.
An IBM Token Ring network with workstations connected to MSAUs and forming one large ring.

Token-passing networks move a small frame—called a token—around the network with possession of the token granting the right to transmit. If a node receiving the token has no information to send, it passes the token to the next end station. Each station can hold the token for a maximum period of time. If a station possessing the token does have information to transmit, it seizes the token and changes one bit of the token.

With this, the token becomes a start-of-frame sequence, appends the information it wants to transmit, and sends this information to the next station on the ring. While the information frame circles the ring, no token remains on the network unless the ring supports early token release. With no token on the network, other stations must wait to transmit and collisions cannot occur. The support of early token release allows the release of a new token at the completion of frame transmission.

The information frame continues to circle the ring until it reaches the intended destination station. At this point, the destination station copies the information for further processing. From there, the information frame continues to circle the ring until removal occurs at the sending station. The sending station can check the returning frame to see whether the destination station accepted and copied the frame. In comparison to CSMA/CD networks, we can refer to token-passing networks as deterministic networks. A deterministic network calculates the maximum time that will pass before any end station can transmit.

Token Ring networks use a sophisticated priority system that permits certain user-designated, high-priority stations to use the network more frequently. Control of these priorities occurs through the priority field and reservation field found within the Token Ring frame. Only stations with a priority equal to or higher than the priority value contained in a token can seize that token.

Once a station seizes the token and changes the token to an information frame, only stations with a priority value higher than that of the transmitting station can reserve the token for the next pass around the network. The generation of the next token includes the higher priority of the reserving station. Stations that raise the priority level of a token must reinstate the previous priority after the completion of the transmission.

Token-passing networks move a small frame—called a token—around the network with possession of the token granting the right to transmit.

Telecommunication Technologies

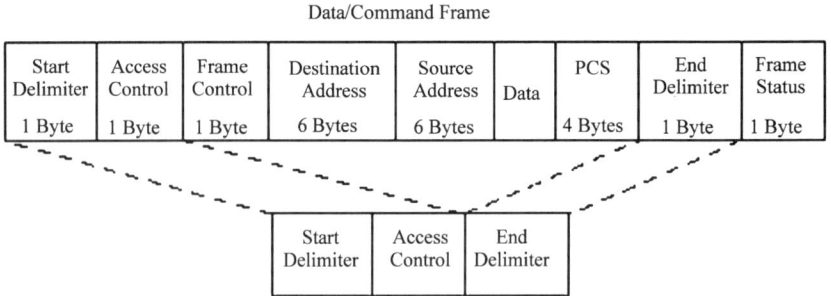

Data/Command Frame

Start Delimiter	Access Control	Frame Control	Destination Address	Source Address	Data	PCS	End Delimiter	Frame Status
1 Byte	1 Byte	1 Byte	6 Bytes	6 Bytes		4 Bytes	1 Byte	1 Byte

Start Delimiter	Access Control	End Delimiter

Figure 4-29.
The three Token Ring frame fields.

The Token Ring standards support two basic frame types called tokens and data/command frames. Tokens have a three-byte length and consist of a start delimiter, an access-control byte, and an end delimiter. Shown in *Figure 4-29*, the three Token Ring frame fields consist of the following information.

- Start Delimiter—Alerts each station of the arrival of a token or data/command frame. This field includes signals that distinguish the byte from the rest of the frame by violating the encoding scheme used elsewhere in the frame.

- Access-Control Byte—Contains the Priority field, or the most significant three bits, and Reservation field, or the least significant three bits, as well as a token bit that differentiates a token from a data/command frame, and a monitor bit. The active monitor uses the monitor bit to determine whether a frame endlessly circled the ring.

- End Delimiter—Signals the end of the token or data/command frame. This field also contains bits to indicate a damaged frame and identify the frame that is the last in a logical sequence.

Data/command frame size depends on the size of the information field. In practice, data frames carry information for upper-layer protocols, while command frames contain control information and have no data for upper-layer protocols. Data/command frames have the same three fields as the Token Ring frame, along with several other fields:

- Frame-Control Bytes—Indicates whether the frame contains data or control information. In control frames, this byte specifies the type of control information.

- Destination and Source Addresses—Two six-byte address fields identify the destination and source station addresses.

- Data—Length of field is limited by the ring token holding time, which defines the maximum time a station can hold the token.

- Frame-Check Sequence (FCS)—Filed by the source station with a calculated value dependent on the frame contents. The destination station recalculates the value to determine whether the frame was damaged in transit. If so, the frame is discarded.

- Frame Status—A one-byte field terminating a command/data frame. The Frame Status field includes the address-recognized indicator and frame-copied indicator.

Token Ring networks employ several methods for detecting and compensating for network faults. One method involves the use of a specific station as an active monitor. The station acts as a centralized source of timing information for other ring stations and performs a variety of ring-maintenance functions such as the removal of continuously circulating frames from the ring. When a sending device fails, a frame may continue to circle the ring, prevent other stations from transmitting frames, and lock the network. The active monitor can detect continuously circulating frames, remove the frames from the ring, and generate a new token.

Reliability also increases through the use of the star topology. Since active MSAUs see all information in a Token Ring network, a network administrator can program the devices to check for problems and—if needed—selectively remove stations from the ring. In addition, a Token Ring algorithm called beaconing detects and attempts to repair certain network faults. Whenever a station detects a serious problem—such as a cable break—within the network, it sends a beacon frame that defines a failure domain. The failure domain includes information about the reporting station, the nearest station in the ring, and other information. In addition to detection, beaconing initiates a process called autoreconfiguration in which nodes within the failure domain automatically perform diagnostics in an attempt to reconfigure the network around the failed areas.

Thin Client Networks

A thin client is a networked desktop device with minimal local resources that optimizes information access and delivery, rather than information processing. Thin client technologies rely on network computers (or a computer with minimal memory), disk storage, and processor power designed to connect to a network. With this in mind, network computers rely on the power of network servers, rather than individual processing power.

Most thin client networks rely on thin client/server software that allows access to existing Windows applications as well as internal custom applications. The networks allow users to run Windows applications remotely through server functions. One of the key criteria for thin client operation is that the server maintains all protocols and applications code. Because of this, organizations can preserve the investment in applications—receiving long-term cost savings from lower maintenance costs and cutting initial equipment purchase costs. *Figure 4-30* shows a diagram of a thin client network that supports both personal computers and network computers. Thin client networks consist of the following components.

- The Windows NT Server base operating system with extensions to NT's memory manager and scheduler that provide NT with true multiuser capabilities.

Figure 4-30.
A diagram of a thin client network that supports both personal computers and network computers.

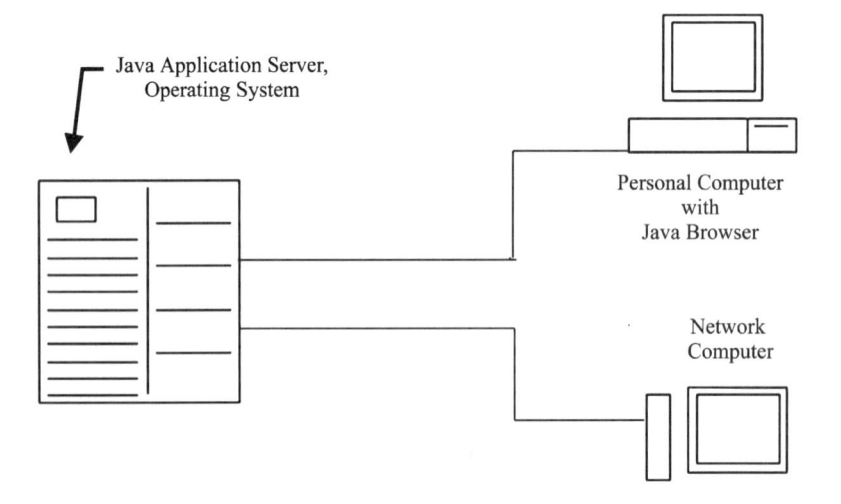

- A display protocol called the Independent Computing Architecture (ICA) that transmits Windows GUI events across a network to connected devices. The ICA allows Windows applications to run on a server, while sending all graphical output to remote software or hardware for display. Working in the opposite direction, the ICA carries remote keystrokes and mouse events back to the application.

- Lightweight client software that communicates with servers through the ICA.

- Miscellaneous server-side components for managing applications and user environments, including load-balancing facilities for distributing client sessions across multiple backend servers.

Virtual LANs

A Virtual LAN (VLAN) solves the scalability problems of large, flat networks by breaking a single broadcast domain into several smaller broadcast domains. As a result, the use of Virtual LANs eases the problems of adapting established network designs to new needs. In a Virtual LAN environment, LAN switches can use VLAN communication to segment networks into logically defined virtual workgroups.

Although logical segmentation provides substantial benefits in LAN administration, security, and management of network broadcast across the enterprise, network designers should consider many components of VLAN solutions prior to large-scale VLAN deployment. Switches control individual VLANs, while routers provide inter-VLAN communication.

Switches remove the physical constraints imposed by a shared hub architecture, because of the logical grouping of users and ports across the enterprise. As a replacement for shared hubs, switches remove the physical barriers imposed within each wiring closet. Routers provide policy-based control, broadcast management, route processing, and distribution. In addition, routers also provide VLAN access to shared resources, such as servers and hosts.

The use of Virtual LANs eases the problems of adapting established network designs to new needs. In a Virtual LAN environment, LAN switches can use VLAN communication to segment networks into logically defined virtual workgroups.

Summary

Chapter 4 provided an overview of popular local-area networking architectures and described applications of Ethernet, Fast Ethernet, and Gigabit Ethernet. As a result, the chapter provided the preliminary foundation for discussions on internetworking found in the following chapters. Just as importantly, Chapter 4 also introduced the reader to the hardware used to build local-area networks and internetworks. With this, the chapter covered network-interface cards, hubs, bridges, routers, and switches.

Chapter 4 provided the preliminary foundation for discussions on internetworking found in the following chapters.

Chapter 5

Wide-Area Networking, Internetworking, and Telephony

Introduction

A wide-area network (WAN) is a data communications network that covers a relatively broad geographic area. WANs can cover a region or can stretch across worldwide boundaries. We can define the connecting of individual local-area networks to create wide-area networks and the connecting of WANs to form even larger WANs as internetworking.

In addition, WANs often take advantage of the transmission facilities and media provided by common carriers such as telephone companies. As with LAN technologies, WAN technologies function at the Physical Layer, the Data Link Layer, and the Network Layer of the OSI model. *Figure 5-1* illustrates the relationship between common WAN technologies and the OSI model.

Internetworking

A network is subject to limits on how far it can extend, how many stations can be connected to it, how fast data can be transmitted between stations, and how much traffic the network can support. If an organization needs to go beyond those limits and link more stations than a LAN can support, it must install another LAN and connect the two together in an internetwork.

WANs often take advantage of the transmission facilities and media provided by common carriers such as telephone companies.

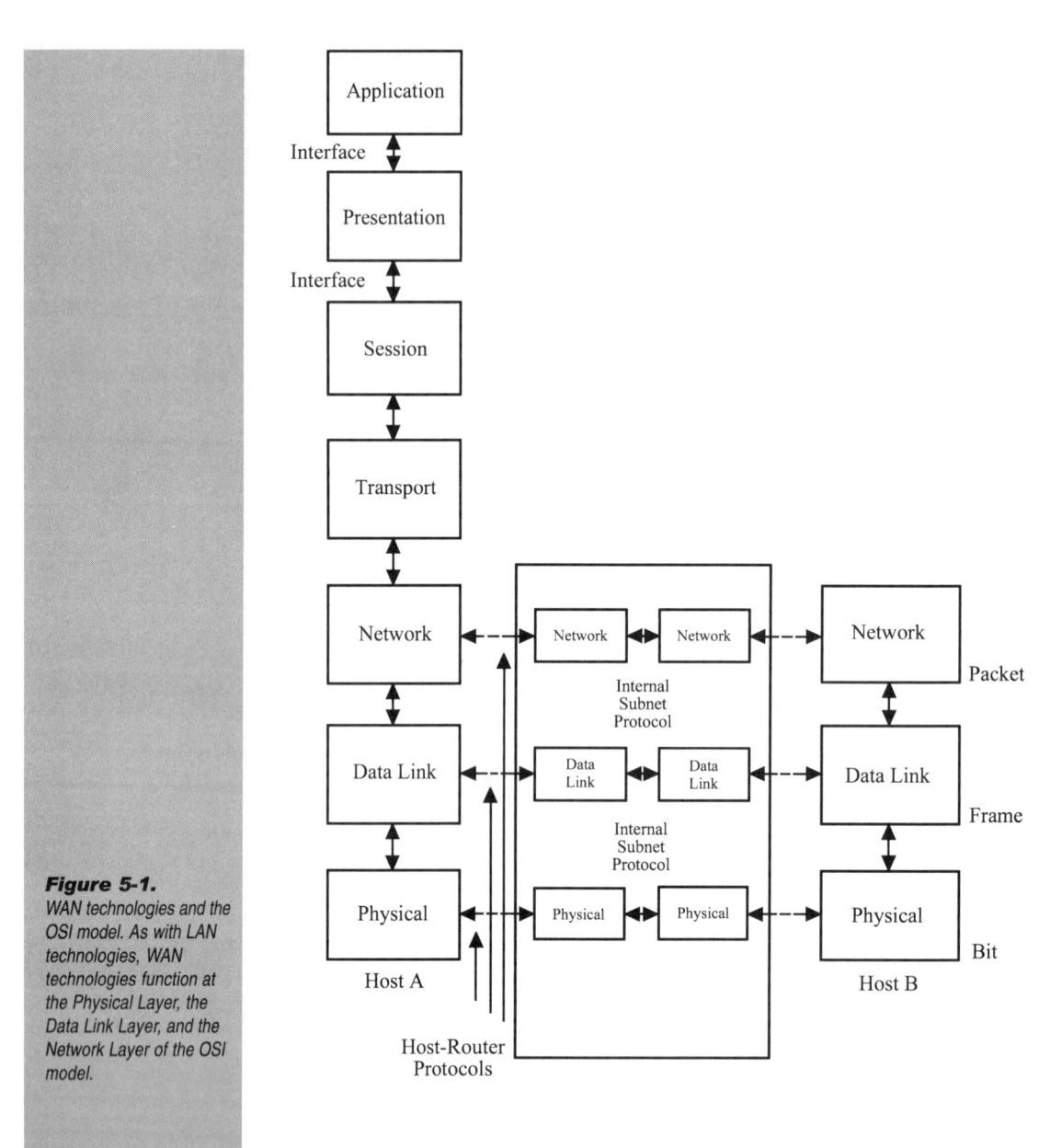

Telecommunication Technologies

Figure 5-1.
WAN technologies and the OSI model. As with LAN technologies, WAN technologies function at the Physical Layer, the Data Link Layer, and the Network Layer of the OSI model.

As *Figure 5-2* shows, the term "internetworking" refers to linking individual LANs together to form a single internetwork. Several reasons exist for implementing multiple networks and then connecting the networks as an internetworking. Workgroup networks on different floors of a building or in separate buildings on a business campus can link together as an internetwork so that all of the computing systems at that site are

Chapter 5: Wide-Area Networking

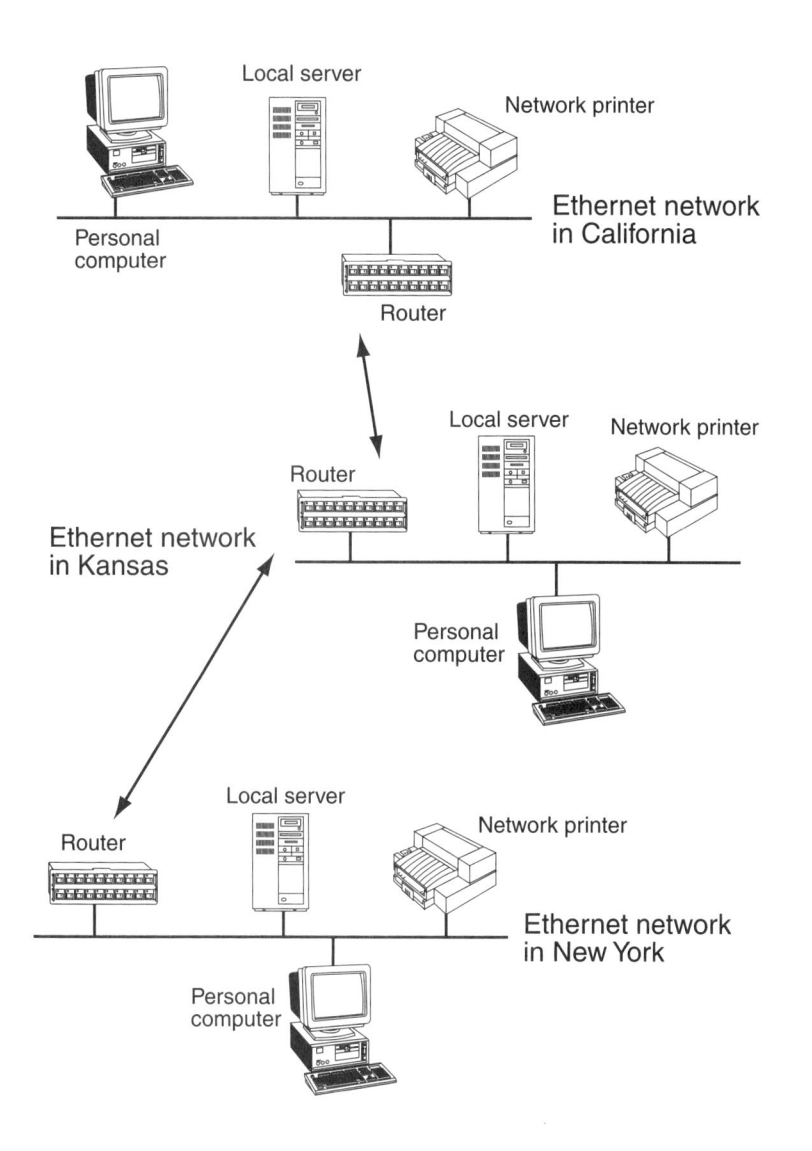

Figure 5-2.
A diagram of an internetwork, which is formed by linking individual LANs together.

interconnected. Geographically distant company sites can also be tied together in the enterprise-wide internetwork. We can also refer to an internetwork as an enterprise network, because it interconnects all of the computer networks throughout the entire enterprise.

As a result, internetworking solves the problem of isolated LANs, ends the duplication of computing and data communications resources at remote sites, and provides a basis for network management. Internetworking also allows the sharing of traffic loads between more than one network. If the load increases beyond the carrying capacity of a network, users will suffer reduced throughput and much of the productivity achieved by installing the network in the first place will be lost. Dividing the traffic between multiple internetworked networks increases network efficiency.

Internetwork Addressing

Internetwork addressing works within either the hierarchical address space or flat address space formats. A hierarchical address space organizes into numerous subgroups, with each subgroup successively narrowing an address until it points to a single device. A flat address space organizes into a single group. Given the differences, hierarchical addressing offers certain advantages over flat-addressing schemes such as address sorting and recall of comparison operations.

Address Assignments

Internetworking devices use either static or dynamic server addressing. Since network administrators assign static addresses according to an internetwork addressing plan, a static address changes only through the action of the network administrator. On the other hand, dynamic addressing occurs when devices attach to a network through a protocol-specific process. With the server assigning the address with each connection and discarding the address with each disconnection, a device using a dynamic address often has a different address each time it connects to the network.

Internetwork Addressing and the OSI Model

Internetwork devices have both a name and an address that represent a logical identifier for internetworking devices. Generally, either a local system administrator or an organization—such as the Internet Assigned Numbers Authority (IANA)—assigns internetwork names and addresses. Internetwork device names have location-independent identities and remain associated with a device despite any change in location.

In contrast, internetwork addresses usually have location-dependent identities and—with the exception of MAC addresses—change with the location of the device. Internetwork addresses also identify devices separately or as members of a group. Addressing schemes vary depending on the protocol family and the OSI layer. During operation, internetworking may rely on data link layer addresses, MAC addresses, and Network-Layer addresses.

Internetworking and Data Link-Layer Addresses.
Since end systems have only one physical network connection, the devices have only one data-link address. But routers and other internetworking devices have multiple physical network connections and, as a result, also have multiple data-link addresses. *Figure 5-3* illustrates the unique identification of each interface on a device by a data-link address.

Internetworking and MAC Addresses.
Media Access Control addresses consist of a subset of Data Link-Layer addresses. During operation, MAC addresses identify network entities in LANs that implement the IEEE MAC addresses of the Data Link Layer. As with most data-link

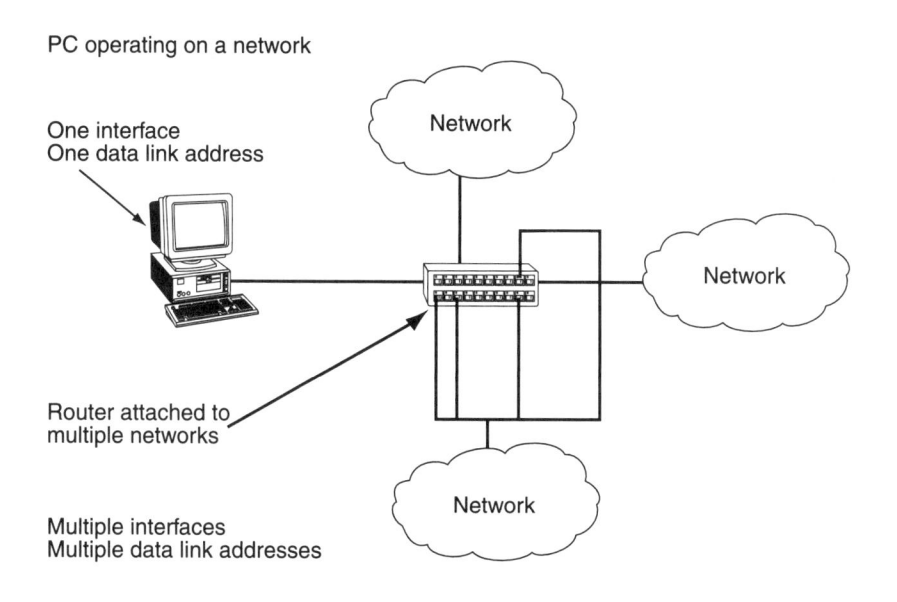

Figure 5-3.
The data-link address identifies the interface on a device.

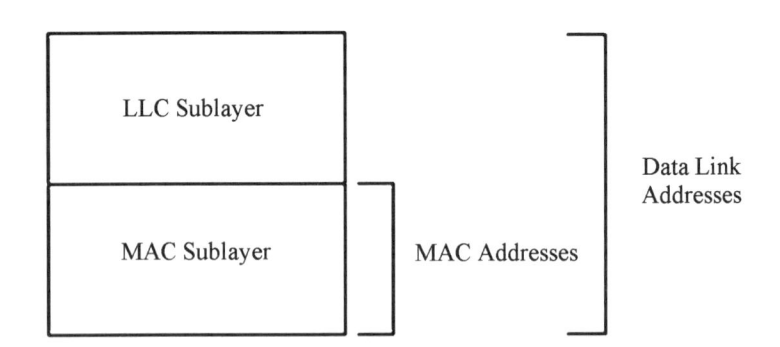

Figure 5-4.
The relationship between MAC address and internetworked devices.

addresses, MAC addresses uniquely identify each LAN interface. *Figure 5-4* illustrates the relationship between MAC addresses and internetworking devices. Different protocol suites may use either the Address Resolution Protocol (ARP), the Hello Protocol, or embedded Network Layer MAC addresses for determining the MAC address of a device.

The Address Resolution Protocol maps network addresses to MAC addresses. This process occurs through the implementation of the Address Resolution Protocol by many protocol suites. When a network address becomes associated with a MAC address, the network device stores the address information in the ARP cache. In turn, the ARP cache enables devices to send traffic to a destination without creating ARP traffic.

Depending on the network environment, the process of address resolution may feature slight differences. For example, address resolution on a single LAN begins when End System A broadcasts an ARP request onto the LAN in an attempt to learn the MAC address of End System B. Although only End System B replies to the ARP request by sending an ARP reply containing its MAC address to End System A, all devices connected to the LAN receive and process the message. End System A receives the reply and saves the MAC network addresses. Whenever End System A must communicate with End System B, the device checks the ARP cache, finds the MAC address of System B, and sends the frame directly without first having to use an ARP request.

All this changes with the interconnection of the source and destination devices found on different LANs through a router. End System Y broadcasts an ARP request onto the LAN in an attempt to learn the MAC address of End System Z. Again, all devices on the LAN—including the router—receive and process the broadcast. The router acts as a proxy for End System Z by checking its routing table and determining that End System Z operates on a different LAN.

In response, the router replies to the ARP request from End System Y and sends an ARP reply containing the router address as if the address belongs to End System Z. From there, End System Y receives the ARP reply and saves the MAC address of the router in its ARP cache within the entry for End System Z. When End System Y must communicate with End System Z, the device checks the ARP cache, finds the MAC address of router, and sends the frame directly without using ARP requests. The router receives the traffic from End System Y and forwards it to End System Z on the other LAN.

Another Network-Layer protocol called the Hello Protocol enables network devices to identify one another and indicate functional status. During operation with the Hello Protocol, a new end system powers up and broadcasts Hello messages onto the network. Devices on the network then return Hello replies. In addition, the devices send Hello messages at specific intervals to indicate the functional status.

Network devices can learn the MAC addresses of other devices by examining Hello-Protocol packets. The Xerox Network Systems (XNS), Novell Internetwork Packet Exchange (IPX), and DECnet Phase IV protocol suites use predictable MAC addresses. The MAC addresses are predictable, because the Network Layer either embeds the MAC address in the Network-Layer address or uses an algorithm to determine the MAC address.

Internetworking and Network-Layer Addresses.

Network addressing identifies a device at the Network Layer and usually exists within a hierarchical address space. We can refer to network addresses as virtual or logical addresses. Since network addresses remain based on either physical network characteristics or on groupings—such as an AppleTalk Zone—that have no physical basis, the relationship between a network address and a device is logical and unfixed. End systems require one Network-Layer address for each supported Network-Layer protocol if the device has only one physical network connection.

Routers and other internetworking devices require one Network-Layer address per physical network connection for each supported Network-Layer protocol support. For example, a router that has three individual interfaces running AppleTalk, TCP/IP, and OSI must have three Network-Layer addresses for each interface. Consequently, the router has nine Network-Layer addresses. *Figure 5-5* shows the assignment of network addresses to an interface that supports multiple protocols.

Internetworking Devices

Internetworking also refers to the industry, products, and procedures that meet the challenge of creating and administering internetworks. *Figure 5-6* depicts different kinds of network technologies that can interconnect through routers and other networking devices to create an internetwork. Switched internetworks have become widely used due to their capability to operate at very high speeds and support such high-bandwidth applications as voice and videoconferencing.

Critical business applications require that users on one network have the ability to reach resources on other networks within the organization. Yet networks must continue to support the filtering of packets so that only those packets that need to pass from one network to another travel across the link. Filtering is based on the destination address of the packet. If a packet's destination is a station on the same segment where it originated, the network doesn't forward the packet onto another network. With filtering, packets sent between two stations on any one network won't cross over onto and congest other networks.

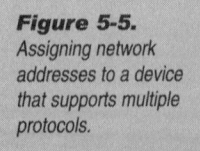

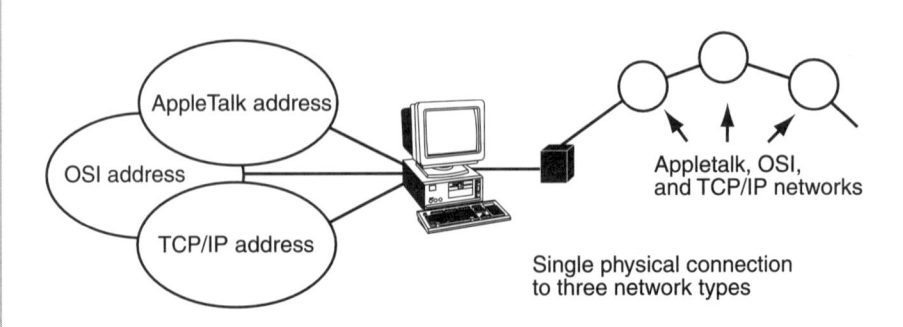

Figure 5-5.
Assigning network addresses to a device that supports multiple protocols.

Chapter 5: Wide-Area Networking

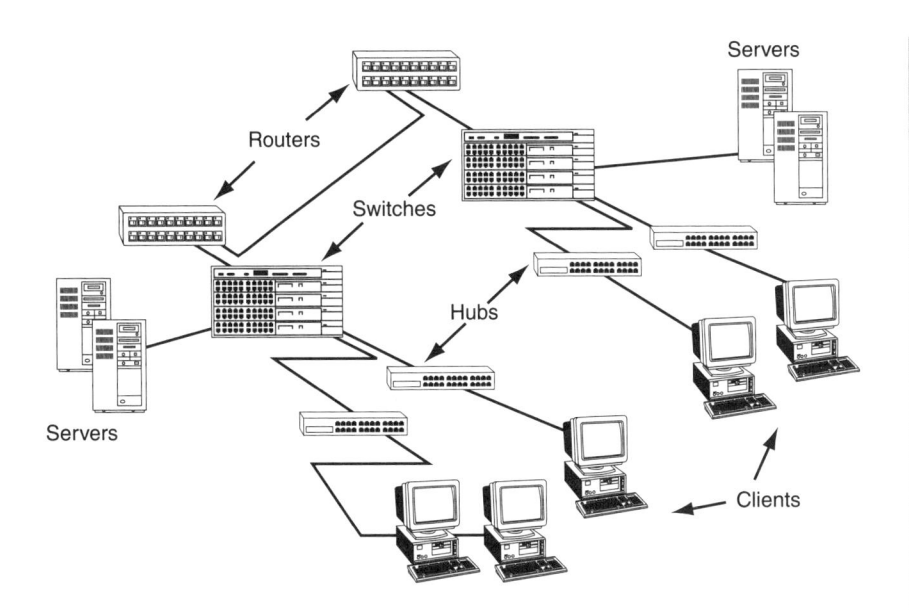

Figure 5-6.
Using routers, switches, and hubs to form an internetwork.

Wide-area networking environments rely on different devices because of specific needs. Many organizations have LANs located at sites that are geographically distant from each other. Both Ethernet and Token Ring specify limits on maximum distances between workstations and hubs, hubs and hubs, and a maximum number of stations that can be connected to a single LAN.

To provide network connectivity for more people, or extend it to cover a larger area, we may need to link two different LANs or LAN segments through routers, switches, and bridges and use access servers, modems, CSU/DSUs, and ISDN terminal adapters. When traffic congestion occurs, users on a single network begin to experience slower response times. Segmenting the network breaks a large network with many users into smaller networks that have fewer users.

WAN Switches

A WAN switch provides multiport internetworking and has applications in carrier networks. In addition, switches offer the higher bandwidth required for video and multimedia communications. Operating in the Data Link Layer

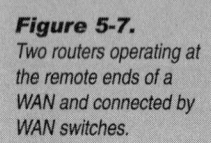

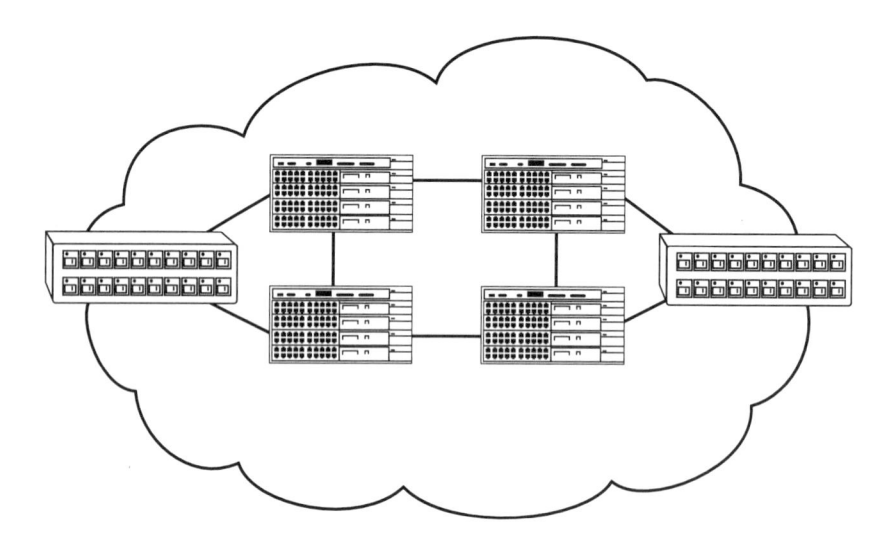

Figure 5-7.
Two routers operating at the remote ends of a WAN and connected by WAN switches.

of the OSI model, WAN switches switch such traffic as ATM, Frame Relay, X.25, and SMDS. *Figure 5-7* shows two routers operating at remote ends of a WAN and connected by WAN switches.

Routers

Today, high-speed routers have become the most commonly used internetworking devices. During operation, routers communicate with each other and share information that allows them to determine the best route through a complex internetwork that links many LANs. All routers use Network Layer Protocol Information found within each packet to route the packet from one network to another.

Due to such reliance on protocol information, a router must have the capability to recognize all of the different Network Layer Protocols used on the networks that it links together. Multiprotocol routers can route data using many different protocols. *Figure 5-8* shows the interconnection of national networks through the use of routers.

Router Benefits. Since routers use structured Layer 3 addresses, the devices can use techniques—such as address summarization—to build networks that maintain performance and responsiveness as growth occurs.

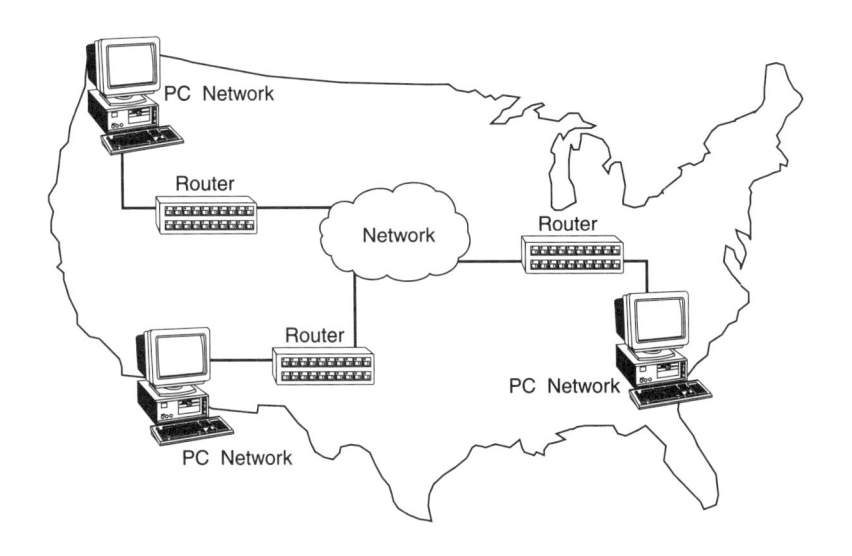

Figure 5-8.
Using routers to interconnect national networks.

In addition, routers offer significant advantages over bridges and switches. Since every data packet must go through the root bridge of a spanning tree, bridging and switching may result in nonoptimal routing of packets. Routers control the routing of packets and utilize optimal paths.

The use of structured addressing allows routers to effectively use redundant paths and determine optimal routes even in a dynamically changing network. While routers ensure network scalability, the devices also provide:

- Broadcast and multicast control
- Broadcast segmentation
- Security
- Quality of service
- Multimedia

Single-Protocol Backbone

With a single-protocol backbone, all routers support a single routing protocol for a single network protocol. As a result, the network ignores all other routing protocols. If multiple protocols pass over the internetwork, then the supported protocol must also encapsulate routing nodes for unsupported protocols. But encapsulation adds traffic on the network. The single-protocol backbone works most effectively if the routing environment supports relatively few other protocols at a limited number of isolated locations.

Multiprotocol Routing Backbone

Router manufacturers recommend support for all the Network Layer protocols in an internetwork with a multiprotocol routing. When multiple Network Layer protocols rout throughout a common backbone and don't utilize encapsulation, a multiprotocol routing backbone exists. We can also refer to multiprotocol routing backbones as native-mode routing. Depending on the routing protocol, a multiprotocol backbone environment can adopt either an integrated routing or a ships-in-the-night strategy.

Integrated routing involves the use of a single routing protocol that determines the least-cost path for different routed protocols. As opposed to integrated routing, another strategy involves the use of a different routing protocol for each network protocol. Each network layer protocol routes independently, has separate routing processes handling traffic, and calculates separate paths.

B/Routers

A B/router combines the functions of a bridge and a router within one device. With this, a B/router bridges frames between segments. In addition, a B/router also routes packets from either segment to another network. Given these functions, a B/router offers great flexibility for internetworking.

CSUs, DSUs, and DTEs

Digital data service (DDS) connections always terminate at a channel service unit (CSU) in the subscriber's building. Pictured in *Figure 5-9*, a CSU actually generates the transmission signals and connects to data terminal equipment (DTE) at the end of the network. In addition to having CSU ports, DTEs also have one or more RS-232 ports that allow the connection to modems. While

Chapter 5: Wide-Area Networking

Figure 5-9.
A photograph and connection diagram of a CSU/DSU. The CSU generates the transmission signals and connects to data terminal equipment (DTE) at the end of the network.

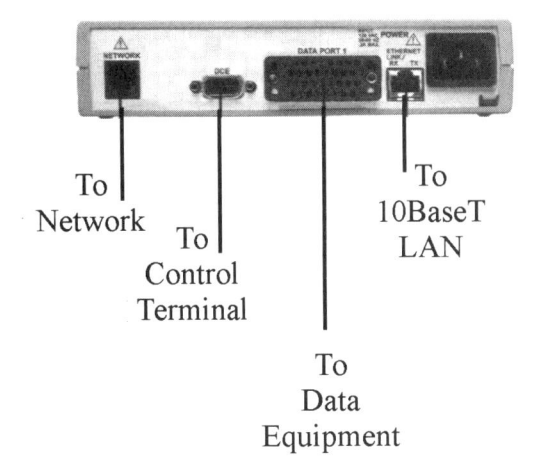

DTEs work as end devices for networks, data circuit-terminating equipment (DCEs) transmits and receives the DTE data.

CSUs also may connect to data service unit (DSU) multiplexers, or to a private automatic branch exchange (PABX). When digital services first became available for subscribers, the telephone companies wouldn't allow subscriber equipment to attach directly to a local loop. As a result, the telephone company provided the CSU, while the subscriber provided the DSU. A PABX provides additional voice services such as call waiting and

141

voice messaging. Since a PABX integrates both voice and data, it can multiplex voice and data onto a T1 carrier and support voice and data communications over a single pair of wires.

Multiplexers

Multiplexers allow multiple signals to travel over the same transmission media. The process of multiplexing combines multiple data channels into a single data or physical channel at the source and works at any of the OSI layers. As a result, multiple traffic sources can share the bandwidth of the physical medium. Demultiplexing separates multiplexed data channels at the destination. *Figure 5-10* illustrates the multiplexing of data from multiple applications into a single lower-layer data packet, while *Figure 5-11* shows a photograph of two multiplexers.

Figure 5-10.
Multiplexing data from multiple applications into a single lower-layer data packet.

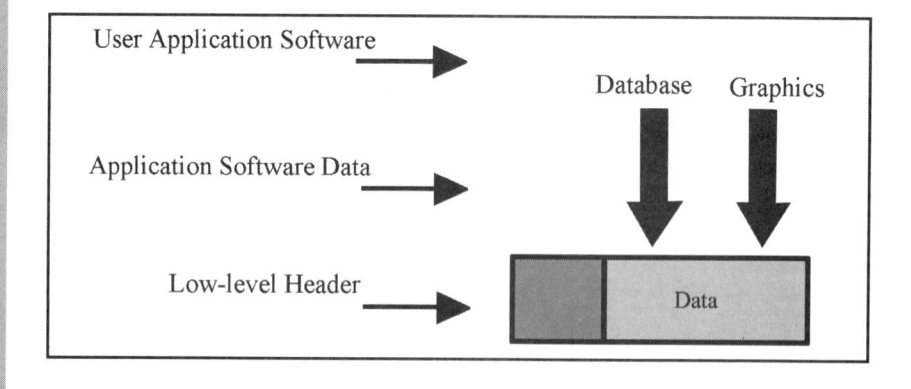

Figure 5-11.
A photograph of two multiplexers.

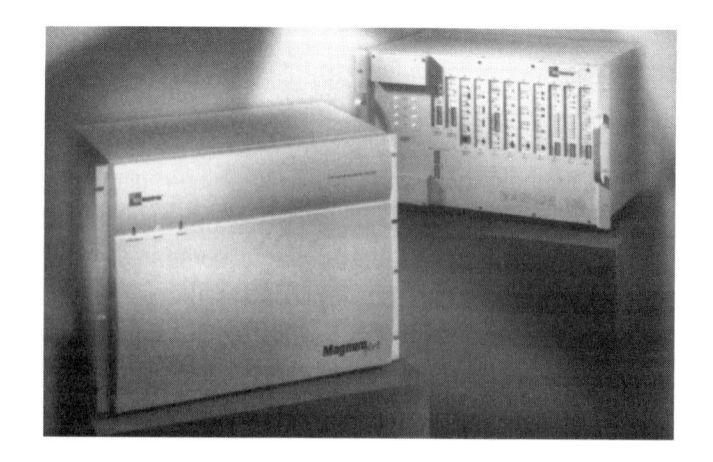

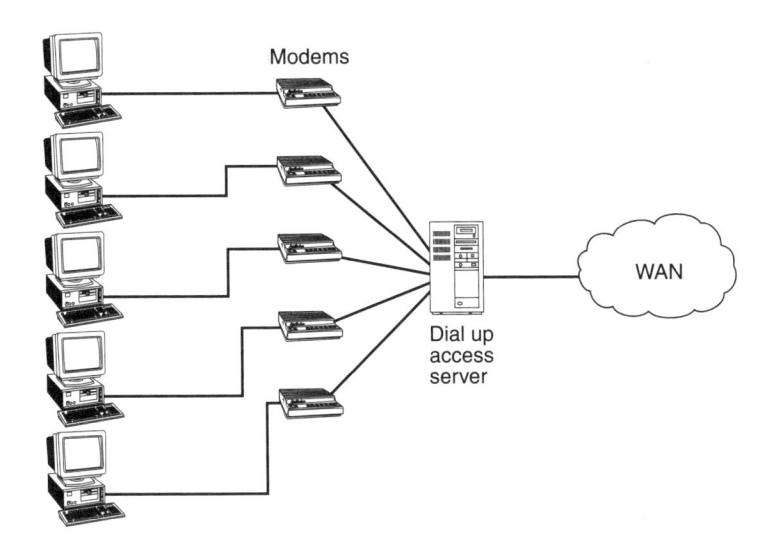

Figure 5-12.
An access server for WAN dial-out connections.

Multiplexing methods include time-division multiplexing (TDM), asynchronous time-division multiplexing (ATDM), frequency-division multiplexing (FDM), and statistical multiplexing. With TDM, the multiplexer allocates bandwidth information from each data channel based on preassigned time slots and has no relationship to the transmission of data.

Access Servers

An access server acts as a concentration point for dial-in and dial-out connections. *Figure 5-12* shows an access server concentrating dial-out connections into a WAN.

Gateways

Bridges and switches cannot interconnect nodes that use different architectures. A gateway allows communication between different architectures and application level protocols. As a result, many different types of gateways exist. For example, one type of gateway called a protocol converter changes the protocol of one architecture to the protocol of another. To accomplish this task, the protocol converter operates above the Data Link Layer. In addition, the operation of the protocol converter must remain transparent to any processes found in the layers located at each end of the connection.

Internetwork Transmission Services

Public transmission facilities rely on two types of services called point-to-point and switched. When considering the two services, point-to-point works in basic terms as a physical connection between the end points of a network. Switched services provide more flexibility than point-to-point services but also have a greater degree of complexity.

> *Public transmission facilities rely on two types of services called point-to-point and switched.*

Point-to-Point Services

With point-to-point services, the physical connections between the end points consist of dedicated dial-up lines, leased lines, digital data services, T1 and fractional T1 lines, and synchronous optical networks. A dial-up line uses a switched telephone network and acts as a circuit when connected between two nodes. In addition, dial-up lines rely on compatible modems at each end point and allow one connection at a time. Given the data-transfer rates of 2.4 Kbps to 56 Kbps, dial-up lines serve as inexpensive solutions for applications—such as the exchange of a short e-mail message—that don't require high-speed data transfers.

WAN Dial-Up Services

Dial-up services exist as cost-effective connectivity across wide-area networks. Popular dial-up services include dial backup and dial-on-demand routing (DDR). Dial backup activates a backup serial line under certain conditions. For example, the secondary serial line can act as a backup link when the primary link fails or as a source of additional bandwidth when the load on the primary link reaches a certain threshold. Dial backup provides protection against WAN performance degradation and downtime.

Dial-on-demand routing relies on a router and dynamically initiates and closes a circuit-switched session as demanded by the transmitting end station. The router has the programmed intelligence to accept some traffic according to protocol while rejecting other traffic. When the router receives acceptable traffic destined for a remote network, dial-on-demand routing establishes a new circuit and transmits the traffic.

If the router receives traffic from a different protocol and has an already-established circuit, the device also transmits that data. The router maintains an idle timer that resets only when receiving acceptable traffic. If the router doesn't receive acceptable traffic before the idle timer expires, the circuit terminates. In addition, the router may receive unacceptable traffic without having an established circuit. In this case, the router drops the traffic. DDR can replace point-to-point links and switched multi-access WAN services.

T1 and T3 Dedicated Circuits

T1 and T3 circuits are dedicated services that connect networks or LANs over extended distances. As an example of T1 and T3 applications, Internet service providers, hospitals, and corporations use T3 lines as a main network line and as a connection to the Internet. A T1 line consists of 24 individual channels with each line supporting a data transmission rate of 64 Kbps for a total bandwidth of 1.544 Mbps.

Each 64-Kbps channel can be configured to carry voice or data traffic. The term "fractional T1" represents the ability of a customer to lease any portion of the 24 channels on a 24-hour basis. In comparison to T1 lines, a T3 line consists of 672 individual channels and supports an even higher transmission bandwidth. As with the T1 circuit, each line of the T3 circuit also supports a data-transfer rate of 64 Kbps.

Switched Services

The emphasis in WAN technologies has shifted away from dedicated networks to switched services. Much of this change has occurred because of the high quality and high reliability offered by switched network services. In addition, switched services have decreased in complexity as usage increased. Switched services consist of circuit-switched networks and packet-switched networks. Dial-up or dedicated connections remain necessary for connecting to the switched network and provide one connection at time.

Packet-Switched Networks

With packet switching, a data packet travels from a source node of a network and through intermediate nodes before arriving at the

The emphasis in WAN technologies has shifted away from dedicated networks to switched services.

destination node. Packet switching provides a method for network devices to share a single point-to-point link to transport packets from a source to a destination across a carrier network. Ethernet networks, Asynchronous Transfer Mode networks, Frame Relay, Switched Multimegabit Data Service, and X.25 serve as examples of packet-switched WAN technologies When the packet travels through the intermediate node—or packet switch—the node switches the packet to the next node in the sequence. We can refer to packet-switched networks as connectionless networks, because no physical connection occurs between the source and destination nodes.

Packet-switched networks that use the X.25 protocol exist on the leased switching. An X.25 network includes:

- The physical hardware and cabling used to make a physical circuit that connects devices together a complete path between two communicating devices used to make a virtual circuit

- A logical connection between the user node and the network

Figure 5-13 illustrates the use of packet switching for network connections between LANs, WANs, public data networks, and leased lines. In the first portion of the figure, access occurs through dial-up modems and a combination of bridges and routers. Physical connections to the public data network and leased line clouds occur through the same bridge/router combination and a combination of DSUs and CSUs.

Figure 5-13.
A diagram illustrating packet switching for network connections between LANs, WANs, public data networks, and leased lines.

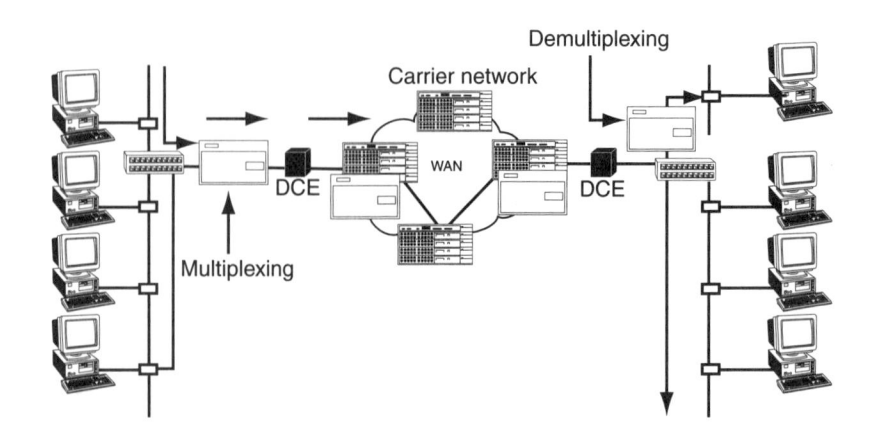

Leased Lines

To serve high-speed applications, telephone companies also establish permanent leased lines that rely on point-to-point T1 connections. Leased lines carry both voice and data traffic and provide consistent, high-quality service. Moreover, leased lines offer security and expanded network control. But increased costs and equipment needs accompany the benefits of high-speed data transfer rates. *Table 5-1* summarizes available options seen with point-to-point services, while *Figure 5-14* represents a wide-area network that uses T1 lines.

Carriers reserve leased lines for the private use of customers as a permanent and fixed path. The links involved with leased line connections accommodate datagram transmissions, or transmissions composed of individually addressed

Service	Link Speed	Equipment Needs
Switched Analog	300 bps - 28.8 bps	Modems
Leased Analog	300 bps - 28.8 bps	Modems
DDS	2.4Kbps - 56 Kbps	DSU and CSU
Fractional T1	64Mbps - 1.544 Mbps	DSU and CSU
T1	1.544 Mbps	Multiplexer
T3	44.736 Mbps	Multiplexer

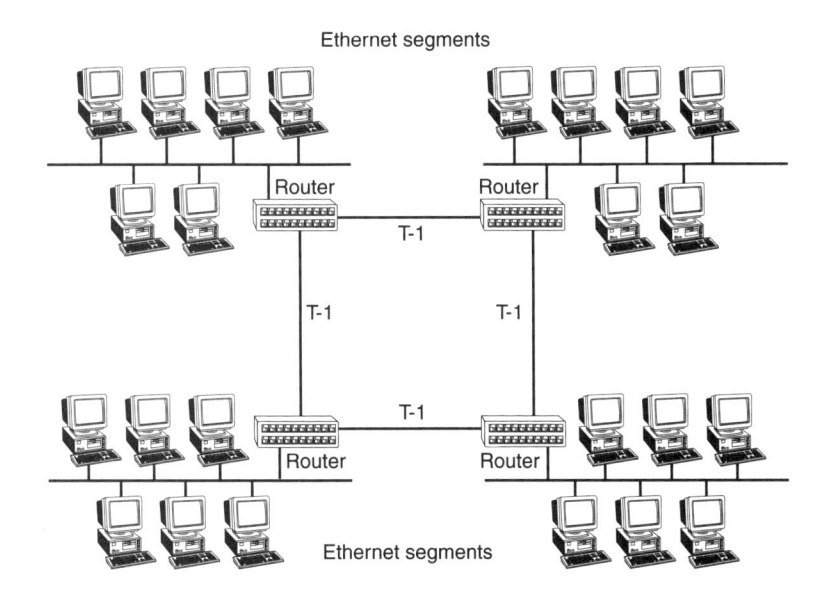

Table 5-1.
Options for point-to-point services.

Figure 5-14.
A wide-area network that uses T1 lines.

Figure 5-15.
A leased line connection. The links involved with leased line connections accommodate datagram and data-stream transmissions.

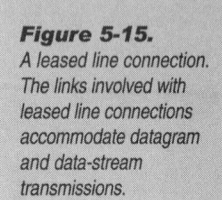

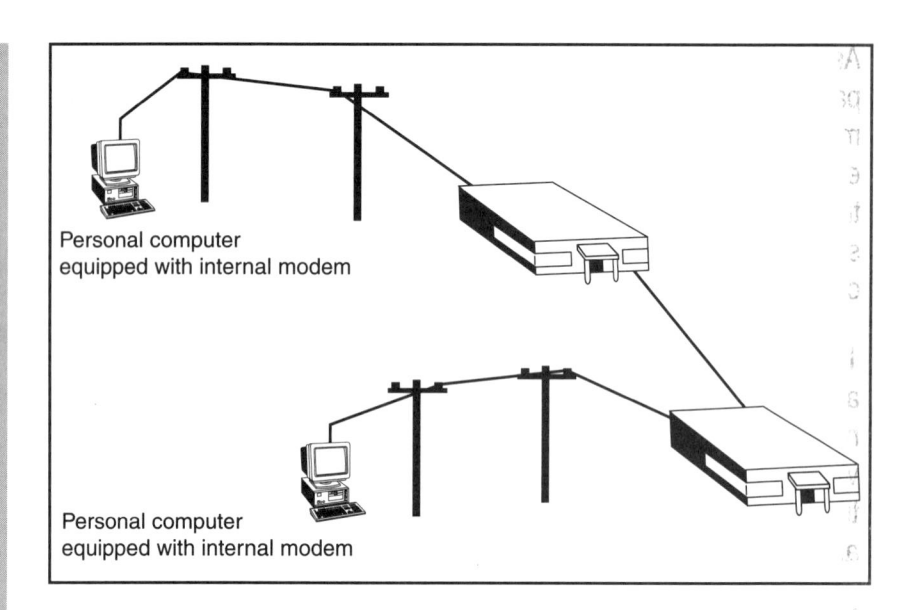

frames, and data-stream transmissions, or transmissions composed of a stream of data for which address checking occurs only once. *Figure 5-15* illustrates a typical leased line connection.

Point-to-point services also make use of the benefits provided by digital data service links. These benefits include more bandwidth and higher reliability. Since a DDS eliminates the conversion of digital data to an analog format and then back to a digital format, the service offers less noise and consistent line quality. A DDS provides permanent connections that have data transfer rates of 2,400 bps, 4,800 bps, 9,600 bps, 19.2 Kbps, or 56 Kbps.

Circuit-Switched Networks

A circuit-switched network establishes a physical connection between source and destination nodes. Circuit switching establishes a dedicated physical circuit maintained, and terminated through a carrier network for each communication session. In addition, circuit switching accommodates datagram transmissions and data-stream transmissions. Used extensively in telephone company networks, circuit switching operates much like a normal telephone call. Integrated Services Digital Network (ISDN) serves an example of a circuit-switched WAN technology.

As with the packet-switched network, intermediate nodes switch the data packet through other nodes or a host computer. With circuit-switched communication, a signaling protocol—or call—establishes a dedicated end-to-end route and transmission capacity that don't change during the connection. Once the completion of the call setup occurs, a call holds the reserved transmission capacity for the entire duration of the call and the capacity is unavailable to other network users.

The circuit established for a call typically passes through several switches and transmission systems. As a consequence, an individual call simultaneously consumes capacity on a sequence of different transmission devices. Circuit switching makes sense for network traffic where the connection time between network nodes is long and congestion-induced delays are unacceptable.

Most voice traffic involves circuit switching. Voice call setup times are typically on the order of a few seconds compared to call holding times on the order of a few hundred seconds. In addition, voice signals must transmit without the introduction of intermittent pauses caused by delay variability. Public-Switched Telephone Networks (PSTNs) throughout the world—as well as most private voice networks in commercial, government, and military organizations—use circuit switching. Moreover, many mobile wireless networks and tasks that require bulk data transfers also use circuit-switched connections.

In recent years, many organizations have started to integrate voice and data traffic on the same network resources in an attempt to reduce network operation and management costs. Such integration occurs at both the transmission and switching levels and in a static or dynamic state. For example, some networks dynamically change the bandwidth of networks as traffic increases.

WAN Virtual Circuits

Wide-area networks also use virtual circuits, or a logical circuit created to ensure reliable communication between two network devices. Switched virtual circuits (SVC) establish dynamically on demand and terminate with the completion of the transmission. Communication over an SVC involves:

- Circuit establishment
- Data transfer
- Circuit termination

The establishment phase creates a virtual circuit between the source and destination devices. Data transfer covers the transmission of data between the devices over the virtual circuit, while the circuit-termination tears down the virtual circuit between the source and destination devices. SVCs operate best when the data transmission between devices has become sporadic, because of the capability to increase bandwidth used during the circuit establishment and termination phases. In addition, the use of an SVC decreases the cost associated with constant virtual circuit availability.

A permanently established virtual circuit (PVC) uses only data transfer. Compared to SVCs, a PVC operates best when the data transfer between devices remains constant. As with switched virtual circuits, PVCs decrease the bandwidth associated with the establishment and termination of virtual circuits. Use of a PVC increases costs, however, due to the constant virtual circuit availability.

Examples of Switched Service Technologies

Common types of switched services include Frame Relay and asynchronous transfer mode (ATM) networks.

Common types of switched services include Frame Relay and asynchronous transfer mode (ATM) networks. With Frame Relay, software runs in an electronic switch that consists of a circuit connected to three or more high-speed links. The Frame Relay switching establishes virtual circuits from one multiplexer to another. Virtual circuit one follows a path from multiplexer A to multiplexer C, while virtual circuit two follows a path from multiplexer A to multiplexer D. Virtual circuit three follows a path from multiplexer A to multiplexer E. Data traveling on all three circuits passes through Frame Relay B.

As an example of Frame Relay operation, if all the data from the three circuits travels into multiplexer A, the multiplexer places the data packets into a frame that stores an address and length with the data. Then the network transfers the data from multiplexer A to multiplexer B. Before sending the data onto the other multiplexers, multiplexer B must remove the information provided by multiplexer A and create new frames for the data.

Telephone Services

A telephone system provides private two-way voice and data communication between any set of locations regardless of distance. Switching circuitry allows a caller to dial and connect with the desired party. The interconnections between telephones occur at a central office and through trunk lines.

Telephone Systems

A telephone system begins with the local loop, or a pair of cables that connect the central office to the user's telephone. Typically, a central office exists for each exchange or area served by the same first three digits of the seven-digit number. A central office supervises the local loop and telephone connected to the loop. In addition, the central office acts as an interface between the individual telephone and the remainder of the system and establishes the physical connection between the telephones.

Local Loop Characteristics

A local loop may extend from several hundred feet to several miles with the distance depending on the locations of the telephone and central office. With the variation in distance, signal levels may vary from one loop to another. Local loops have frequency and time-delay characteristics. Frequency characteristics include a passband that has an approximate 10 dB variation from 300 to 2500 Hz. Signal attenuation outside the passband reaches almost 30 dB at 100 Hz and 3500 Hz.

Time-delay characteristics, or envelope delay, introduced by a phone line cause distortion of the overall waveform for different spectrum components. While time delay may not affect voice transmission to a great extent, it degrades the transmission of digital information over a telephone line. *Figure 5-16* compares the frequency and time-delay characteristics for an unconditioned line and a conditioned line.

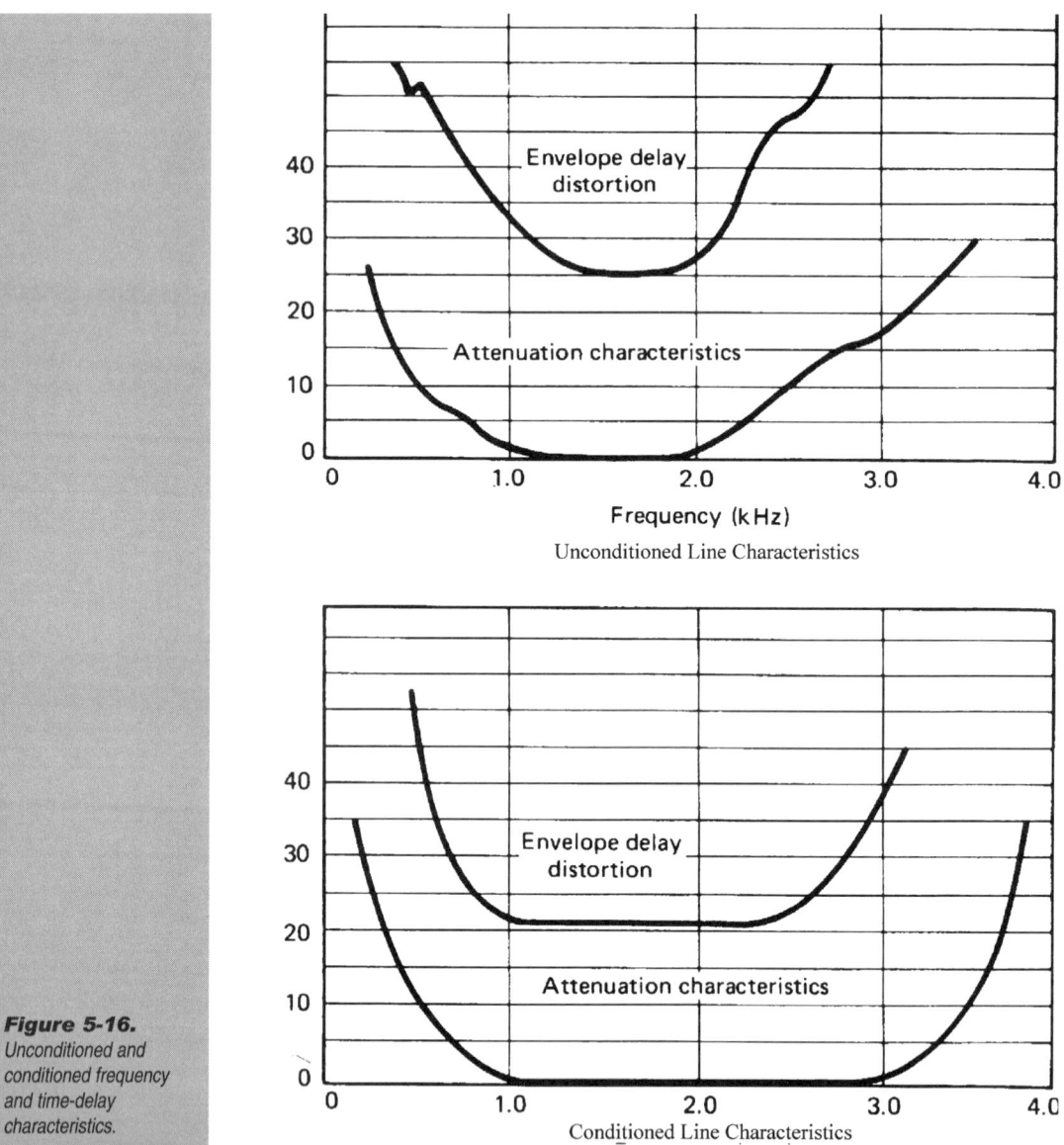

Figure 5-16.
Unconditioned and conditioned frequency and time-delay characteristics.

Unconditioned Line Characteristics

Conditioned Line Characteristics

Trunks

Central offices connect to other exchanges through a special four-line link called a trunk that may consist of copper cable, fiber-optic cable, a radio link, or a satellite connection. While calls between adjacent cities may

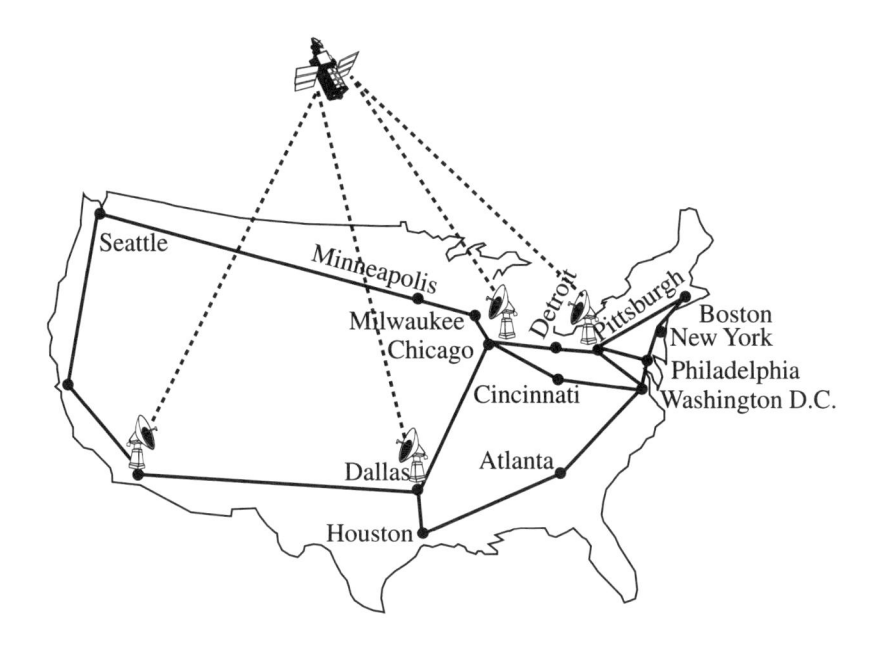

Figure 5-17.
Trunks and central offices form a network of interconnected nodes.

involve only a single trunk between two exchanges, long-distance calls may rout through several trunks and central offices. As *Figure 5-17* shows, the combination of central offices and trunks form a network of interconnected nodes.

Supertrunks collect many trunk lines at regional switching centers and multiplex the lines together. A tandem switch, or connecting node for a supertrunk, connects only to other central offices but not to any local loops. Only long-distance calls travel through a tandem switch.

Trunk Characteristics. As with the local loop, trunk characteristics cover frequency response, envelope delay, and noise. A trunk controls the characteristics tightly, however, due to more efficient design and performance criteria. Compared to the local loop that traverses different types of equipment, a trunk connects only between central offices and serves as the major link in the telephone system.

Due to the performance requirements seen with a trunk and the impact of any signal degradation on the total system, trunks rely on digital regenerative amplifiers spaced at intervals to boost signal levels, reduce noise, and correct frequency response. The signals between central offices multiplex together into a single wider-bandwidth signal that transmits over a single physical link or carrier. At the receiving central office, equipment demultiplexes the signal into its original separate components and switches the signals to the called telephone loop or another trunk.

Telephone System Hierarchy

The small-, medium-, and large-sized switching centers that combine with trunks and local loops to make up a telephone system form a carefully planned hierarchy. Shown in *Figure 5-18*, a center at one level of the hierarchy can connect to another center at a different level. Call routing doesn't follow any rule for going up to the top of one hierarchy and then down to the final destination on another. The variety of network connections and options gives the telephone system the needed flexibility to route calls through the shortest route or to route the call through an alternative link when an overload condition exists.

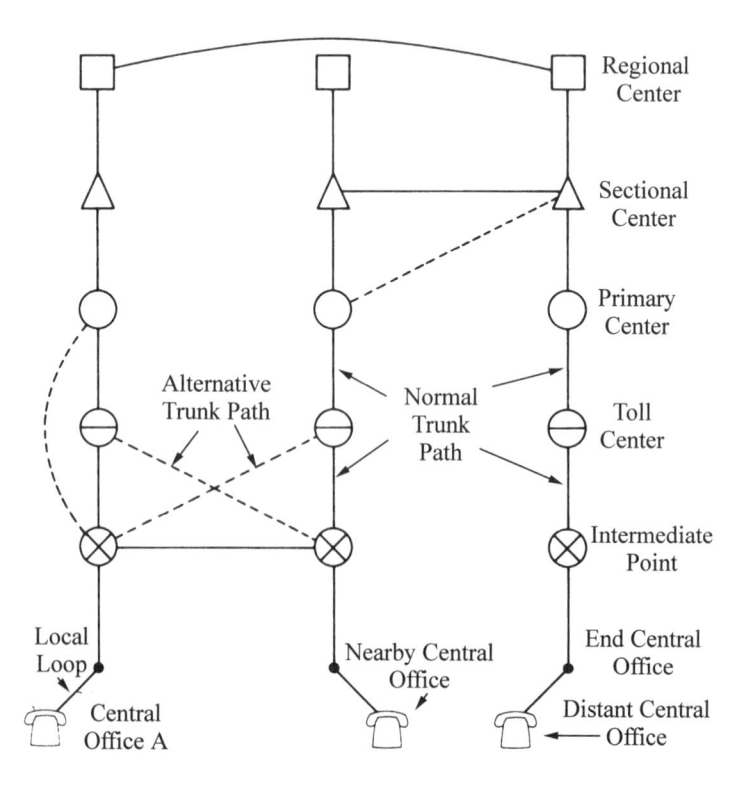

Figure 5-18.
A hierarchy of telephony switching centers. A center at one level of the hierarchy can connect to another center at a different level.

POTS

Plain Old Telephone Service (POTS) represents the basic form of telephone service provided between the central office, the local loop, and a telephone. When on-hook, the telephone appears as an open circuit on the loop and doesn't draw current. When off-hook, the telephone draws 20 mA of current and completes the local loop. The central office detects the current flow. In addition, the central office applies an ac ringing voltage that activates the ringer located within the telephone. *Figure 5-19* depicts the signals found on the local loop during different phases of use.

Signals traveling along the local loop must have enough level for use but remain low enough to prevent overloading of the telephone or amplifiers found in the system. For example, the outgoing voice signal has a power level of approximately –10 and 0dBm into a 600-ohm load. Since an incoming voice signal travels through trunks, links, and switches, the power levels for

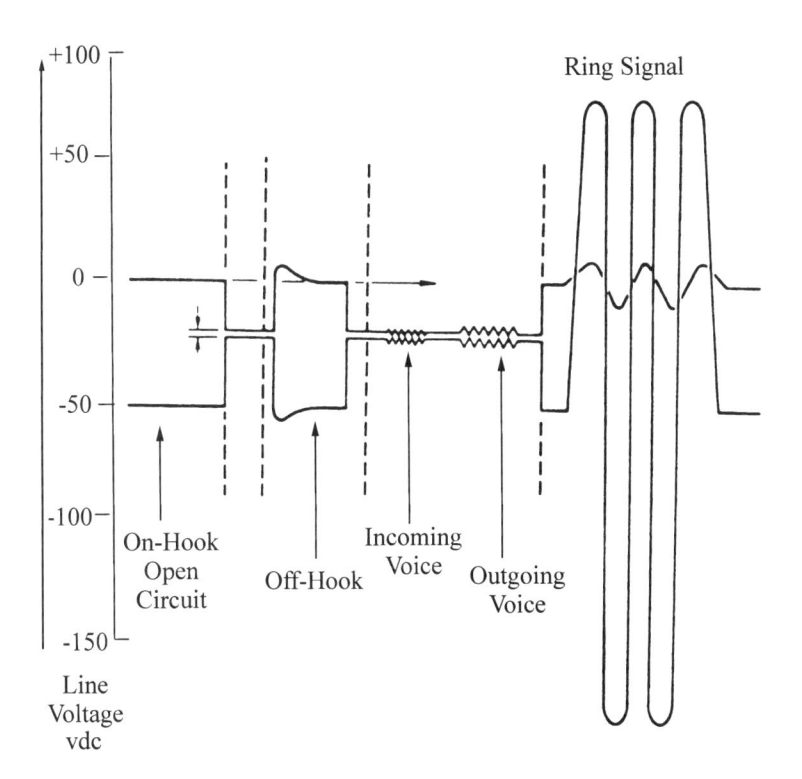

Figure 5-19.
The signals at the local loop during different phases of use.

incoming signals vary. The significant spread in acceptable levels seen with outgoing and incoming telephone signals requires that equipment used within the system operate over a wide range.

Dialing

Telephone systems rely on older pulse dialing and the widely used tone dialing methods. With pulse dialing, the loop circuit opens and closes a number of times with respect to the dialed digit, at a rate of 10 make/break cycles, and with a 50% duty cycle. Central office equipment senses and counts the make/break sequences when recognizing the pulse-dialed digit. The inter-digit time, or the time required to move a person's finger back to dial another number, notifies the system that following make/break pulses belong to a new digit rather than the preceding digit. Time-out circuitry located at the central office ensures that the break portion of the dialing pulse doesn't signify a hangup.

Pulse dialing uses electromechanical relay or an electronic equivalent to generate the make/break signals. Because of this, pulse dialing lacks the controlling signals needed for special customer services. In addition, the combination of pulse dialing and the use of a rotary telephone requires longer time to complete the dialing routine.

In comparison, tone dialing uses a unique pair of tones to represent each digit. Low-group tones correspond to each row, while high-group tones correspond to each column. The selection of a digit causes the phone to generate a mix of high- and low-group tones that represent the digit. Given the use of tones rather than make/break pulses, tone dialing offers a much shorter dialing time.

Dual-Tone Multifrequency Dialing

Dual-tone multifrequency dialing—or the use of low- and high-group tones to represent a digit—offers additional benefits along with dialing time savings. The circuitry used to support DTMF has a lower cost than the circuitry used to support pulse dialing. Moreover, the use of tones permits the application of special telephone services for individual customers. Since the tones correspond to the voice band of frequencies, or in-band signals, the telephone system can ignore any signals in the voice band that follow the initial dialing sequence.

Switching

The path created between two telephone loops must allow the signal voltage to pass. Central office switching systems make the connection through a matrix that aligns incoming caller lines along the vertical side and outgoing caller lines along the horizontal side. Solid-state switches use a single IC and digital logic to point to the designated matrix connection point and switch the connection on or off. Digital recognition circuitry in the central office produces a complete number of the desired called telephone by converting the number into a binary value that refers to the physical location of the telephone. Decoding logic uses the binary value to set the connection points of the initiating and receiving locations while operating across central office boundaries.

Solid-state switches use a single IC and digital logic to point to the designated matrix connection point and switch the connection on or off.

Rather than use fixed function logic, switching systems depend on a computer that operates as a controller, or an electronic switching system (ESS). The ESS provides additional capabilities along with less expense, less power consumption, flexibility, and higher reliability. An ESS can monitor its own performance and auto-configure when it detects any signal degradation. The controlling computer maintains a data base of all connected telephones and trunks while listing services—such as call forwarding, call blocking, and speed dialing—and available paths.

An ESS uses physical numbers and logical numbers. Physical numbers refer to the actual location of the local loop termination. Logical numbers consist of the actual telephone numbers associated with the physical numbers.

Common Channel Signaling

Common channel signaling (CCS) separates the signal-carrying lines from the voice information-carrying lines in a trunk. Illustrated in *Figure 5-20*, a common channel carries only signaling and supervisory information. The

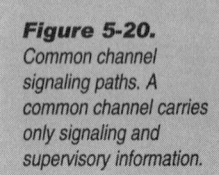

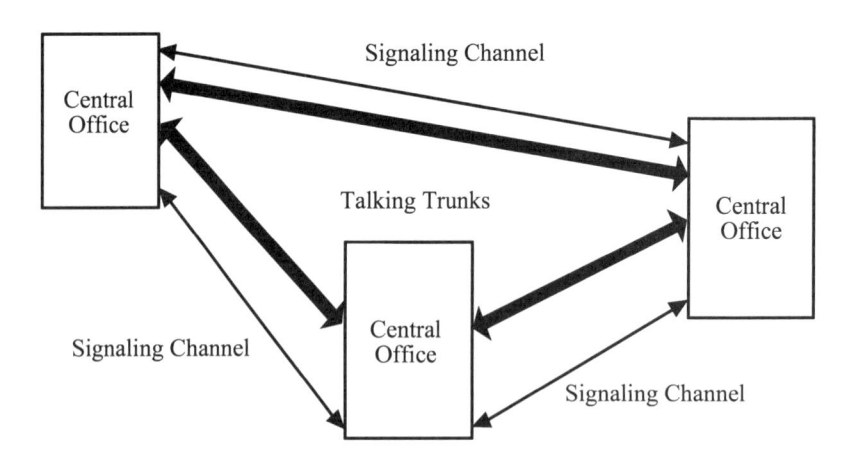

Figure 5-20.
Common channel signaling paths. A common channel carries only signaling and supervisory information.

central office passes the signaling and supervisory information to control actions or determine the status of system lines. Since common channel signaling separates voice information from signaling information, we can refer to CCS as out-of-band signaling. CCS provides detailed information about central office activity and phone-line status to other central offices, while extending services across central office boundaries.

Digital Telephony

Although traditional telephone systems operate with analog signals, new trunks and central offices use digital signals that may originate from digital sources or the analog-to-digital conversion of signals. Due to incompatibilities between some telephone systems and digital signals, an interface modem converts the signals to usable data. Tandem switching centers provide the analog-to-digital conversion for those systems. Other systems operate only with digital signals. In either case, protocols manage the operation of the system.

Trunk lines for digital signals must transmit the signals without excessive signal attenuation or noise in the lines and amplifiers will corrupt the signals and cause bit errors. Capacitance found within the lines may cause intersymbol interference that causes bits to spread out in time. Amplifiers used for digital signals must judge and restore small, noisy signals to original values. A digital regeneration amplifier uses a comparator to perfectly re-create a digital signal without noise or attenuation.

LECs, LATAs, and Carriers

When AT&T was divested of its local phone companies on January 1, 1984, the order legally defined two distinct types of telephone companies. The first type is the local exchange carriers (LECs), consisting of the 23 Bell Operating Companies (BOCs) created by the divestiture; the former independent telephone companies such as GTE; and about 1,500 small-town telephone companies. The second type is the interexchange carriers (IECs), consisting of long-distance carriers such as the former AT&T Long Lines organization and other carriers, such as MCI and Sprint.

In addition, the U.S. was partitioned into Local Access and Transport Areas (LATAs). LATA boundaries conform more or less to the standard metropolitan statistical areas defined by the U.S. Department of Commerce. For example, two area codes represented in *Figure 5-21*—212 and 787—also represent two LATAs. Regulations prohibit LECs from carrying inter-LATA calls, or calls between two LATAs. Previously, IECs couldn't carry

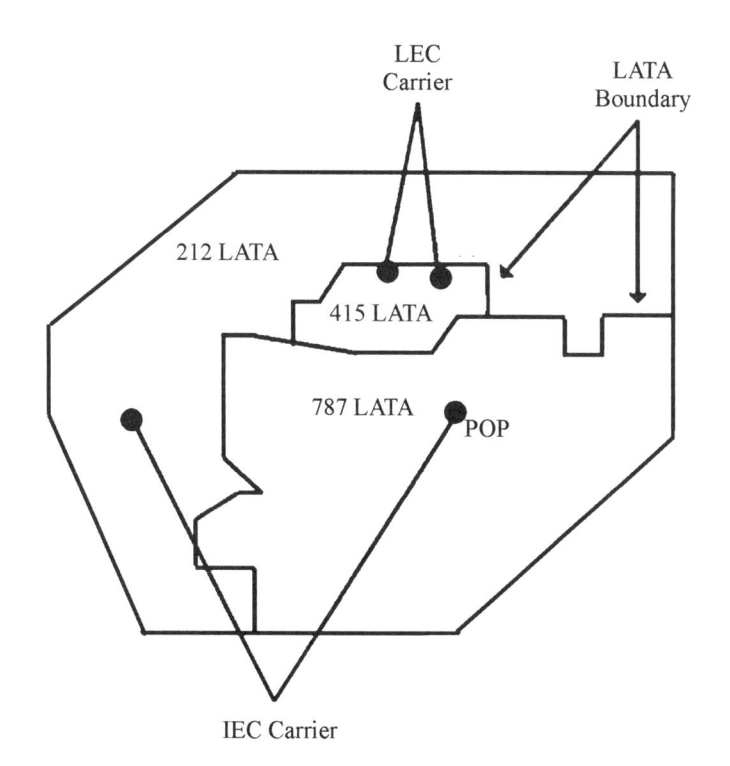

Figure 5-21.
A diagram of LATAs, LECs, and IECs.

intra-LATA calls, or calls within one LATA. Now, however, phone customers in most areas can designate which company will carry intra-LATA long-distance calls. In some instances, an LEC covers two LATAs. Even with that coverage, however, the LEC must route the call through an IEC. Looking at the figure, we see that the Point of Presence (POP) represents the IEC. Subscribers can connect to an IEC facility only at a POP. Typically, the connection occurs through a line provided by a Local Exchange Carrier from the subscriber's premises to the POP. The providers of inter-LATA services fall into two classes:

1) Long-Distance Carriers—Companies that provide long-distance private lines, virtual private networks, or switched lines that the subscriber uses in the same manner as dedicated private lines.

2) Packet Carriers—Companies that provide packet-switching services. These companies often use the facilities of the long distance carriers to construct their networks.

Long-Distance Carriers

Long-distance carriers provide: (1) voice-grade service through leased private analog lines; (2) digital data services that provide leased digital lines used for data only at speeds ranging from 2400 bps to 19.2 Kbps or, in some cases, 64 Kbps; (3) fractional T1 and full T1 access; (4) and T3 access.

In addition, long-distance carriers also offer international private lines (IPLs) of the various types provided by some carriers to international locations through certain gateway cities. With this, a subscriber must lease one line to the gateway and one line from the gateway to the international location. Finally, long-distance carriers sell switched data services (SDS), often referred to as software-defined networks or virtual private networks (VPNs). Switched data services allow the subscriber to make use of the circuit-switching capabilities of the carrier's long-distance facilities to control and monitor their private network.

Packet Carriers

Packet carriers offer economical network access for subscribers who have relatively low volumes of message traffic or message volumes that fluctuate

Long-distance carriers offer international private lines (IPLs) of the various types provided by some carriers to international locations through certain gateway cities.

widely. By allowing subscribers to share transmission facilities and to pay for usage based on message traffic and connect time, the carriers provide an economical alternative to acquiring leased or circuit-switched lines from the long-distance carriers. Along with economical network access, packet carriers also provide value-added services such as electronic mail. Services provided by packet carriers separate because of these factors:

- Type of Access—Most offer dedicated access, or a leased line from the subscriber's premises to the nearest point of access to the network. Many also offer public or private dial access. With private access, only one subscriber uses the access lines.

- Access Speed—Most offer 9,600-bps dial lines and 64-Kbps dedicated lines. Packet movement within the network occurs at 64 Kbps.

- Protocols—Virtually all support X.25. A wide variety of other protocols are supported, including SNA 3270, and X.400. Often these protocols are linked to value-added services offered. Recently, several of the vendors have used X.400 to interconnect their electronic mail services.

- Dial Availability—These services are only economical for dial users if there is a local number to dial. The number of such access points varies a great deal from provider to provider.

The Internet

The Internet is a global network connecting millions of computers. Currently, the Internet serves more than 70 million users worldwide and links more than 100 countries into exchanges of data, news, and opinions. Since the Internet is decentralized by design, each Internet computer is independent. Host operators can choose which Internet services to provide to local users and which local services to make available to the global Internet community.

An Internet service provider (ISP) is a company that provides access to the Internet for a monthly fee. After receiving a subscription, the ISP provides a software package, user name, password, and access phone number for individual access to the Internet. In addition to serving individuals, ISPs also serve large companies and provide a direct connection from the company's networks to the Internet. ISPs connect to one another through network access points (NAPs).

In comparison to an ISP, an on-line service—such as America On-line or CompuServe—is a business that provides its subscribers with a wide variety of data transmitted over telecommunications lines. On-line services provide an infrastructure so that subscribers can communicate with one another through e-mail messages or through on-line conferences. In addition, an on-line service can connect users with an almost unlimited number of third-party information providers. With this, subscribers can get up-to-date stock quotes, news stories hot off the wire, articles from many magazines and journals, and almost any information that has been put in electronic form.

The World Wide Web

The World Wide Web is a system of Internet servers that support documents formatted in a language that supports links to other documents, graphics, audio, and video files. Applications called web browsers allow easy access to the World Wide Web. Each web site, or location on the World Wide Web, usually contains a home page, additional documents, and files. Due to the relative independence given through the Web, each site is owned and managed by an individual, company, or organization. A home page is the main document of the web site and serves as an index or table of contents to other documents stored at the site. Web pages are documents found on the World Wide Web and are identified by a unique address.

Uniform resource locators (URLs) signify the global address of documents and other resources on the World Wide Web. When considering the makeup of a URL, the first part of the address indicates what protocol to use, while the second part specifies the IP address or the domain name where the resource is located. The first specifies an executable file that should be fetched using the File Transfer Protocol (FTP); the second specifies a web page that should be fetched using the HyperText Transfer Protocol (HTTP).

Uniform resource locators (URLs) signify the global address of documents and other resources on the World Wide Web.

Every web site is serviced through a web server, a computer that delivers web pages. While every web server has an Internet Protocol (IP) address, many also use domain names.

For example, if we enter the URL http://www.goofy.com/index.html in your browser, this sends a request to the server that has the domain name "goofy.com". The server then fetches the page named "index.html" and sends it to your browser. Any computer can operate as a web server through the installation of server software and the connection of the machine to the Internet. Web server software applications include commercial packages from Microsoft, Netscape, and other manufacturers.

As the process unfolds, a client in a network sends a request for a web page to a web server by addressing a frame to a router. The information travels from the NIC installed in the client to a hub and then to a switch. Basing its action on the destination address of the frame, the switch forwards the frame to the appropriate router. As with the client computer, the router contains an NIC with a specific address. The router sends the packet across the WAN link by placing the packet in a new frame. At the destination, another router looks at the packet address and forwards the data to the web server. The web server responds to the data request from the client computer by sending a packet containing web page information back to the client.

The suffix html found on the index.html page stands for Hypertext Markup Language, the authoring language used to create documents on the World Wide Web. The hypertext language is a special type of database system that links objects to one another. Each object may consist of text, pictures, music, or programs. When a user selects an object, the other linked objects are also displayed. In addition, a user can move from one object to another even though the objects may have very different forms.

Short for web browser, a browser is a software application used to locate and display web pages. Two of the most popular browsers are Netscape Navigator and Microsoft Internet Explorer. Each of those browsers can display graphics as well as text. In addition to graphics, most browsers can present information in the form of sound and video.

A search engine is a program that searches documents for specified keywords and then returns a list of the documents where the keywords were found. Although search engines are a general class of programs, the term often describes systems such as Yahoo, WebCrawler, Alta Vista, and Excite. Each of these services allows individuals to search for documents on the World Wide Web.

Search engines operate by sending out a program called a spider to fetch as many documents as possible. Another program, called an indexer, reads these documents and creates an index based on the words contained in each document. Each search engine uses a proprietary algorithm to create its indices so that only meaningful results are returned for each query.

Summary

Chapter 5 began by defining the benefits of expanded networking technologies. As with other chapters, Chapter 5 used the OSI reference model to illustrate specific characteristics and functions. The analysis of those characteristics and functions led to definitions of protocol routing and descriptions of router operation. As with Chapter 4, Chapter 5 described a wide variety of network devices such as CSUs, DSUs, Multiplexers, and Gateways.

Chapter 5 also moved into the world of telephony by illustrating private branch exchange technologies and applications. From that starting point, the chapter progressed to definitions of dial-up services, dedicated services, and switched services. All this creates a better understanding of the telephone system technologies described in the latter portions of the chapter. In addition, Chapter 5 discussed digital telephony and set the tone for later chapters.

Search engines operate by sending out a program called a spider to fetch as many documents as possible. Another program, called an indexer, reads these documents and creates an index based on the words contained in each document.

Chapter 6
Analog Modems

Introduction

Modems function as an extension of a computer and use ordinary telephone lines to carry digital computer data. Combined with the appropriate software, modems allow low-cost, individual access to on-line services, electronic mail, print services, and file transfers from a home or an office. Thanks to adherence to standards, any modem—given the correct cable and settings—will work with any computer.

Analog-to-Digital Conversion

During operation, analog modems use processes called analog-to-digital and digital-to-analog signal conversion to change digital information transmitted from a computer into analog audible tones that can flow over ordinary phone calls. Digital-to-analog conversion changes the digital signals to an analog format. ADC and DAC circuits perform those operations, while acknowledging requests from microprocessor controllers and other circuits. In addition to the analog/digital conversion, those circuits also require the conversion between digital formats and the translation of digital data to a display by encoders and decoders. Analog-to-digital-conversion (ADC) circuits take an analog signal and convert the signal to a digital number that corresponds to the value of the analog signal. Digital-to-analog conversion (DAC) circuits produce a dc output voltage that corresponds to a binary code and convert digital properties to analog voltages.

Thanks to adherence to standards, any modem—given the correct cable and settings—will work with any computer.

Modulation

The transferal of the information from computer to modem also involves a process called modulation. Modulation involves the encoding of a carrier wave with another signal or signals that represent some type of intelligence. With modulation, an audio frequency signal affects the frequency or amplitude of radio frequency waves so that the waves represent communicated information. The carrier wave is the sinusoidal component of a modulated wave and has a frequency independent of the modulating wave. As the name suggests, the carrier wave carries the transmitted signal.

Radio waves can only carry signal information when modulated by another signal. While a perfect unmodulated carrier has zero bandwidth and contains no information, the modulated signal occupies a bandwidth least comparable to the modulating signal. Modulation combines the waveforms of the combined signals and yields different combinations of those signals.

Several different types of modulation methods exist and may be seen in various types of communications equipment. For example, AM (amplitude modulation) and FM (frequency modulation) are used to transmit the picture and sound information in television systems. Along with the oft used AM and FM methods, some communication systems use phase modulation and single-sideband modulation.

Amplitude Modulation

With amplitude modulation, the amplitude of the RF carrier changes, while its frequency remains constant. The changes in the carrier amplitude vary proportionately with the changes in the frequency of the modulating signal. Commercial AM radio stations operate with broadcast frequencies in the 550- to 1650-kHz range. When considering television signals, the video signal is amplitude modulated. In addition, many amateur radio operators use amplitude modulation for their signal transmissions.

A very good representation of amplitude modulation is seen with super heterodyne radio. With this, an incoming radio frequency (RF) signal combines with the frequency of the local oscillator found in the radio. For example, an AM radio local oscillator will have a frequency 455 kHz higher than the incoming RF signal. *Figure 6-1* is a block diagram of the RF signal entering the radio and modulating with the local oscillator frequency at the mixer stage.

Chapter 6: Analog Modems

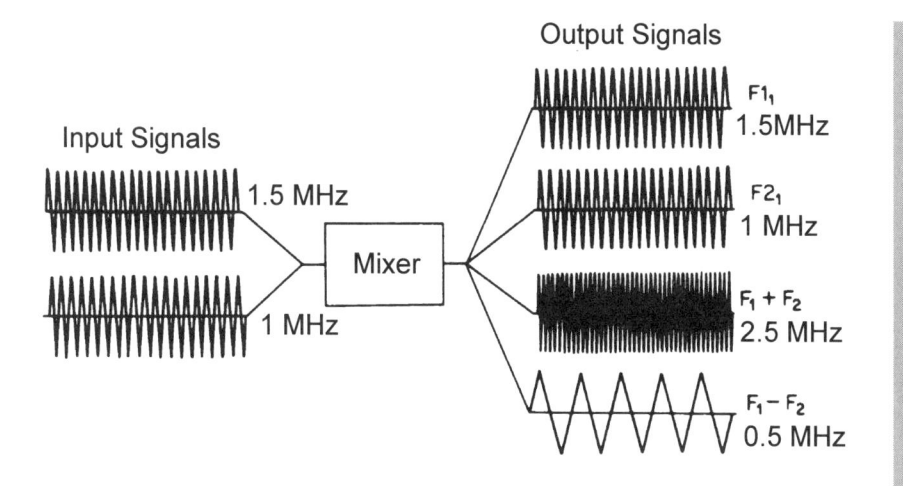

Figure 6-1.
A representation of heterodyning. The RF signal enters the radio and modulates with the local oscillator frequency at the mixer stage.

The constant amplitude carrier wave (CW) doesn't contain any modulating information. Referring to *Figure 6-2*, we see that the thin outline drawn along the peaks of the modulated carrier wave represents the modulation envelope. Exact reproduction of the transmitted signal at the receiver requires that the modulation envelope produced at the transmitter has the same waveform as the modulating signal.

Frequency Modulation

With frequency modulation, the frequency of the carrier changes, while the amplitude remains constant. As the modulating signal increases to a maximum positive value, the carrier frequency also changes. When the modulating frequency drops to zero, the carrier frequency decreases to its original value.

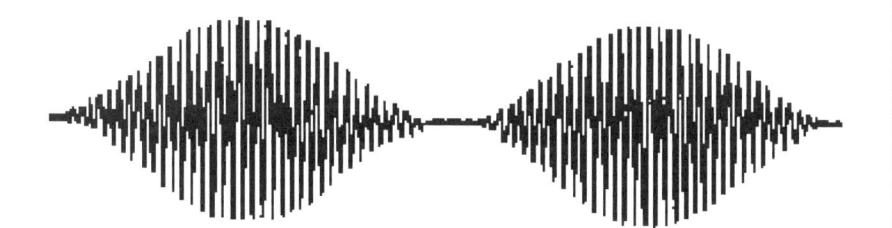

Figure 6-2.
Waveform of an amplitude modulated signal. The thin outline drawn along the peaks of the modulated carrier wave represents the modulation envelope.

Figure 6-3.
Waveform of a frequency modulated carrier. The positive peak of the sine wave coincides with an increase in oscillator frequency.

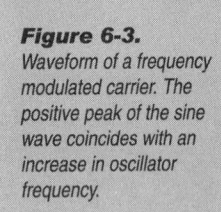

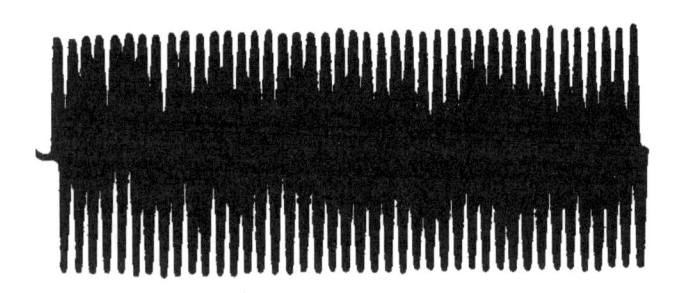

As the modulating frequency increases to its maximum negative value, the carrier frequency decreases. Looking at *Figure 6-3*, we see that the positive peak of the sine wave coincides with an increase in oscillator frequency. As a result, the change in frequency of the frequency-modulated carrier wave corresponds with a change in the amplitude of the input signal. The rate of the frequency changes corresponds with the modulating frequency. As opposed to amplitude modulation, the amplitude of the frequency-modulated waveform in the figure remains constant.

Figure 6-3 also shows that the carrier frequency varies above and below a center frequency. We refer to the amount of positive or negative change in the carrier frequency as deviation. While the amplitude of the modulating signal determines the deviation, the frequency of the modulating signal determines the rate that the carrier varies above and below the center frequency.

Phase and Single-Sideband Modulation

Both the phase and single-sideband modulation methods suppress the carrier and one sideband, while transmitting one sideband from the original sideband signal. Since the upper and lower sidebands also include the modulating information, a typical AM demodulation scheme suppresses the carrier wave and one sideband and transmits the remaining sideband. This type of transmission is referred to as single-sideband transmission (SSB).

Pulse Modulation

Digital communication systems employ another method called pulse modulation; the system converts the intelligence held within the modulating signal into a pulse. After the conversion occurs, the system pulses the RF signal for the type of pulse modulation used. The modulating pulses may control the amplitude, frequency, on time, or phase of the carrier. We refer to the pulse modulation methods as pulse-amplitude modulation, frequency shift keying, pulse-width modulation, and pulse-phase modulation.

Demodulation

On the receiving end, a modem demodulates the analog signals arriving over the phone line into electrical signals, which are then fed to the computer. Demodulators, or detectors, decode and recover the intelligence from the carrier signal. When considering the demodulation of amplitude-modulated signals, the demodulator circuits detect the envelope that corresponds with the modulating signal and eliminate the carrier wave.

With the simple diode detector circuit shown in *Figure 6-4*, the modulated RF signal travels from the secondary of the transformer and encounters the diode. The nonlinear diode detector allows only half of the modulated RF waveform to pass.

Since the filter capacitor cannot follow the RF signal, it passes only the information-carrying envelope of the waveform. The detection of a frequency-modulated signal for the recovery of audio intelligence is more difficult than the detection of an amplitude-modulated signal. Due to the characteristics of frequency modulation, a circuit called a frequency

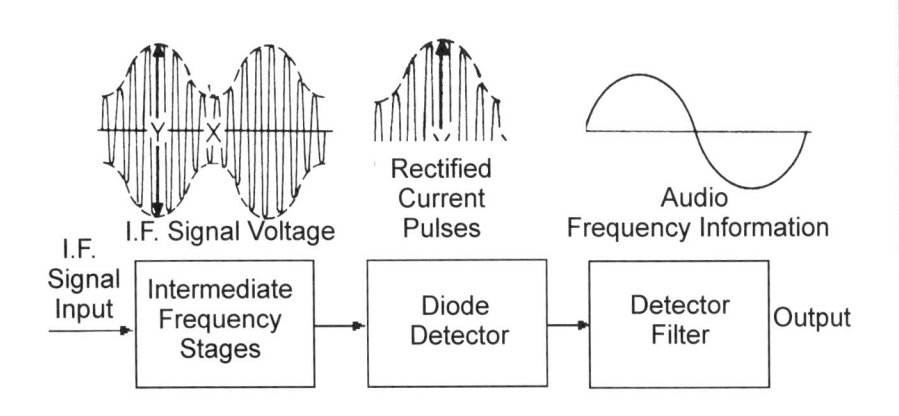

Figure 6-4.
A block diagram of a simple diode detector circuit. The modulated RF signal travels from the secondary of the transformer and encounters the diode.

discriminator combines with a limiting circuit to ensure that the amplitude of the signals remains constant. In those circuits, changes in frequency cause a change in the output voltage.

Serial Communications

Serial communications transmit data one bit at a time and occur in either a synchronous or asynchronous format. Synchronous communications involve the use of clock pulses to synchronize the transfer of data. Asynchronous communications transmit data intermittently rather than in a steady stream. Both synchronous and asynchronous communications rely on sync bits that acquire and maintain the synchronization between the sending and receiving devices.

Synchronous Communications

With synchronous communications, a constant flow of data allows each data to remain ready for a data transmission. Each character represents either actual data or an idle character. Since synchronous communications don't mark the beginning and end of each data byte, faster data-transfer rates occur. Before modems begin to transmit and receive data, the modems use a set of synchronized clock pulses to set the internal timing of clock circuits.

At the beginning of the established connection, each modem transmits a burst of bits that has a specific length. During the sending of the data, the transmitting modem places a one or zero on the line at established intervals. The receiving modem samples the line on the same timetable and transmits the condition of the line to the other modem. The modems must remain synchronized so that communication can occur.

Asynchronous Communications

Compared to synchronous communications, asynchronous communications don't rely on an established timetable for the transmission of data. Instead, asynchronous communications require the use of a start and stop bit that identifies the beginning and end of data. As an example of asynchronous communications, both parties involved in a telephone conversation can speak at any time during the conversation.

Transmitting the start bit indicates the start of each character, while the stop bit indicates the end of the character. Without the use of a start and stop bit, the receiving system would have difficulty separating the data from noise. Even if the modem clocks don't have exact synchronization, the data transfer remains successful. During transmission, the modems need to stay synchronized only for the length of time needed to send eight bits of data.

DCE and DTE

A modem operates as data communications equipment (DCE), while the computer attached to the modem operates as data terminal equipment (DTE). Each of these terms also defines the pinouts for connectors located on a device or a cable and establish the direction of the signals. Referring to *Figure 6-5* and *Tables 6-1* and *6-2*, we see that DTE devices use male connectors, while DCE devices rely on female connectors.

While the attachment of a DTE device to a DCE device can occur through straight-through connections, the linking of a DTE device to another DTE device (or a DCE device to another DCE device) requires the use of a null modem cable. Shown in *Figure 6-6*, a null modem crosses the transmit and receives lines in the cable.

DB-9 Female Connector Used for DCE Devices

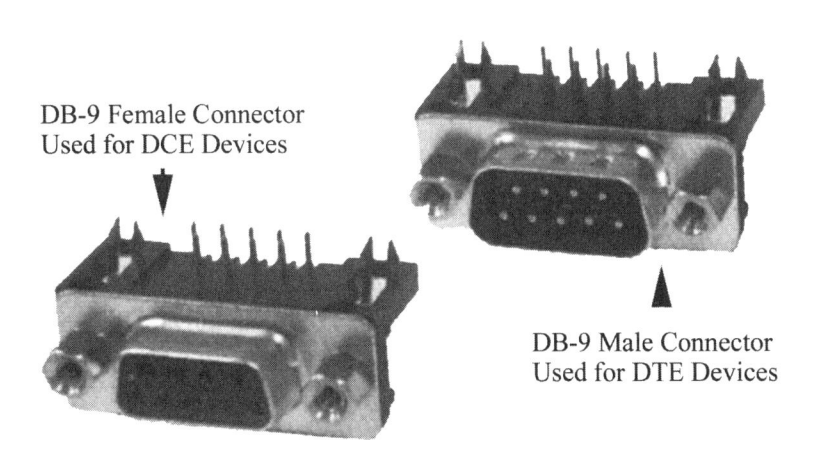

DB-9 Male Connector Used for DTE Devices

Figure 6-5.
DB-9 connectors. DTE devices use male connectors, while DCE devices use female connectors.

Table 6-1.
25-pin connector on a DTE device.

Pin	Name	Direction of signal
1	Protective Ground	
2	Transmitted Data (TD)	Outgoing Data (from a DTE to a DCE)
3	Received Data (RD)	Incoming Data (from a DCE to a DTE)
4	Request To Send (RTS)	Outgoing flow control signal controlled by DTE
5	Clear To Send (CTS)	Incoming flow control signal controlled by DCE
6	Data Set Ready (DSR)	Incoming handshaking signal controlled by DCE
7	Signal Ground Common reference voltage	
8	Carrier Detect (CD)	Incoming signal from a modem
20	Data Terminal Ready (DTR)	Outgoing handshaking signal controlled by DTE
22	Ring Indicator (RI)	Incoming signal from a modem

Table 6-2.
9-pin connector on a DTE device.

Pin	Name	Direction of signal
1	Carrier Detect (CD)	Incoming signal from a modem (from DCE)
2	Received Data (RD)	Incoming Data from a DCE
3	Transmitted Data (TD)	Outgoing Data to a DCE
4	Data Terminal Ready (DTR)	Outgoing handshaking signal
5	Signal Ground Common reference voltage	
6	Data Set Ready (DSR)	Incoming handshaking signal
7	Request To Send (RTS)	Outgoing flow control signal
8	Clear To Send (CTS)	Incoming flow control signal
9	Ring Indicator (RI)	Incoming signal from a modem (from DCE)

External Modems

External modems attach to computers through the use of an RS-232C cable, while internal modems install into an expansion slot found on the motherboard of a computer. Looking back to Figure 6-5, we see that the male (or DTE) connector plugs into the computer, while the female (or DCE) connector plugs into the modem.

Table 6-3 supplements Figure 6-5 by showing the pin assignments for both connectors. At the DCE connector, the data seems to travel in reverse, because of the relationship to the DTE connector.

Chapter 6: Analog Modems

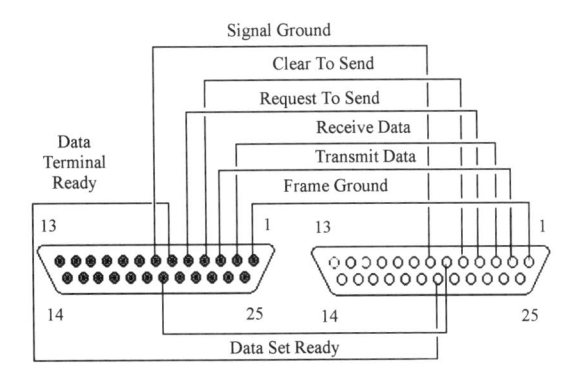

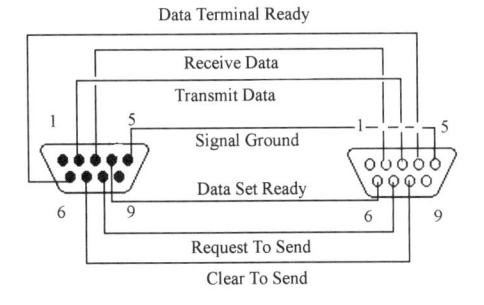

Figure 6-6.
Null modem connectors. A null modem crosses the transmit and receives lines in the cable.

Table 6-3.
RS-232C pin assignments and signal definitions.

- Data Set Ready (DSR)—This signal is controlled by the DCE. It tells the DTE that the Equipment (Modem) is available and functioning to receive data.
- Ready to Send (RTS)—This signal is also controlled by the DCE. It tells the DCE that the Equipment is ready to accept data for transmission over the line, or to the command processor.
- Carrier Detect (CD)—This signal, controlled by the DCE, tells the DTE that there is a carrier, which means there is a DCE-DCE connection.
- Ring Indicator (RI)—The DTE will use this signal to let the DCE know that it hears a RING on the line, i.e. there is an incoming call.
- Data Terminal Ready (DTR)—This signal is controlled by the DTE. It tells the DCE that the computer is available and functioning to receive data.
- Clear to Send (CTS)—This signal is also controlled by the DTE. It tells the DCE that the equipment is ready to accept data for reception from the DCE. All these signals, as well as the standard way of sending and receiving data, are grouped together in a protocol that we know as RS-232. The computers we use today use all RS-232 signaling with a slightly different voltage level for Signals. Their interfaces are referred to as RS-232C. All these signals are electrical levels on a set of Pins.
- Signal Ground (GND)—This is the electrical Zero for all signals on the serial port.
- Transmit Data (TX)—This is where a device transmits data to the other device.
- Receive Data (RX)—This is where all data comes in from the other side.

173

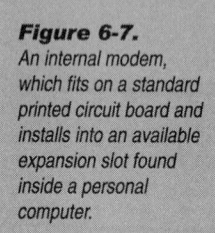

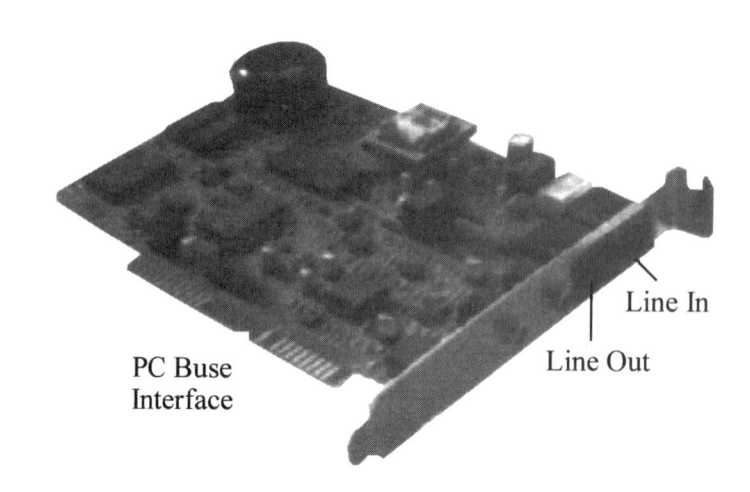

Figure 6-7.
An internal modem, which fits on a standard printed circuit board and installs into an available expansion slot found inside a personal computer.

PC Buse Interface

Line In

Line Out

Internal Modems

Shown in *Figure 6-7*, an internal modem fits on a standard printed circuit board and installs into an available expansion slot found inside a personal computer. As a result, internal modems conform to the architecture of the computer. Internal modems provide their own serial port and do not require a separate power adapter or a modem cable.

Dipswitch settings on any internal modem must correspond to the used serial/com port. Each serial/com port is assigned a name, address, and interrupt. Interrupts are signals to the CPU that it needs attending to. In this case, when the serial/com port has data to transfer, its IRQ signals the CPU to act upon it. *Table 6-4* lists the COM ports along with the corresponding addresses and interrupts.

AT Commands

Communication software issues commands that cause a modem to complete functions such as dialing, answering, or hanging up. The method

Table 6-4.
COM ports, addresses, and interrupts.

COM 1	3F8	IRQ4
COM 2	2F8	IRQ3
COM 3	3E8	IRQ4
COM 4	2E8	IRQ3

for issuing these commands takes the form of two-letter AT commands that conform to a standard set of rules. An AT command asks the modem for attention. The TIA/EIA/602, 578, and 592 standards list hundreds of AT commands for modems and fax machines. Nearly all modem manuals contain the complete list of AT commands for their particular modems. The basic AT commands used for troubleshooting modems include:

- AT [enter], for basic serial/com port to modem tests
- ATE1 [enter], to make the letters appear on the screen as they are typed
- ATE0 [enter], to make the letters disappear (these are echo commands)
- ATH1 [enter], makes the modem pick up the receiver
- ATH0 [enter], makes the modem hang up the receiver
- ATM1 [enter], turns the modem speaker on
- ATM0 [enter], turns the modem speaker off
- ATM2 [enter], sets modem speaker medium range
- ATDXXXXXXX [enter], dials a number
- ATZ [enter], resets the preferred profile that you have previously chosen

Modem Command Codes

Computers and modems combine to provide fully programmable communications functions. Rather than rely on possibly incompatible signals, modem manufacturers use a standard set of in-band signaling commands to control the modem. When placed in command mode, the modem will read every line prefaced with the letters AT. After issuing the AT command, the software can show another set of functional commands. *Table 6-5* lists functional commands, a definition for each command, and the actions resulting from the command.

Command	Default	Command Definition	Command Action
A		Answer incoming call	When a call RINGS, the word RING will appear on the screen.
AT		Attention	
D		Dial a telephone number	This is followed by the number to call.

Table 6-5.
Modem commands.

Telecommunication Technologies

Table 6-5 (cont.)
Modem commands.

Command	Default	Command Definition	Command Action
T		Dial Tone dial	
P		Dial with Pulse dial	
W		Wait	Wait for a second dialtone.
Ex	E1	Echo	Echo Characters in Command Mode.
Hx		Hook	Hook Switch Control.
Kx	K1	Break	Allows or prevents retrain breaks during Error Control situations.
Mx	M1	Speaker Mode	Controls when the sound on the phone line is passed through in the internal modem speaker: M0 Speaker always off M1 Speaker On until Carrier Detect M2 Speaker Always On
Qx	Q0	Quiet Mode	When set to Quiet mode, a modem doesn't return responses to commands given to it. Q1 enables Quiet mode.
Vx	V1	Verbose Mode	When set to V1, all responses to commands will be sent as verbal messages, such as OK and ERROR. When set to V0, only a number will be returned.
Xn	X5	Response and Dialing Mode	This command controls what responses will be given when dialing other modems and what checks will be performed while dialing and connecting:

176

Command	Default	Command Definition	Command Action
			X0 Basic Response, do not check for Dialtone or Busy
			X1 Extended response, do not check for Dialtone or Busy
			X2 Extende response, check Dialtone, do not check Busy
			X3 Extended response, do not check Dialtone, do check Busy
			X4 Extended response, Dialtone and Busy detection
			X5 Basic Response, Dialtone and Busy detection
Z			Reset modem to NVRAM defaults
&Bx	&B0	DSR Behavior	Controls how DSR behaves on the modem:
			&B0 DSR always ON
			&B1 DSR does what DTR does. The modem equipment is Ready
			&B2 DSR signals normal RS-232C behavior
&Cx	&C1	Carrier Detect Behavior	Informs the software about the presence of a carrier:
			&C0 DCD Always ON
			&C1 DCD follows RS-232C standards
			&C2 DCD Wink; always on but 'winks' off when carrier lost
&Dx	&D2	Response to DTR	This setting controls what the modem does when DTR is toggled:
			&D2 Causes the modem to disconnect a soon as the DTR line is dropped

Telecommunication Technologies

Table 6-5 (cont.)
Modem commands.

Command	Default	Command Definition	Command Action
			&D3 Drop connection and reset to NVRAM settings after DTR drop
&F		Reset to Factory Settings	Will load the modem's factory setting from the ROM in the modem. Factory Settings are usually Full Flow Control, All Error Correction.
S0=x	0		Numbers of Rings to Auto Answer.
S7=n	60		Wait time for remote carrier.
S9-n	5		Carrier Detect Time required for Recognition.
S10=n	60		Wait on Loss of Carrier.

Looking Inside a Modem

Universal Asynchronous Receiver and Transmitter IC

Modems and serial ports include an integrated circuit called a Universal Asynchronous Receiver and Transmitter (UART) that provides conversion between synchronous and asynchronous data while notifying the modem that a data transmission can occur. While modems communicate through a synchronous format that includes error correction and data compression, a UART must translate the asynchronous data from the computer into the synchronous data for transmission and the synchronous data back into asynchronous data for the computer.

When the transmission of data begins, the UART appends a start bit to the desired information as a notification about the presence of the data. At the end of the transmission, the UART adds a stop bit so that the sending and receiving modems will recognize that the data transmission has ended. At the receiving modem, the UART strips the start and stop bits from the desired information. With that, the modem relays only the desired information to the computer.

While the UART controls all out-of-band signaling, the software commands listed in the next section control in-band signaling. As shown in *Figure 6-8*, UARTs have an internal clock used to control the serial port and baud generator speed. As a result, the UART controls the speed and timing of the serial data transfer. An oscillator sets the clock speed by oscillating at a specific frequency.

Parity Checks

A parity check defines whether the sum of all bits in a data packet occurs as an even or an odd number. For example, the data packet 1111110 has a sum of six and stands as even, while the sum of all bits for 1111111 is seven or an odd number. Changing one of the bits in either data packet changes sum of the data packet. Computer systems add one extra bit—a parity bit—to each set of data.

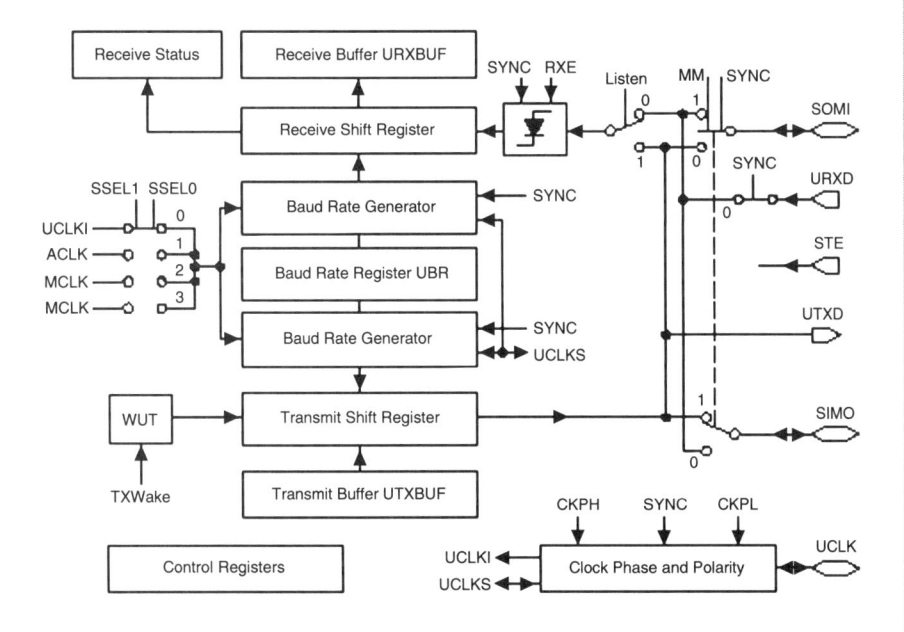

Figure 6-8.
A block diagram of a UART. UARTs have an internal clock used to control the serial port and baud generator speed.

During operation, the parity bit remains set or unset and ensures that the data transfer will pass the parity test. If the data transfer requires even parity, sending the data packet 1111111 requires a parity bit that has a value of one. Since 1111111 yields an odd parity, the addition of the parity bit provides an even parity.

Data Flow Control

Since the flow of data from DCE and DTE occurs at different rates, modems use flow control to synchronize the stop and start bits within the data blocks. The use of flow control prevents both modems from transmitting at the same time. As such, flow control defines the origin and timing of data communication between modems. With flow control properly set, a receiving computer can signal a sending computer for a pause or resumption in the transfer of data.

Modem manufacturers may rely on one of several techniques to accomplish flow control. Those techniques include in-band signaling, hardware flow control, and out-of-band signaling. In-band signaling involves the sending of a signal within the data stream to indicate the completion of the data transmission. Hardware flow control involves the sending of signals through a separate channel to indicate the completion of the data transmission. Out-of-band signaling dictates the type of flow control used between the two modems. Local flow control specifies the method for establishing flow control between a serial port and a modem.

In-band signaling inserts special characters into the transmitted data stream. The improper configuration of flow control settings can cause slow data transfer rates, the aborting of downloads, or may prevent a connection from occurring. In addition, both the modem and the communication software must have the same flow control settings.

In-Band Signaling

The insertion of stop and start characters into the data stream for the purposes of flow control is called in-band signaling. In-band signaling offers disadvantages, because of the need to continuously monitor the data stream. The monitoring

The insertion of stop and start characters into the data stream for the purposes of flow control is called in-band signaling. In-band signaling offers disadvantages, because of the need to continuously monitor the data stream.

provides a method for finding the signaling characters and must occur at high speed. Because modem transmission speeds have increased, in-band signaling has become less desirable. Faster modem and computer speeds combined with higher DTE speeds complicate the process of acquiring the in-band signals within the needed time.

Hardware Flow Control

In addition to software configuration settings, flow control also refers to the link between the personal computer and the attached modem. Hardware flow control uses separate circuits in the modem cable to signal stop and start requests. Activation occurs through increased signal voltages, while decreasing the signal voltages stops the transmission. Combinations of voltage changes on these lines not only cause the ready state for the modem but also cause the actual data transfer to start and stop. Flow control and call control signals travel back and forth between the modem and computer and establish the off and on conditions.

Out-of-Band Signaling

Either the error-control standards of the modem or the file control protocol used by the communication program determines out-of-band signaling. In most cases, software allows the DCE and DTE functions to alternate between modem and computer. As a result, eight signals exist for the application of out-of-band signaling with serial communications. Out-of-band signaling allows call signaling to occur separate from the data channels.

Frequency Response and Line Speeds

Frequency response defines the lowest and highest sounds that a line will reproduce. An analog telephone line has a frequency response ranging from about 150 Hz to 6 kHz and has a peak performance within the 1500-300 Hz range. Noise distortion appears as clicks, pops, and hum at low frequencies and hisses, whistles, and static at high frequencies. For the most part, the highest quality signal occurs in the middle of the frequency band.

As modem speeds and the number of frequencies on a line increase, the chances of noise interference also increase. An error caused by noise interference can result in the loss of bits during a data transfer. To lower the chances for line errors, modem manufacturers first used line equalization and then compression. Line equalization occurs through the use of circuitry in a modem, measures the capability of a telephone line to respond to certain frequencies, and allows the modem to utilize or not utilize frequencies.

Compression technologies analyze a stream of data for patterns and condense the patterns into a shorter representation of the data. For example, the data stream 0101 0101 0101 0101 0101 0101 0101 0101 could appear as 8 times 0101. Protocols such as MNP-5 and V.42Bis maintain error correction but also compress the data. During operation, the compression of data occurs before transfer occurs at the standard speed. At the receiving modem, decompression of data occurs and the computer linked to the modem receives the data at a higher speed.

Error-Correction Protocols

Line noise can cause errors as data either becomes garbled or drops out during transit. All high-speed modems use some type of error correction to detect and correct data errors. The Microcom Networking Protocol (MNP) ensures error-free transmission of data. Five MNP error-correction levels exist with the fifth also providing compression.

Error-correction protocols such as MNP-3 calculate a checksum for a block of data that reflects changes in the block. During operation, a modem uses asynchronous error correction to calculate the data checksums during the data transfer. In addition, the modem passes the checksum numbers to the receiving modem within every 128 bytes of the transfer.

From there, the receiving modem generates a checksum and then compares the two checksums for each completed data block. Any error that occurs during the transfer results in a request for the resending of the data. Newer error correction protocols such as MNP-4 and V.42 send data synchronously and place synchronization bits on the line.

MNP-5

MNP-5 has two different modes of operation called adaptive frequency encoding (AFE) and run-length encoding (RLE). If a single character occurs more than three times in sequence, MNP-5 switches to run-length encoding from adaptive frequency encoding. With AFE, the modem tracks the occurrence of every possible character. The modem uses this information to assign the shortest code words to the most often occurring characters. For example, the modem could receive the following byte stream.

AARTUEIODADAEAAFEERIAEAA

Looking at the stream, we see that "A" occurs as the most frequent character. The MNP-5 compression splits the stream into two sections that include a three-bit header and a body consisting of one to seven bits. The three header bits present the option of eight different code sets. Sets one and two use header bits 000 and 001 and indicate a one-bit body.

Going back to the example stream, the "A" would be coded as:

000 0 (three-bit header, one-bit symbol)

and the next most-frequent character—the letter E—codes as:

000 1 (three-bit header, one-bit symbol)

The next header sequence, 010, indicates a two-bit body. All this leads to the coding listed in *Table 6-6*.

Header	Body	# of Possible Code Words
000	1 bit	2
001	1 bit	2
010	2 bit	4
011	3 bit	8
100	4 bit	16
101	5 bit	32
110	6 bit	64
111	7 bit	128

Table 6-6.
MNP-5 coding table.

Referring to Table 6-6, we see that the MNP-5/AFE scheme offers 256 possible code words. But only one quarter of the code words is shorter than the original eight bits. As a result, AFE compression gains very little and may make matters worse in some cases.

As mentioned, MNP-5 also relies on run-length encoding as a method for compressing data. Whenever a character appears more than three times in sequence, the modem will no longer send the AFE code words to the remote. Instead, the modem circuitry begins to count the repetitions. When the pattern changes or the counter reaches a maximum number of 250, the sending modem transmits the repetition count to the receiving modem. If the data contains byte patterns with high repetition counts, RLE provides an efficient compression method.

Modern Compression

MNP-5 was quite useful back in those days when text and ANSI graphics were the only traffic. Nowadays, where most of the transmitted data is already compressed using CPUs and algorithms far more powerful than anything ever built into a modem, I'd say MNP-5 has outlived its usefulness. WWW traffic, with its nicely compressible HTML code, would be a good place to use AFE. Unfortunately, web traffic tends to contain more and more images, and even a small GIF icon will disrupt the AFE's frequency table and severely impair the coding efficiency.

Modem Standards

Four parameters define the effectiveness and quality of a modem link. Most manufacturers list the modem speed in bits per second, the parity, and the number of stop bits. For example, the configuration of a modem may show 33,600 bits per second, seven data bits, even parity, one stop bit or 33,600 bps 7E1. In this example, the 33,600 bps 7E1 separates into:

Modem Speed	Number of Data Bits	Parity	Number of Stop Bits
33,600	7	E	1

MNP-5 was quite useful back in those days when text and ANSI graphics were the only traffic... I'd say MNP-5 has outlived its usefulness.

Modem specifications may list either the baud rate or the data-transfer rate in bits per second. The baud rate of a line indicates how many changes in modem signal frequencies occur per second. Since each transition can carry multiple bits of information, the baud rate does not equal the measurement of bits per second. For example, a 300-baud line has 300 tone changes per second, while a 2,400-baud line that features four frequencies has 2,400 tone changes per second. With higher speed modems, the number of frequencies sent simultaneously in each direction increases, while the speed of frequency change remains the same.

In contrast, the bits-per-second (bps) rate shows the quantity of bits that can travel over the line within one second. With the early 300-baud and 2,400-baud modems, one bit corresponded with one baud. That is, a 2,400-baud modem transfers data at a rate of 2,400 bits per second. As modem technologies improved, the number of bits per baud began to increase. In addition, as modem speeds increased, the number of different and simultaneous frequencies on the line also increased.

Table 6-7 lists the standards that have evolved as modems have become common tools for communication. As the first section of the table shows, the standards have grown from covering 300-bps communication to the present-day standards that cover 56K communication. The table doesn't list the V.90 standard.

Table 6-7.
Modem communication standards.

- V.21—Standard for modem communications at 300 bps.
- V.22—Standard for modem communications at 1200 bps.
- V.22bis—Standard for modem communications at 2400 bps.
- V.23—Standard in the U.K. for modem communications at 1200 bps with a 75-bps back channel.
- V.25—Overseas 2100-Hz answering tone, different from the Bell tone used in the U.S. and Canada.
- V.25bis—Standard for synchronous communications between the host and modem using HDLC protocol.
- V.32—Standard for modem communications at 9600 bps, as well as 4800 bps.
- V.32bis—Extends the V.32 connection range to 14.4K and adds fallback speeds of 7200 and 12K.
- V.42—Standard that defines a two-stage process of detection and negotiation for LAPM error control, and also incorporates MNP levels 1-4.
- V.42bis—Extension of V.42 that defines a specific data-compression scheme.

Table 6-7 (cont.)
Modem communication standards.

- V.FC—Proprietary standard developed by Rockwell and Hayes that allows a top modem speed of 28,800 bps with fallback to 26.6K, 24K, 21.6K, and 19.2K as well as other lower speeds.
- V.turbo—Proprietary standard advanced by AT&T that specifies a top modem speed of 19,200 bps with fallback to 16.8K, 14.4K, and lower speeds as well. USR incorporates V.turbo in their Courier modem.
- V.34—Standard that supports connections up to 28,800 bps.
- V.34Bis—Standard that specifies a top modem speed of 33,600 bps.

The V.90 Standard

During the mid-1990s, Lucent/Rockwell and U.S. Robotics introduced independent modem protocols that allowed modem data transmission at speeds of 57,600 bits per second on standard phone lines. X2 Protocol for one-way 56-kbps data transmission was proposed by U.S. Robotics in 1996 and submitted as a proposed standard to the ITU. It takes advantage of a fully digital interface at one side of the connection to overcome signal decay in one direction, allowing an appropriately configured service provider to transmit data at 56 kbps to its subscribers; however, data from the subscribers to the provider will only travel at a 33.6-kbps rate.

The difference between modem standards and protocols became a confusing issue for many consumers with the introduction of the X2 and Kflex protocols. While the manufacturers proposed the protocols as standards, basic incompatibilities prevented true 56K connections from occurring between an X2-based modem and a Kflex-based modem. When a modem utilizing the X2 protocol dialed into a provider that used modems based on the K56flex protocol, a V.34bis 33.6-kbps connection resulted.

Due to this problem, the International Telecommunications Union (ITU) established the V.90 standard. With this standard, the X2 and Kflex protocols merge together and provide maximum modem performance. Since FCC restrictions limited the transfer speeds to 56,000 bps, modems using the protocols became known as 56K modems. With the V.90 standard recognized by modems at both the transmitting and receiving ends, the type of protocol used by a 56K modem will not harm performance. The standard only remains useful, however, if any V.90 client modem can reliably connect to any manufacturer's V.90-compliant modem.

V.90 modem technologies assume that one end of the modem session has a pure digital connection to a phone network and takes advantage of the high-speed digital connection. Seeing the public switched telephone network as a digital network, V.90 technology accelerates data downstream from the Internet to your computer at speeds of up to 56 Kbps. In this way, V.90 technology works differently than other standards, because it digitally encodes downstream data rather than modulating the data as analog modems do.

The V.90 data transfer operates asymmetrically and causes lower-bandwidth upstream transmissions from a computer to flow at the conventional rates of up to 33.6 Kbps. The lower data-transfer rate occurs because data sent from an individual modem takes the form of an analog transmission and mirrors the V.34 standard. The downstream data transfer that may occur from an ISP takes advantage of the higher speed V.90 rates and allows faster downloading of large files or World Wide Web pages that include sound and video. *Table 6-8* compares the data-transfer rate of different modem standards.

Analog Modem Operation

Figure 6-9 depicts a modem-to-modem communication from one computer to another. In the diagram, a computer user located in one state wishes to send an electronic mail message to a computer user located in another

Connection	Bits/Sec.	Bytes/Sec.	KB/Min.	MB/Hour	MinSec/MB
Modem	9,600	1200	70	4	14m 33s
Modem	14,400	1800	106	6	9m 42s
V.34 Modem	28,800	3600	211	12	4m 51s
Modem	33,600	4200	246	14	4m 09s
V.90 Modem	42,000	5250	308	18	3m 19s
V.90 Modem	50,000	6250	366	22	2m 48s

Table 6-8.
Data-transfer rate of different modem standards.

Figure 6-9.
A modem-to-modem communication from one computer to another.

state. Both send and receive messages through different Internet service providers. As the communication begins, the modem found in the sender's computer responds to a software command and dials the number of the local ISP. Most Internet service providers employ banks of modems that handle incoming calls and transfer the information from those calls to other servers. Within the modem bank, one modem responds to the call from the computer user's modem and allows the computer user to connect with the ISP server.

Carrier Detect

Once the ISP modem answers the call, it also responds with a set frequency tone that notifies the sender modem that contact has occurred with another modem. In turn, the sender modem acknowledges the tone and recognizes it as a modem response. After sending the guard tone, the ISP modem transmits an unmodulated carrier wave that occurs as a continuous tone. The frequency of the tone depends on the data-transfer speed of the modem.

With the tone in place, the sender modem waits for the frequency that corresponds with the modem data-transfer rate and then begins to send a carrier signal. In response, the ISP modem sends a receiving carrier signal that has a slightly different frequency. If the sender modem attempts to connect with the ISP modem and the ISP modem doesn't respond, then the sender will send the result code: NO CARRIER to the computer. The CARRIER DETECT result code indicates that the modem has detected the carrier of the second modem.

As long as the modem connection exists, the carrier signals remain. The absence of the carrier signals ends the connection. Referring to *Figure 6-10*, we see that an external modem has a carrier detect (CD) indicator

Figure 6-10.
An external modem, which has a carrier detect (CD) indicator that lights with the sending of the carrier.

that lights with the sending of the carrier. As the modems query about data-transfer capabilities that include error correction, protocols, and standards, the receiving device adds carriers that correspond with the transmission speed, while the sending modem answers with additional carrier frequencies.

Carriers that remain unanswered drop from the signal transaction. For example, 33.6 modem dialing into a 56K modem will recognize and answer only the 33.6 carriers. The 56K modem will also drop carriers and the data transfer will occur at 33,600 bps. In some protocols, the exchange of carriers appears as a series of beeps going from low to high or as another burst of static. At this point, the modem has emitted:

- Dialing tones
- Ringing tones
- A longer tone that varies in frequency
- A higher continuous frequency
- Static

Data Transmission

Once the modems begin "handshaking" and establish a connection, the data transmission can begin. As a result, each modem begins to accept data from the local computer. Each stream of digital information traveling across the phone line causes the frequencies of the send/receive tones to constantly change.

Adoptive Equalization

As the modems begin to communicate, the adoptive equalization stage occurs. With adoptive equalization, the modems measure the quality of the line and adjust transmission/reception to compensate for line imperfections. In terms of audible tones, adoptive equalization includes the sequence of a short low buzz, a lower hiss, another beep and a higher hiss, another beep, and then a normal sounding hiss.

Retrain, Fall Back, and Fall Forward

During the data transmission, the modems continuously monitor the status of the line, the quality of received signals, the number of corrected errors, and the validity of data on each of the carriers. If the modems sense excessive problems, a reevaluation of the line occurs through the interruption of the carriers and the sending of a low burst tone on the line. With this, the modems retrain, or again work through the processes of sending out carriers and adoptive line equalization.

Usually, the initial retrain will result in a lower speed connection that uses only a limited number of carriers and has a higher chance for successful connection. The use of a lower speed connection for these purposes is called Automatic Fall Back. If the modem sees only a few errors, Fall Forward occurs and the device initiates another retrain that uses higher speeds and more carriers. When the connection with multiple carriers occurs successfully, the data-transfer rate also increases.

Summary

Chapter 6 introduced analog modem technology and discussed the principles of heterodyning, modulation, and demodulation. In addition, the chapter covered the internal operation of modems by illustrating the operation of a simple diode detector and showing the pin connections for DTE and DCE devices. The chapter built from this discussion with a detailed look at UART ICs. As the chapter progressed, it also considered the differences between internal and external modems while discussing modem-to-modem communication in a WAN environment.

While Chapter 6 didn't cover digital modems, the overall consideration of analog modems laid the foundation for later chapters in this book. Digital subscriber line and cable transmission technologies rely on digital modems. With each of those, as well as with other technologies, the modems rely on the same principles discussed in this chapter.

While Chapter 6 didn't cover digital modems, the overall consideration of analog modems laid the foundation for later chapters in this book.

Chapter 7

ISDN and Frame Relay

Introduction

An integrated services digital network (ISDN) provides an on-demand digital telephony and data-transport service between two points. In brief, the ISDN technology digitizes the telephone network and permits the transmission of voice, data, text, graphic images, music, video images, and other information over existing telephone connections to devices attached to local-area networks. *Figure 7-1* shows the connection of an ISDN to a 10Base2 Ethernet network.

To accomplish this, an ISDN works from the same conceptual model that we considered with public switched telephone networks. Indeed, designers had intended the circuit-switched ISDN as a replacement for those networks and as a method for transporting digital information over existing copper telephone lines. Regional telephone carriers as well as local telephone operating companies offer ISDN services over networks that have high-speed capabilities and support packet switching.

ISDN also provides end-to-end transmission paths according to user calls and includes signaling and a variety of optional features. The end-to-end digital connectivity supports the high-speed transmission of voice and nonvoice applications.

Contrasted with public switched telephone networks, ISDN features advantages with an all-digital system, a common user interface, and the use of advanced networking methods. In addition, ISDN supports faster connections on a dial-up basis for data applications, support for simultaneous multiple applications, worldwide connectivity based on international standards, enhanced management functions, and decreasing

> *Contrasted with public switched telephone networks, ISDN features advantages with an all-digital system, a common user interface, and the use of advanced networking methods.*

Figure 7-1.
ISDN connected beween a PSTN and a LAN.

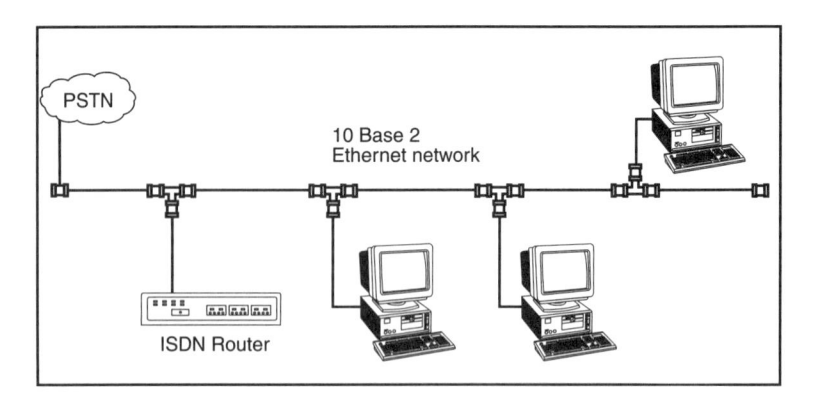

costs. With this, ISDN provides a well-defined, efficient, standardized method for transmitting and receiving digital information.

ISDN Services

We can categorize ISDN services as "bearer services" that use lower layer transport facilities or as "tele-services" that have distributed application capabilities. Bearer services allow a user to transfer information to another user without restriction on the type or format of the data. ISDN tele-services include telephony, facsimile, interactive videotex and videophone and offer guaranteed end-to-end compatibility regardless of the terminal equipment.

In addition, we can define two types of ISDN. Narrowband ISDN (N-ISDN) covers the ISDN BRI and PRI formats discussed in this chapter. We will consider Broadband ISDN (B-ISDN) in the next chapter. B-ISDN provides for very high-speed transmission using Asynchronous Transfer Mode technology and hasn't gained wide use due to the lack of an infrastructure and high implementation costs.

Single-point ISDN connections called the basic rate interface (BRI) use a standard two-wire telephone loop and offer two data channels. ISDN provides great flexibility for high bandwidth applications due to the capability to offer 23 data channels with the Primary Rate Interface (PRI). As a result, the emergence of ISDN represents an effort to standardize subscriber services, interfaces, and network and internetwork capabilities.

Moreover, ISDN provides superior data handling capabilities through the use of multiple data channels.

As we have seen, an ISDN requires 64-Kbps digital data channels to carry information. A digital control channel sends control information that manages the flow of information between the user and the network. X.25 data packets travel through the control channel at data transfer speeds that depend on the type of interface used for the transmission.

Physical BRI and PRI interfaces provide connectivity to an ISDN according to application need and traffic engineering. In terms of traffic engineering, an application may require either multiple BRI services or PRI services at some sites. A single BRI or PRI interface provides a multiplexed bundle of B and D channels. While the B channel provides bearer services such as high bandwidth data of up to 64 Kbps per B channel or voice services, the D channel provides the signaling and control channel and can also be used for low-bandwidth data applications. The selection of BRI or PRI interfaces becomes the first step in the ISDN design process.

ISDN BRI

The ISDN basic rate interface service offers two B channels and one D channel, or a configuration called 2B+D. While the BRI B-channels operate at 64 kbps and carry user data, the BRI D-channel service operates at 16 kbps and carries control and signaling information. One B-channel cannot carry voice and data simultaneously. Since ISDN BRI also provides framing control, the total bit rate climbs to 192 kbps.

Service Profile Identifiers

Every ISDN carrier establishes a service profile identifier (SPID) that identifies the line configuration of the BRI service. A SPID allows multiple voice and data ISDN devices to share the local loop. Since SPID numbers have no set format, the numbers vary according to switch vendor and the carrier.

Each SPID points to line setup and configuration information. When a device attempts to connect to the ISDN network, the device performs a D channel

The ISDN basic rate interface service offers two B channels and one D channel, or a configuration called 2B+D.

Layer 2 initialization process that causes a TEI to be assigned to the device. From there, the device attempts D channel Layer 3 initialization. Incorrect configuration or no configuration of necessary SPIDs causes the failure of the Layer 3 initialization and makes the use of ISDN services impossible. A typical SPID configuration appears as:

- interface BRI0
- isdn spid1 0835866201 8358662
- isdn spid2 0835866401 8358664

Part of the SPID command specifies the seven-digit local directory number (LDN) assigned by the service provider and used for call routing. The LDN becomes necessary for the reception of incoming calls on B channel 2 and when two SPIDs are used for switching. Configuring the LDN causes incoming calls to B channel 2 to be answered properly. If the LDN is not configured, incoming calls to B channel 2 may fail. Each SPID associates with an LDN.

ISDN PRI

Offered as a high-speed alternative to ISDN BRI, the ISDN PRI service operates on traditional T1 leased lines between the customer premise equipment and the ISDN switch. Rather than terminating at an end user's terminal equipment, a PRI line serves as a trunk between customer-based switching equipment and an ISDN termination at the central office. In North America and Japan, T1-based ISDN PRI provides 23 B channels and one D channel (23B+D), or a total bit rate of 1.544 Mbps. In other parts of the world, E1-based PRI provides 30 64 Kbps B channels and one 64 Kbps D channel (30B+D) for a total bit rate for 2.048 Mbps.

ISDN Components

ISDN components include terminals, terminal adapters, network-termination devices, line-termination equipment, and exchange-termination equipment. We can refer to specialized ISDN terminals as terminal equipment type 1 (or TE1) and non-ISDN terminals as terminal equipment type 2 (or TE2). Specialized ISDN terminals connect to the ISDN network through a four-wire, twisted-pair digital link.

Chapter 7: ISDN and Frame Relay

In contrast, non-ISDN terminals connect to the ISDN through a terminal adapter that generates traffic on the ISDN line. Terminal adapters include devices such as ISDN phones and ISDN modems. Pictured in *Figure 7-2*, an ISDN terminal adapter works as either a stand-alone device or a board inside the terminal. A single computer user can have direct control of the initiation and release of an ISDN session through the PC terminal adapter. If the non-ISDN terminal operates as a stand-alone device, it connects to the terminal adapter through a standard physical-layer interface such as an RS-232-C cable. The addition and removal of a secondary B channel requires the use of automated mechanisms. *Figure 7-3* depicts an ISDN network configured for both ISDN and non-ISDN terminal devices.

Network termination type 1 (NT1) or network termination type 2 (NT2) devices connect the four-wire subscriber wiring to the conventional two-wire local telephone loop. In North America, the NT1 operates as a customer

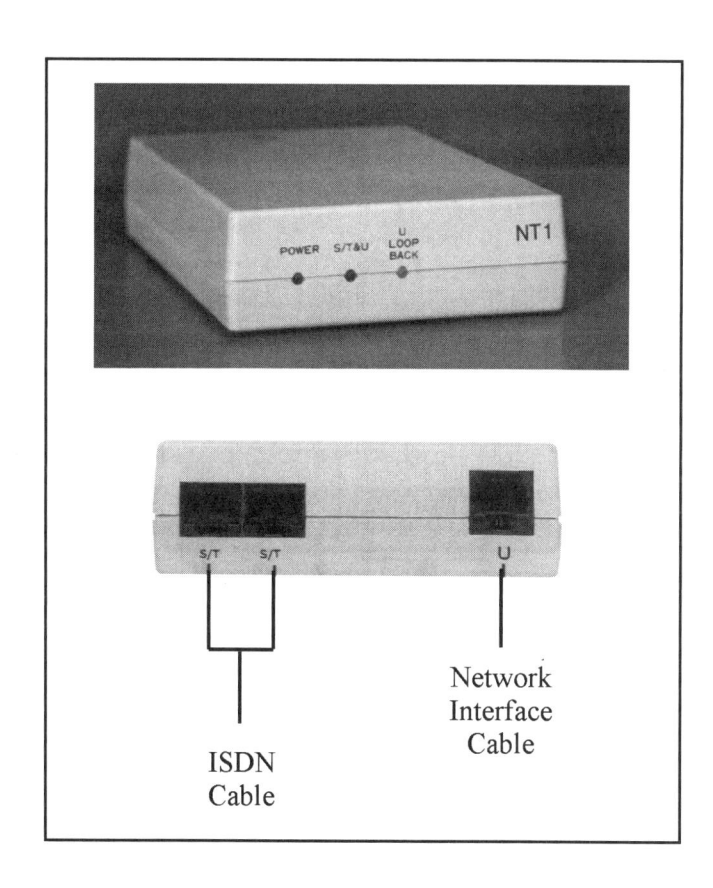

Figure 7-2.
A photograph of an ISDN terminal adapter, which works as either a stand-alone device or a board inside the terminal.

Figure 7-3.
An ISDN network configured for both ISDN and non-ISDN terminal devices.

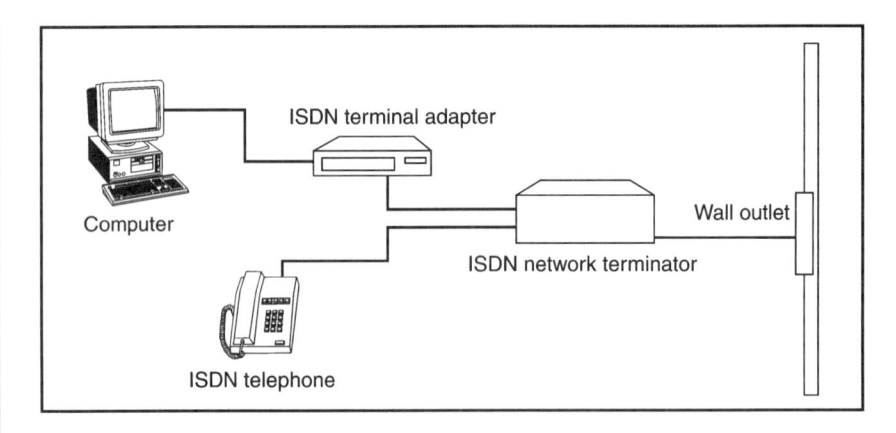

premises equipment (CPE) device. The NT1 operates as part of the network provided by the carrier in other parts of the world.

The NT2 provides more complicated functionality than the NT1 and operates within a digital private branch exchange. Network termination type 2 devices provide switching, perform Layer 2 and 3 protocol functions and concentration services, and include devices such as PABXs, routers, concentrators, and multiplexers. In addition to the NT1 and NT2 devices, a hybrid device called the NT1/2 combines the functions of an NT1 and an NT2. *Figure 7-4* shows an ISDN router.

Termination of the BRI local loop occurs at an NT1 found at the customer premise. The ISDN standard uses reference points that define logical interfaces between functional groupings such as terminal adapters and

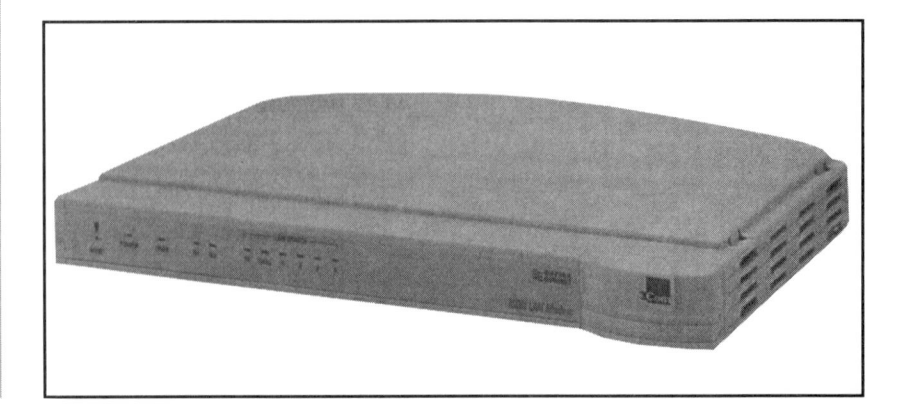

Figure 7-4.
A photograph of an ISDN router.

network termination type 1 devices. Moving to *Figure 7-5*, we see that ISDN reference points include:

- R—The reference point between non-ISDN equipment and a Terminal Adapter. Any R Interface is manufacturer dependent. The use of an R interface allows a non-ISDN device to appear as if it were ISDN-compliant.

- S—The reference point between user terminals and the NT2. In practice, the S interface connects between the network termination equipment and ISDN terminal equipment such as a Terminal Adapter.

- T—The reference point that connects switching equipment to an NT1.

- U—The reference point between line-termination equipment and the carrier network. The U (United States) reference point is relevant only in North America, where the NT1 function isn't provided by the carrier network.

Figure 7-5 also shows the configuration of a sample ISDN with three devices attached to an ISDN switch at the central office. Two ISDN-compatible devices attach through an S reference point to NT2 devices. A third standard, non-ISDN telephone device attaches through the reference point to a terminal adapter. Any of these devices could attach to an NT1/2 that

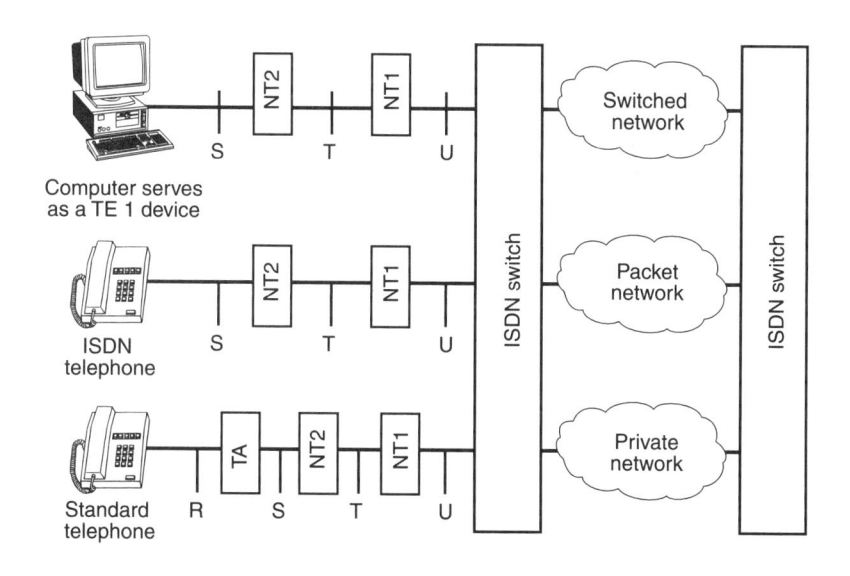

Figure 7-5.
ISDN reference points and the configuration of a sample ISDN.

replaces both the NT1 and the NT2. In addition, similar user stations attach to the far-right ISDN switch.

At the local exchange, ISDN central office switches provide local termination and exchange termination. Local termination operates with the transmission facility and termination of the local loop. The exchange termination function deals with the switching portion of the local exchange. During these operations, the exchange termination function demultiplexes information on the B and D channels. From there, the B channel information routes to the first stage of the circuit switch, while D channel packets routes to D channel packet separation circuitry.

The ISDN Physical Layer

The ISDN physical layer relies on different frame formats for inbound and outbound frames. An inbound frame travels from the network to a terminal, while an outbound frame moves from a terminal to the network. ISDN frames have a length of 48 bits, with 36 of the bits representing data. Bits found within the ISDN physical layer provide:

- Synchronization
- Adjustment of the average bit value
- Assurance of contention resolution when several terminals on a passive bus contend for a channel
- Device activation
- Handling of user data

Multiple ISDN user devices can physically attach to one circuit. With this configuration, collisions can result if two terminals simultaneously transmit data. As a result, the ISDN provides features that establish contention for links. Terminals cannot transmit into the D channel unless the device first detects a specific number of ones that indicate a "no signal" condition that corresponds to a preestablished priority. By assigning priorities, this technique allows only one terminal to transmit its D channel message at a time.

After a successful D-message transmission, the contention rules reduce the priority of the terminal by requiring the device to detect more continuous ones before transmitting. Terminals cannot raise their priority until all other devices on the same line have had an opportunity to send a D message. Telephone connections have higher priority than all other services, and signaling information has a higher priority than nonsignaling information.

ISDN Signaling

The signaling system employed by ISDN controls activities and actions of the network. Layer 2 of the ISDN consists of the Link Access Procedure D channel (LAPD) and provides a signaling protocol for the network. With an all-digital network such as that seen with ISDN, a method must exist for controlling the routing of calls. Along with this requirement, digital networks also require a method for transmitting the control data down the network to ensure that the call reaches the correct destination.

Using supervisory, information, and unnumbered frames, layer 2 works across the D channel and ensures the proper flow and reception of control and signaling information. Depending on the condition of the first address byte, the LAPD address field may have a length of either 1 or 2 bytes. A set first byte enables a one-byte address, while the opposite condition enables a two-byte address.

The first address-field byte contains an identifier service access-point identifier (SAPI) that identifies the portal for providing LAPD services to Layer 3. Within the SAPI, one bit indicates whether the frame contains a command or a response. The terminal endpoint identifier (TEI) field identifies the use of either a single terminal or multiple terminals, while a TEI consisting of all ones indicates a broadcast.

In-channel Signaling. Control information can travel along with call data through in-channel signaling or common-channel signaling. In-channel signaling allows the control information to travel with the call data along the same path. Within in-channel signaling, the in-band signaling technique places control signals in the same channel as the voice data and uses the same transmission bandwidth. Unfortunately, the control information must either precede or follow the call data—that is, control information only flows during intervals when call data isn't moving.

Unfortunately, the control information must either precede or follow the call data—that is, control information only flows during intervals when call data isn't moving.

Again considering in-channel signaling, the out-of-band signaling technique takes advantage of the unused portion of the voice bandwidth for the transmission of the control information. Since a channel has a 4,000-Hz bandwidth and with voice data consuming only 300 Hz to 3,400 Hz of that bandwidth, space remains for the transmission of the control information. As a result, the two different types of information can arrive at the destination simultaneously. But the use of out-of-band signaling limits the bandwidth available for the control information.

Common-channel Signaling. Common-channel signaling allows the control information to travel along a separate path from the call data. This occurs through the establishment of a new channel dedicated to the transport of control information for all voice channels. The control channel remains common to all voice channels. Given the use of the separate control channel, higher bandwidth becomes available for the transmission of control signal information.

ISDN technologies typically use common-channel signaling, except in situations where the network doesn't feature the full implementation of ISDN. The D, or data control, channel transmits control information at a rate of 16 Kbps with the BRI service and at a rate of 64 Kbps with the PRI service. The control information consists of call information that describes the transmitting and destination nodes as well as ISDN control information.

Signaling System Number 7

As mentioned, ISDN technologies use common-channel signaling. More specifically, though, an ISDN relies on the common-channel signaling found with the Signaling System Number 7 (SS7) standard. Controls given through the signaling system affect interaction between the user and the network and occur through the protocols found on the D channel called Digital Subscriber Signaling System No. 1 (DSS-1).

SS7 consists of four protocol layers that operate in terms of message transfer and network service. Rather than routing packets through the network, SS7 increases network flexibility by using a datagram approach. With this, SS7 treats each packet individually and allows individual nodes to make the best decision about routing. In short, the connectionless method used with the

Controls given through the signaling system affect interaction between the user and the network and occur through the protocols found on the D channel called Digital Subscriber Signaling System No. 1 (DSS-1).

datagram approach allows the packets to follow any pathway in the network. Messages contained within the datagram contain the ISDN number of the called party, the type of required connection, and type of circuitry found at the end point of the sending node.

Signaling System 7 also meets the requirements of typical telephone systems that use packet-switched networks. To accomplish this, SS7 provides telephone switches with signaling capabilities for telephony switch-to-switch DS0 connections. For these applications, SS7 offers benefits such as reduced call setup time, 64-Kbps data paths, caller-ID, and dialed number information (DNIS). In addition, the setup message includes a bearer capability information element (IE) that indicates the type of application carried on the B channel. All this allows the application to better utilize the ISDN connection.

Prior to use of the SS7 standard, in-band signaling provided the only solution for end-to-end call management and operated by subtracting bits from the DS0 trunks. Switch-to-switch signaling used the eighth and least significant bit of each voice byte and didn't detract from the quality of the voice transmission. The available data channel, however, had a maximum transfer rate of 56 Kbps. End-to-end out-of-band signaling through SS7 and PRI/BRI D channels allows the placing of data calls through ISDN networks using the full 64-Kbps DS0 trunk.

SS7 Layers. Earlier in this section, we saw that SS7 uses four distinct layers that divide into message delivery and network service functions. The SS7 layers operate from the concept that a network has signaling points, or points that can handle control messages, and signal-transfer points, or points that route control messages. Because of this relationship, the signaling data-link layer corresponds to the Physical Layer of the OSI model and specifies the physical and electronic characteristics of connections between the network points.

The signaling data-link layer provides full-duplex data transmission at a transfer rate of 64 Kbps. A single PRI D channel can support caller and call-control information for as many as 20 separate ISDN connections. As a result, an ISDN could allocate transmission content to one PRI facility, while designating signaling information for another channel. With this, the ISDN provides nonfacility associated signaling.

Telecommunication Technologies

> *The SS7 signaling link layer uses three different types of signal units called the message signal unit, the link status signal unit, and the fill-in signal unit.*

Moving from the signaling data-link layer, the signaling link layer establishes error-free transmission of the data. In addition, the link layer ensures that data arrives at the destination in the same order as seen at the transmission point. Finally, the signaling link layer allows the receiver to control the flow of data from the sender.

The SS7 signaling link layer uses three different types of signal units called the message signal unit, the link status signal unit, and the fill-in signal unit. While the MSU carries the message data, the LSSU carries control information. In addition, the MSU also carries a forward-sequence number and a backward-sequence number that maintain the order of the signal units and a length-indicator that shows the length of the MSU. Message signal units also include service and signaling information, a message type indicator, a service type indicator, and a series of status indicators. The FISU pads a short message so that the message fills the designated message space.

SS7 also includes a signaling network layer that supports message handling and network management. With message handling, the signaling network layer decides whether a message has arrived at the proper destination or whether the message should forward to the next node. The signaling network layer also determines the best method for failure recovery and has the capability to dynamically reroute messages if the network overloads or fails.

Layer 3 of SS7 has a relationship with the OSI-model Network Layer, but does not provide a complete set of Network Layer functions. Because of this, SS7 adds a signaling connection control part (SCCP) that provides high-level services to the ISDN user part (ISUP) and the operations, administration, and maintenance (OA&M) part. The ISUP consists of a set of software messages that establish, control, and terminate an ISDN session. As a result, ISDN application software relies on the ISUP.

The software messages found within ISUP set up calls to other ISDN subscribers and include the:

- forward address that designates outgoing lines for the transmission of calling part information

204

- general setup that establishes requests or sends additional call-related information

- backward setup that confirms that the network has received call-routing information from the originating ISDN node

- call supervision that indicates the answering of a call, requests assistance from an international operator, or reports to the caller that the call setup didn't occur

- circuit supervision that releases circuits, holds circuits open during a network hiatus, indicates the release of circuits, and controls the circuit when a pause occurs at a user terminal

- circuit group supervision that controls a block of circuits called a circuit group

- in-call modification that allows a sending or receiving party to change the characteristics of a call from voice to data or from data to voice

- node-to-node messages that transfer information between two network nodes

- user-to-user messages that have no current definition

The ISDN Connection Layer

Two Layer 3 protocols support user-to-user, circuit-switched, and packet-switched connections. The protocols use a variety of call-establishment, call-termination, information, and miscellaneous messages that include SETUP, CONNECT, RELEASE, USER INFORMATION, CANCEL, STATUS, and DISCONNECT. These messages compare functionally with the messages provided by the X.25 protocol.

Datagram Encapsulation

The creation of an end-to-end path over the ISDN requires some method of datagram encapsulation so that connectivity can occur. While an ISDN can use the PPP, HDLC, X.25, and V.120 protocols for datagram encapsulation and delivery over the D channel, most internetworking designs use PPP as the encapsulation format. The point-to-point protocol provides a powerful and modular peer-to-peer mechanism for establishing data links, providing security, and encapsulating data traffic. Since internetworking peers negotiate a PPP when establishing a new connection, the protocol becomes useful for setting internetwork connectivity with network protocols such as IP and IPX. To support greater bandwidth needs, Multilink PPP can provide greater throughput for internetworking applications.

PPP Framing. An ISDN router may use synchronous PPP, while an ISDN terminal adapter connected to a PC serial port and the same network may use asynchronous PPP. When this occurs, two methods exist for providing framing compatibility. With the first, synchronous to asynchronous PPP frame conversion occurs at the ISDN terminal adapter. The second and lesser-preferred method uses the V.120 protocol to encapsulate the asynchronous PPP frames for transport across the ISDN.

PPP Link Control Protocol. The PPP Link Control Protocol (LCP) provides a method of establishing, configuring, maintaining, and terminating the point-to-point connection. Before the exchange of any network-layer datagrams can occur, the LCP must open the connection and negotiate the configuration parameters. At the completion of this phase, the LCP sends and receives a configuration acknowledgment frame.

PPP Authentication Protocols. After the link control protocol has established the PPP connection, PPP can employ an optional authentication protocol before proceeding to the negotiation and establishment of the Network Control Protocols. Authentication must occur as an option before the LCP establishment phase can exist as either bidirectional or unidirectional. With the bidirectional phase, both sides authenticate one another; with the unidirectional phase, the called side authenticates the other side. Most ISDN designs require the called device to authenticate the calling device. Aside from providing security benefits, authentication also establishes a foundation for DDR and Multilink PPP bundling.

PPP Network Control Protocols. PPP allows the simultaneous use of multiple network-layer protocols. Before this can occur, however, PPP uses a family of Network Control Protocols (NCPs) for establishing and configuring the different network-layer protocols. After LCP has established the connection and authentication has passed, the PPP nodes send NCP frames to negotiate and establish connectivity for one or more network-layer protocols. For example, PPP negotiates and establishes IPCP to support IP over a PPP connection. The successful implementation of IPCP allows the transmission of IP datagrams over the PPP connection.

Multilink PPP. Multilink PPP (MP) provides a standard for aggregating multiple PPP links and allows for multivendor interoperability. To accomplish this task, Multilink PPP defines a method for sequencing and transmitting packets over multiple physical interfaces. Using MP, BRI devices can double the available connection bandwidth across the link from 56/64 Kbps to 112/128 Kbps.

In addition, MP also defines a method of fragmenting and reassembling large packets. When an NCP packet arrives at a Multilink PPP master-interface for transmitting and has a length of more than 30 bytes, the packet can become fragmented and sent on each physical link in the MLP bundle. When MP packet fragments arrive on PPP destination, Multilink reassembles the original packets and sequences them correctly in the data stream. *Figure 7-6* provides a conceptual view of MP in action.

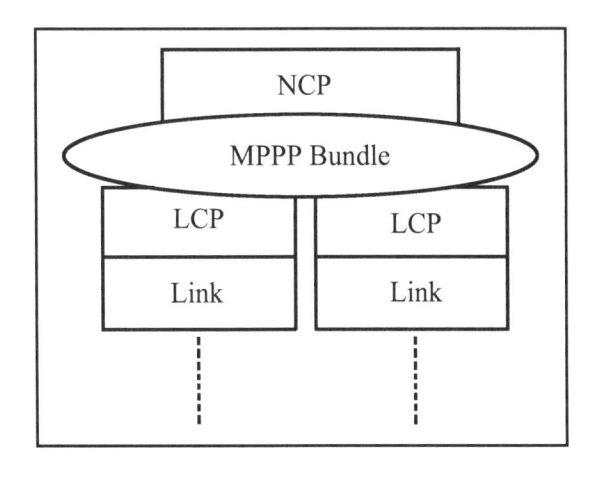

Figure 7-6.
A block diagram of Multilink PPP (MP) in action.

PPP Compression Control Protocol. The PPP Compression Control Protocol (CCP) defines a method for negotiating data compression over PPP links. These links can occur through leased lines or circuit-switched WAN links including ISDN. Compression increases throughput and shortens file transfer times.

Dial-on-Demand Routing

Along with those applications, ISDN also provides an ideal transport medium for Dial-on-Demand Routing and Dial Backup solutions for remote office/branch office connectivity. When building internetworking applications, designers must determine methods for initiating, maintaining, and releasing ISDN connections. Dial-on-Demand Routing (DDR) intelligently establishes and releases circuit switched connections as needed by internetworking traffic.

DDR begins with the arrival of qualified packets at a Dialer interface. During operation, DDR can create the illusion of full-time connectivity over circuit-switched connections by mimicking internetwork routing and directory services. From there, connectivity over the ISDN occurs until the network encounters set period of inactivity and disconnects. Dial-on-Demand Routing adds and removes additional ISDN B channels from the Multilink PPP bundles through the use of configurable thresholds. *Figure 7-7* depicts the use of DDR for internetworking between ISDN connected sites.

Frame Relay

Frame Relay provides a high-performance, packet-switched WAN protocol that operates at the physical and data link layers of the OSI reference model. Originally designed for use across ISDN interfaces,

Figure 7-7.
Using DDR for internetworking between ISDN connected sites.

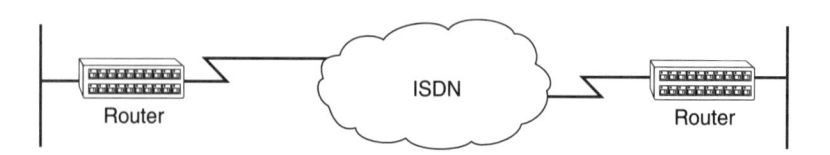

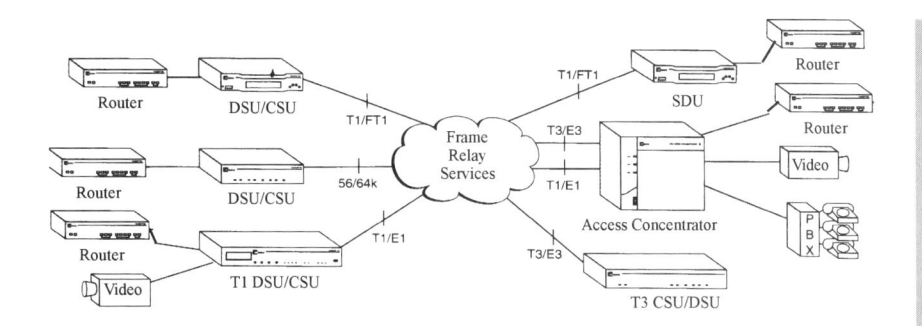

Figure 7-8.
Frame Relay applications, which may interconnect with other types of network services.

Frame Relay has become useful over a variety of other network interfaces. As *Figure 7-8* shows, Frame Relay services may interconnect with other types of network services and provide high-bandwidth connectivity over a variety of interfaces.

Frame Relay offers fewer of the robust capabilities—such as windowing and retransmission of lost data—seen with the X.25 protocol, because it operates over WAN facilities that offer more reliable connection services and a higher degree of reliability. Since Frame Relay operates as a layer 2 protocol suite, it offers higher performance and greater transmission efficiency than X.25 and works well for WAN applications such as interconnecting LANs through DTEs and DCEs.

Frame Relay Virtual Circuits

Frame Relay provides connection-oriented data-link layer communication through a Frame Relay virtual circuit. Operating as either a switched virtual circuit or permanent virtual circuit, Frame Relay virtual circuits establish a logical connection created between devices across a Frame Relay packet-switched network. In addition, virtual circuits provide a bidirectional communications path from one DTE device to another and are uniquely identified by a data-link connection identifier (DLCI). Since a number of virtual circuits can multiplex into a single physical circuit for transmission across the network, Frame Relay can reduce the equipment and network complexity required for connecting multiple DTE devices.

Most—if not all—Frame Relay virtual circuit applications use the permanently established connections and consistent data transfers between

Frame Relay reduces network overhead by implementing simple congestion-notification mechanisms rather than explicit, per-virtual-circuit flow control.

DTE devices that occur through permanent virtual circuits. Communication across a PVC doesn't require the call setup and termination states seen with switched virtual circuits. Permanent virtual circuits always operate with either a data transfer where DTE devices transmit data between one another over the virtual circuit, or in an idle condition where the devices remain active but no data transfer occurs.

Data-Link Connection Identifier. Data-link connection identifiers (DLCIs) identify Frame Relay virtual circuits. Service providers such as telephone companies assign DLCI values that have local significance—that is, the DCLI values don't remain unique to the Frame Relay WAN. For example, two DTE devices connected through a virtual circuit may use different DLCI values when referring to the same network connection.

Congestion-Control Through Frame Relay. Frame Relay reduces network overhead by implementing simple congestion-notification mechanisms rather than explicit, per-virtual-circuit flow control. Since Frame Relay becomes implemented on reliable network media and leaves flow control to higher-layer protocols, the network doesn't sacrifice data integrity. Frame Relay implements forward-explicit congestion notification (FECN) and backward-explicit congestion notification (BECN).

A single bit contained in the address field of the Frame Relay frame header controls FECN and BECN. In addition, the Frame Relay frame header contains a discard eligibility (DE) bit that identifies less important traffic that the system can drop during periods of congestion. Frame Relay initiates the FECN mechanism when a DTE device sends Frame Relay frames into the network. If the network congestion exists, DCE switches set the value of the FECN bit to one.

When the frames reach the destination DTE device, the address field containing the set FECN bit indicates that the frame experienced congestion in the path from source to destination. The DTE device can relay this information to a higher-layer protocol for processing. Depending on the implementation, Frame Relay may initiate flow-control or ignore the indication. As with FECN, the BECN bit is part of the address field in the Frame Relay frame header. DCE devices set the value of the BECN bit to one in frames traveling in the opposite direction of frames with a set FECN bit. This informs the receiving DTE device that a particular path through the network has become congested. Then, the DTE device can relay this information to a higher-layer protocol for processing. Again, as we saw with FECN, Frame Relay may initiate flow control or ignore the indication.

Frame Relay Discard Eligibility. Frame Relay uses a discard eligibility (DE) bit to indicate that a frame has lower importance than other frames. The DE bit is part of the Address field in the Frame Relay frame header. DTE devices can set the value of the DE bit of a frame to 1 to indicate that the frame has lower importance than other frames. When the network becomes congested, DCE devices will discard frames with the set DE bit before discarding other frames. The use of the DE bit reduces the likelihood of the Frame Relay DCE devices dropping critical data during periods of congestion.

Frame Relay Error Checking. Frame Relay uses a common error-checking mechanism known as the cyclic redundancy check (CRC). The CRC compares two calculated values when determining whether errors occurred during the transmission of data from source to destination. Implementing error checking rather than error correction reduces network overhead. Error correction occurs at higher-layer protocols.

Frame Relay Local Management Interface. Cisco Systems, StrataCom, Northern Telecom, and Digital Equipment Corporation built the Local Management Interface (LMI) as a set of enhancements to the basic Frame Relay specification. The LMI offers a number of extensions for managing complex internetworks. LMI extensions include global addressing, virtual-circuit status messages, and multicasting. In addition, the LMI global addressing extension gives Frame Relay DLCI values global rather than local significance.

The global addressing extension adds functionality and manageability to Frame Relay internetworks. For example, the Frame Relay can identify individual network interfaces and the end nodes attached to them by using standard address-resolution and discovery techniques. In addition, the entire Frame Relay network appears as a typical LAN to routers outside the network.

LMI virtual circuit status messages provide communication and synchronization between Frame Relay DTE and DCE devices. These messages periodically report on the status of permanent virtual circuits and prevent the sending of data to nonexistent virtual circuits. The LMI multicasting extension allows the assignment of multicast groups and saves bandwidth by allowing the sending of routing updates and address-resolution messages only to specific groups of routers. The extension also transmits reports on the status of multicast groups in update messages.

Frame Relay Frame Formats. As shown in *Figure 7-9*, flags indicate the beginning and end of the frame. A header and address area, the user-data portion, and the frame-check sequence (FCS) make up the frame. The two-byte long address area consists of 10 bits representing the actual circuit identifier and 6 bits of fields related to congestion management. *Table 7-1* describes the parts of the frame.

Virtual Remote Nodes

By using network address translations (NAT) features, a remote site can appear to the ISDN NAS as a single remote node IP address. Virtual Remote Nodes can alleviate IP address consumption problems and the routing design complexity often associated with large-scale ISDN DDR deployment. As a result, the ISDN can maintain support of LAN and DDR-based connectivity from the remote site. National Address Translations use the IP address received from the NAS during IPCP negotiation. All packets routed between the LAN and the PPP link have IP address and UDP/TCP port addresses translated to appear as a single IP address. The ISDN uses the port number translation to determine which packets need to return to which IP addresses on the LAN.

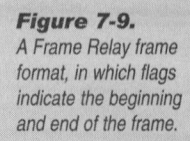

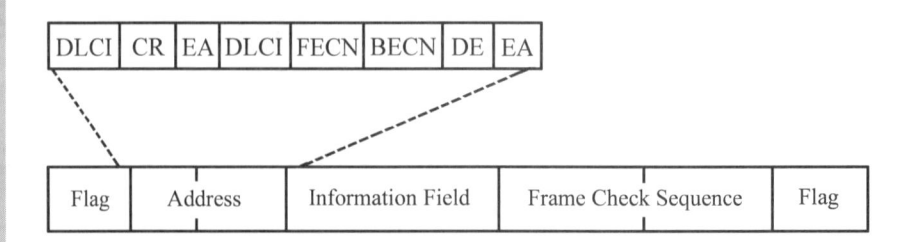

Figure 7-9.
A Frame Relay frame format, in which flags indicate the beginning and end of the frame.

| DLCI | CR | EA | DLCI | FECN | BECN | DE | EA |

| Flag | Address | Information Field | Frame Check Sequence | Flag |

Frame Portion	Purpose
Flag	Delimits the beginning and end of the frame. The value of this field is always the same and is represented either as the hexadecimal number 7E or the binary number 01111110.
Address	Contains the DLCI, Extended Address, C/R Bit, and congestion-control mechanisms. The Extended Address (EA) is used to indicate whether the byte in which the EA value is 1 is the last addressing field. If the value is 1, then the current byte is determined to be the last DLCI octet. The C/R is the bit that follows the most significant DLCI byte in the Address field. The C/R bit is not currently defined.
Data	Contains encapsulated upper-layer data. Each frame in this variable-length field includes a user data or payload field that will vary in length up to 16,000 octets. This field serves to transport the higher-layer protocol packet (PDU) through a Frame Relay network.
Frame Check Sequence	Ensures the integrity of transmitted data. This value is computed by the source device and verified by the receiver to ensure integrity of transmission.

Table 7-1.
Parts of the Frame Relay frame.

ISDN Providers and Customers

From one perspective, an ISDN consists of providers such as telephone companies or private enterprises. ISDN providers offer the networks that combine to form the worldwide-integrated digital network fabric. We can also describe ISDN networks from a customer perspective where a user sees a communications "line" on the wall. For example, the small office/home office (SOHO) market takes advantage of the bandwidth and data transfer speeds seen with ISDN. The high bandwidth and data transfer speeds given through ISDN have prompted many SOHO sites to use ISDN BRI services. SOHO designs typically involve dial-up only and can take advantage of emerging address translation technology to simplify design and support. Using these features, the SOHO site can support multiple devices and appear as a single IP address.

ISDN Dial Backup

With Dial Backup, ISDN operates as a backup service for a leased-line connection between the remote and central offices. If the primary connectivity breaks down, Dial Backup establishes an ISDN circuit-switched connection and reroutes traffic over the ISDN. Once the primary link reinitializes, Dial Backup redirects the traffic to the leased and releases the ISDN. Dial Backup operates either with floating static routes and DDR or through the use of interface backup commands. Some ISDN Dial Backup option involves configuring the network on the basis of traffic thresholds as a dedicated primary link. If the traffic load exceeds a user-defined value on the primary link, the ISDN link activates and increases bandwidth between the two sites.

ISDN Security

Since internetwork devices can connect through the PSTN, internetworking design considerations must also include robust security. Because ISDN relies on the SS7 standard for signaling, the ISDN can deliver end-to-end Information Elements such as caller-ID and dialed number information service (DNIS). Along with PPP authentication, all this information can provide additional security for ISDN solutions.

Limiting ISDN Costs

High implementation costs have limited ISDN usage prior to the last few years. With ISDN operating as a circuit-switched connection, customers pay for ISDN services according usage. As a result, one of the primary goals when selecting ISDN for an internetwork becomes avoiding the cost of full-time data services. With this in mind, an evaluation of data traffic profiles and the monitoring of ISDN usage patterns to ensure the control of WAN costs rise in importance.

Most ISDN solutions can remain cost effective only if the ISDN B channels remain idle for most of the day. Because of this, Frame Relay offers a more cost-effective solution. Some ISDN service providers charge a per-connection and per-minute charge even for local calls. When considering selecting DDR design and parameters, consider local and long-distance tariff charges. ISDN Callback can centralize long-distance charges and—in turn—significantly reduce administrative overhead and provide opportunities for reduced rate structures.

Most ISDN solutions can remain cost effective only if the ISDN B channels remain idle for most of the day. Because of this, Frame Relay offers a more cost-effective solution.

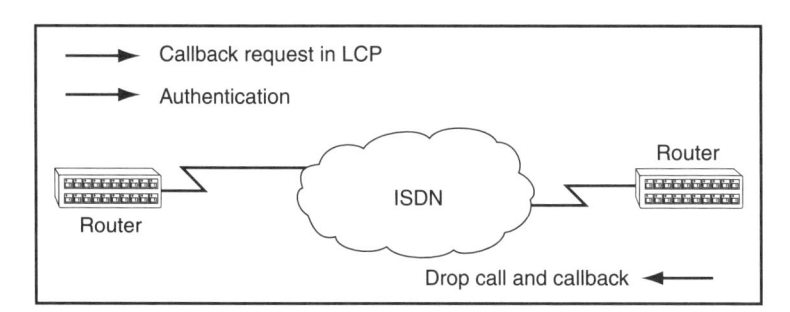

Figure 7-10.
The ISDN Callback.

Callback uses PPP and allows a router to initiate a circuit-switched WAN link to another device and request that device to call back. The router responds to the callback request by calling the device that made the initial call. Callback provides centralized billing for synchronous dial-up services and allows the customer to gain an advantage with tariff disparities on both a national and international basis. Since callback requires the establishment of a circuit-switched connection before passing the callback request, the router initiating the call that requests a callback incurs a small charge.

Moving to *Figure 7-10*, we see that the ISDN Callback follows these steps:

1) Router A brings up a circuit-switched connection to Router B.
2) Routers A and B negotiate PPP Link Control Protocol. Router A can request a callback, or Router B can initiate a callback.
3) Router A authenticates itself to Router B. Router B can optionally authenticate itself to Router A.
4) Both routers drop the circuit-switched connection.
5) Router B brings up a circuit-switched connection to Router A.

To provide total control over when the DDR connections are made, the network designer must carefully consider:

- Sites that initiate connections based on traffic
- Dial-out requirements
- Sites that can terminate connections based on idle links
- Support for directory services and routing tables across an idle connection
- Support for applications over DDR connections
- The number of users

Using SNMP

The Simple Network Management Protocol uses management information bases (MIBs) to store information about network events. With these MIBs, SNMP-compliant management platforms such as HP OpenView or SunNet Manager can query ISDN routers for statistics about interfaces and network information. For example, the Call History MIB stores call information for accounting purposes and provides historical data about an ISDN interface. The information includes the number of placed calls and call length.

ISDN and Internetworking

To gain the full benefits from an ISDN, the calling party, the receiving party, and the provider's network must use ISDN-compatible equipment. Many of the issues that slowed ISDN deployment involve the need for interconnection to legacy applications and equipment. In most instances, a need exists for attaching non-ISDN devices to an ISDN line. *Figure 7-11* shows an ISDN connected as part of an internetwork.

The ability to link ISDN to older services such as the telephone network also allows a gradual migration of technologies. When used for ISDN to ISDN, ISDN to PSTN, ISDN to CSPDN, or ISDN to Frame Relay connections, the ISDN network serves as a gateway for internetworking and facilitates the operating of older protocols on top of ISDN.

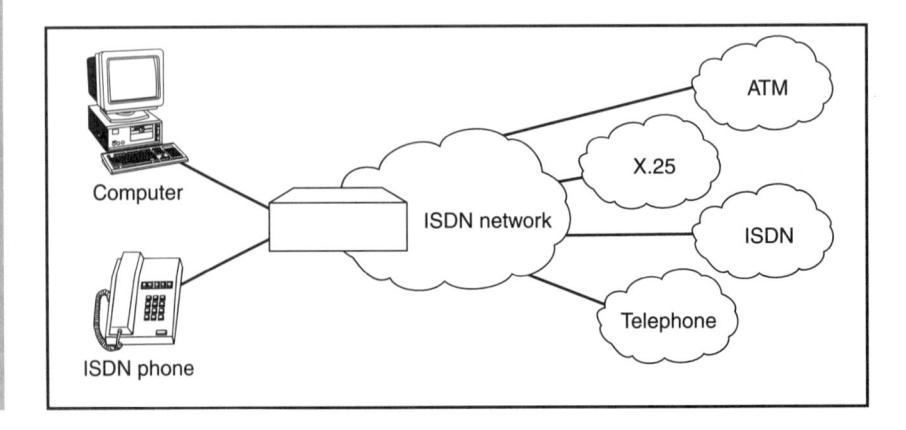

Figure 7-11.
ISDN functioning as part of an internetwork.

Integrating multiple independent ISDN networks stands as an essential requirement for international operation of ISDN services. As ISDN implementation grows across international borders, no single supplier can provide universal coverage. Connection between ISDN and the PSTN becomes essential for the consistent implementation of voice applications are involved. Internetworking of ISDN to PSTN involves interconnecting voice circuits and the standardization of signaling.

The connection of Circuit-Switched Public Data Networks to an ISDN may also require the conversion of signaling protocols and rate adaptation. For connections to a Packet-Switched Public Data Network through the B Channel, a terminal adapter provides the conversion between LAPB to LAPD. An X.25 packet level protocol can also operate using ISDN as the physical network.

Non-ISDN services make the most use of Frame Relay. With an ISDN-to-Frame Relay connection, the international ISDN number—a variable length number of up to 17 digits—uniquely identifies every subscriber. The ITU I.330 series of standards define the use of ISDN numbering and other network addresses. Subaddresses provide an additional addressing capability outside of the ISDN numbering plan and additional addressing within a private network accessed through an ISDN.

The ISDN Marketplace

ISDN connectivity offers solutions for many of today's business requirements, because of support for multiple high-bandwidth applications, decreasing implementation costs, and interoperability with other network services. Due to these factors, application demands for ISDN connectivity have grown. For example, the demand for Internet connectivity has also fueled demand for videoconferencing across the World Wide Web. Customer applications—such as medical imaging—that use extensive file transfers, secure on-line transactions, or utilize graphics or imaging require the bandwidth capacity of ISDN.

The use of ISDN has also grown due to a global commitment to a universal ISDN Infrastructure. Numerous inter-carrier agreements have prompted

Telecommunication Technologies

moves toward the use of ISDN as an international connectivity standard. In North America, the adoption of National ISDN-1 (NI-1) allowed interconnection among regional Bell Operating Companies and long distance carriers. The National ISDN-1 standard defines how ISDN-compatible devices signal their status to the carrier switches. As a result, ISDN has become available in more than 200 metropolitan areas.

Since the implementation of an integrated services digital network requires ISDN-specific equipment at the customer site, many customers have opted for other, less costly networking methods. But costs for devices such as ISDN terminal adapters have dropped as demand has increased. In addition, new technologies that combine NT1 functionality with voice/data adapters have dropped ISDN device costs down into the range of high-end analog modems.

Summary

ISDN technologies have become a cost-effective alternative for needs that don't require a point-to-point leased line. Applications for ISDN technologies include the interconnection of multiple LANs, videoconferencing, file-sharing applications between individuals or groups at two or more sites, and providing overflow capacity at high transfer rates. Chapter 7 provided information about implementing those applications by describing ISDN services.

Costs for ISDN terminal adapters have dropped as demand has increased. And new technologies that combine NT1 functionality with voice/data adapters have dropped ISDN device costs down into the range of high-end analog modems.

Chapter 8

ATM Networks, Frame Relay, and SMDS

Introduction

In Chapter 7, we explored ISDN technologies and found that two different types of ISDN exist—narrowband and broadband. Chapter 7 emphasized the characteristics and protocols that make up narrowband ISDN. Asynchronous Transfer Mode (ATM) technologies work from a standard developed by the International Telecommunication Union–Telecommunication Standardization Sector for cell relay technologies and from the format devised for broadband ISDN.

As a transport vehicle, ATM networks allow ISDN to deliver promised capabilities. When considering the transmission of voice information, data, and video information, each type of transmission places a different demand on the network. A circuit-switched public telephone system network can transfer a 64 Kbps voice data stream with no problem. Local-area networks route data packets over packet-switched networks at a transfer rate of 1.4 Mbps. The transfer of video information requires a data-transmission rate of 30 Mbps.

In addition to data-transfer rate differences, the three streams of data also differ in the type of transmission. While data moving between computers travels in short bursts, voice and video information require transmission at a constant bit rate. All this points us toward another key difference between the transmission of the three data types: the software that supports the transfer of each type relies on different protocols. ATM networks match the potential of ISDN by having the capability to support the simultaneous transfer of voice, data, and video information. Within the past several years, ATM has become the strategic technology for WAN managers across the world. As equipment

Asynchronous Transfer Mode (ATM) networks match the potential of ISDN by having the capability to support the simultaneous transfer of voice, data, and video information.

costs have decreased and new features have become available, new customer demand has risen. For larger enterprise networks, ATM delivers at Optical Channel rates that include OC-3 and OC-12. For midsize networks, the capability of ATM to provide multiplexed services that can aggregate several T1 circuits or fractional T3 circuits delivers data transfer speeds not seen previously at that level.

Looking Inside ATM Networks

ATM networks use cell-switching and multiplexing technologies that combine benefits of circuit switching and packet switching. Originally intended for public networks, the connection-oriented ATM has gained usage in both public and private networks. With cell relay, the network conveys voice, video, or data information in small, fixed-size cells.

The use of circuit switching in an ATM network provides guaranteed capacity and transmission delay, while the use of packet switching provides flexibility and efficiency for intermittent traffic. As a whole, ATM networks offer scalable bandwidth that ranges from a few megabits per second to many gigabits per second.

As the name shows, ATM networks work in an asynchronous world. Given this characteristic, ATM offers greater efficiency than that seen with synchronous technologies. For example, a network that relies on time-division multiplexing assigns a time slot to each user. No other station can send data in that time slot. If a station has data to send on a TDM-based network, the station must wait for the correct time slot before sending the data. If the station doesn't send data during the designated time slot, the time slot remains unused. The asynchronous nature of ATM uses time slots on demand.

ATM Cells

ATM transfers information in fixed-size units called cells. Each cell consists of 53 bytes, with the first five bytes containing cell-header information and the remaining 48 containing user information. The header contains information identifying the source of the transmission. When compared to the use of data packets, the use of small fixed-length cells provides benefits in the transfer of voice and video traffic. The downloading of a large data packet causes delays in data traffic.

An ATM cell header may use either a User Node Interface or a Network Node Interface format. The UNI header establishes communication between ATM endpoints and ATM switches in private ATM networks. Given these tasks, the cell header with the UNI format contains a Generic Flow Control (GFC) field. In comparison, the NNI header establishes communication between public ATM switches. Rather than relying on a GFC field for those tasks, the cell header with the NNI format has a Virtual Path Identifier (VPI) field that occupies the first 12 bits of the field and allows for larger trunks between the switches. *Figure 8-1* depicts the ATM cell as well as ATM cells with a UNI cell header and an NNI cell header. *Table 8-1* provides a listing and description of all the ATM cell-header fields.

ATM Layers

The ATM architecture corresponds to the Physical Layer and part of the Data Link Layer of the OSI reference model. With this, the ATM architecture breaks down into an ATM physical layer, the ATM layer, and the ATM adaptation layer. While the ATM physical layer provides tasks associated with the OSI-model physical layer, the ATM layer and ATM

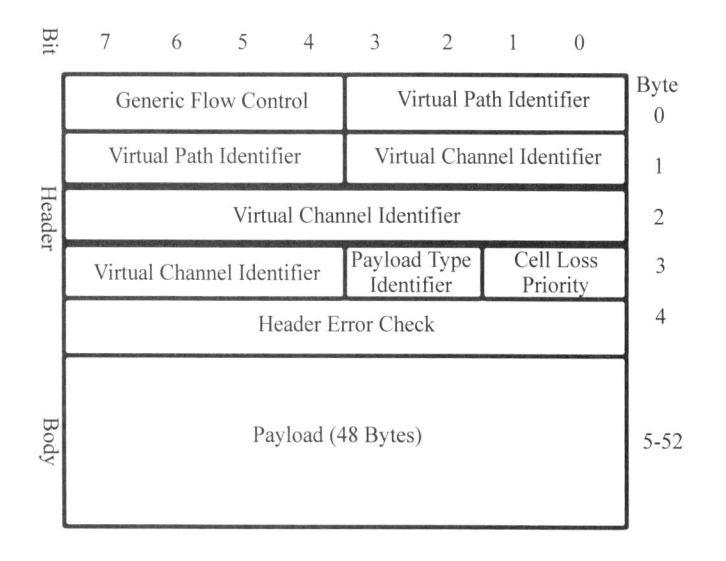

Figure 8-1.
The ATM cell format.

Telecommunication Technologies

Table 8-1.
Listing and description of ATM cell-header fields.

Field Type	Field Purpose
Generic Flow Control (UNI Only)	Provides local functions such as identifying multiple stations that share a single ATM interface. This field is set to its default value.
Virtual Path Identifier (NNI Only)	Combined with the VCI, this field identifies the next destination of a cell as it passes through a series of ATM switches on the way to its destination.
Virtual Channel Identifier (NNI Only)	Combined with the VPI, this field identifies the next destination of a cell as it passes through a series of ATM switches on the way to its destination.
Payload Type (PT)	Indicates in the first bit whether the cell contains user data or control data. If the cell contains user data, the second bit indicates congestion, and the third bit indicates whether the cell is the last in a series of cells that represent a single AAL5 frame.
Congestion Loss Priority (CLP)	Indicates whether the cell should be discarded if it encounters extreme congestion as it moves through the network. If the CLP bit equals 1, the cell should be discarded in preference to cells with the CLP bit equal to 0.
Header Error Control (HEC)	Calculates checksum only on the header itself.

adaptation layer provide data-link layer functions. Higher layers residing above the ATM adaptation layer accept user data, arrange the data into packets, and hand the data to the adaptation layer. *Figure 8-2* illustrates the ATM reference model.

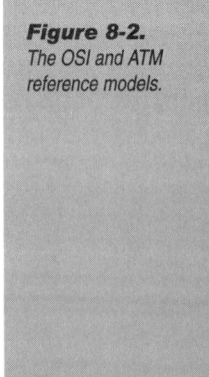

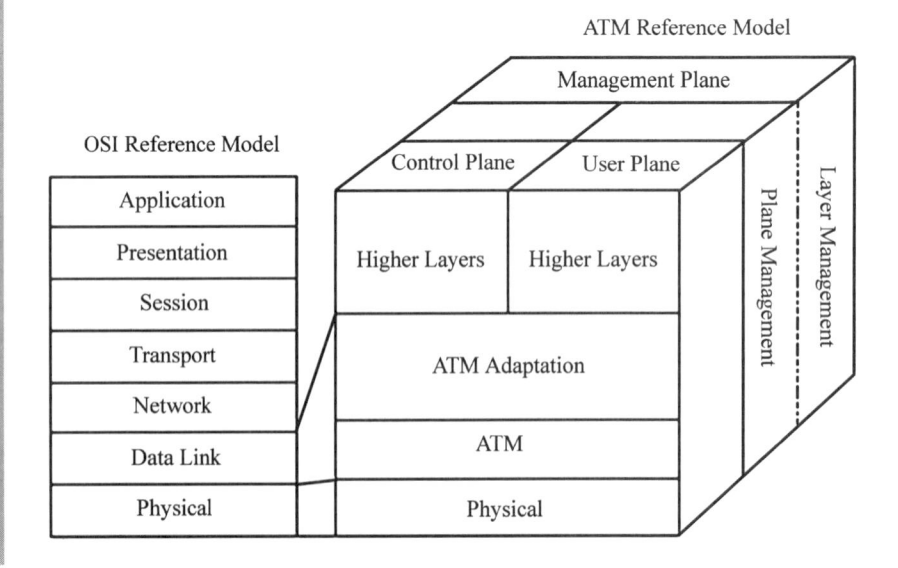

Figure 8-2.
The OSI and ATM reference models.

In addition to the layered architecture, the ATM reference model consists of the control, user, and management planes that span all layers. While the control plane generates and manages signaling requests, the user plane manages the transfer of data. The management plane manages layer-specific functions such as the detection of failures and protocol problems along with managing functions related to the entire system.

The ATM Physical Layer

Consisting of two sublayers, the ATM physical layer converts bits into cells, controls the transmission and receipt of bits on the physical medium, tracks ATM cell boundaries, and packages cells into the appropriate types of frames for the physical medium. The ATM physical layer divides into the physical medium-dependent (PMD) sublayer and the transmission-convergence (TC) sublayer. The PMD sublayer synchronizes transmission and reception by sending and receiving a continuous flow of bits with associated timing information.

In addition, the PMD sublayer specifies the type of physical media for including connector types and cable. ATM networks may use SONET/SDH media, multimode fiber, coaxial cable, or shielded twisted-pair cabling. Each of the media types supports a data transfer rate of 155 Mbps. The latter two types of physical media require the use of an 8B/10B-encoding scheme. As a result, many ATM service providers have taken the opportunity to move into lower-speed and lower-cost ATM access options such as xDSL, T1, and fractional DS-3.

Customers can obtain ATM services through both 1.5-Mbps and 45-Mbps ATM ports. The port speed defines the maximum rate at which the customer premises equipment can transmit and receive data from the ATM network. In many instances, businesses can attain OC-3 speeds through suitable transmission facilities. Interexchange carriers, local exchange carriers, or alternate vendors offer access to ATM interexchange carrier ports with data transfer speeds ranging from 64 Kbps to 2 Mbps. Given this connectivity and range of data transfer speeds, customers can connect X.25 and Frame Relay to ATM endpoints without technology upgrades.

Resuming our conceptual discussion of the ATM physical layer, the TC sublayer provides cell delineation, header error-control (HEC), sequence generation and verification, cell-rate decoupling, and transmission-frame adaptation. Cell delineation function maintains ATM cell boundaries and allows devices to locate cells within a stream of bits. HEC sequence generation and verification generates and checks the header error-control code to ensure the transmission of valid data. Cell-rate decoupling maintains synchronization and inserts or suppresses unassigned ATM cells so that the rate of valid ATM cells adapts to the payload capacity of the transmission system. Transmission frame adaptation packages ATM cells into frames acceptable to the particular physical-layer implementation.

ATM Adaptation Layer 1

The ATM Adaptation Layer 1 (AAL1) provides a connection-oriented service that handles circuit-emulation applications such as voice and video conferencing. In addition, AAL1 establishes a circuit-emulation service that accommodates the attachment of equipment currently using leased lines to an ATM backbone network. Since AAL1 requires timing synchronization between the source and destination, the layer depends on a medium—such as SONET—that supports clocking.

The AAL1 process prepares a cell for transmission in three steps. At the outset, the protocol layer inserts synchronous samples into the Payload field. For example, the insertion might involve one byte of data at a sampling rate of 125 microseconds. Second, the protocol layer adds Sequence Number and Sequence Number Protection fields to provide information that the receiving AAL1 uses to verify that the reception of cells in the correct order. Finally, AAL1 fills the remainder of the Payload field with enough single bytes to equal 48 bytes. *Figure 8-3* illustrates how the ATM Adaptation Layer 1 prepares a cell for transmission.

ATM Adaptation Layers 3 and 4

ATM Adaptation Layers 3 and 4 (AAL3/4) support both connection-oriented and connectionless data. The layers accommodate network service providers and align with the Switched Multimegabit Data Service (SMDS). AAL3/4 allows the transmission of SMDS packets over an ATM network.

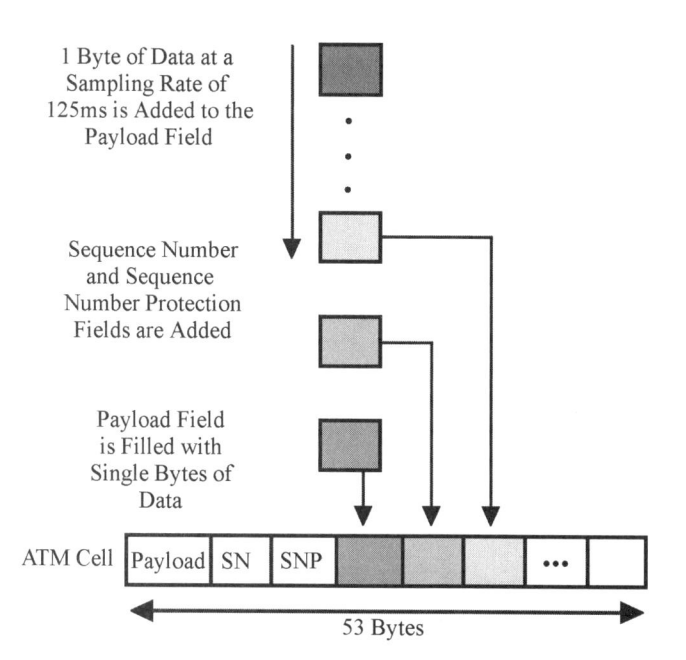

Figure 8-3.
The ATM Adaptation Layer preparing a cell for transmission.

As the layers prepare a cell for transmission, the convergence sublayer (CS) creates a protocol data unit (PDU) by attaching a beginning/end tag header and length field trailer to the frame. Next, the segmentation and reassembly (SAR) sublayer fragments and attaches a header to the PDU. Then the SAR sublayer appends a CRC-10 trailer to each PDU fragment for error control.

The completed SAR PDU then becomes Payload field of an ATM cell and receives a standard ATM header. An AAL 3/4 SAR PDU header consists of Type, Sequence Number, and Multiplexing Identifier fields. Type fields identify the placement of a cell within the message. A cell may occur at the beginning, as a continuation, or at the end of a message. Sequence number fields identify the reassembly order of cells. The Multiplexing Identifier field determines which cells from different traffic sources should interleave on the same virtual circuit connection and ensures the reassembly of the correct cells at the destination.

ATM Interconnect

The ATM Interconnect integrates all of the required functionality to transport ATM cells across a backplane architecture with high-speed cell traffic exceeding 1.5 gigabits per second to a maximum of 32 destinations. In addition, the device manages multiple service classes and monitors ATM traffic performance. *Figure 8-4* shows a diagram of the ATM Interconnect.

ATM Concentrator

The ATM Access Concentrator shown in *Figure 8-5* includes the system controller, two control ports, a PCMCIA slot, an ac power supply, a T1 network interface, and two user module slots, and one or two V.35 user ports. The flexible architecture and RISC processor found within the concentrator allows the device to allocate bandwidth to applications as needed. As a result, the use of the concentrator maximizes data, video, and voice WAN access.

Switched Multimegabit Data Service

In the last section, we found that AAL4/5 aligns with the Switched Multimegabit Data Service. SMDS is a high-speed, packet-switched,

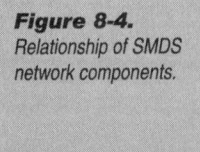

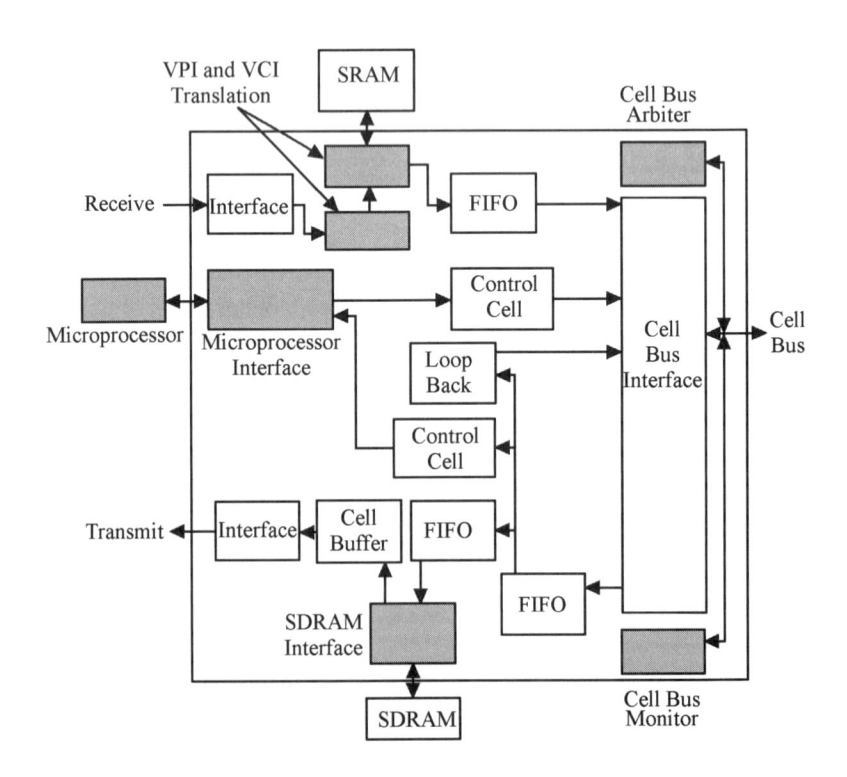

Figure 8-4.
Relationship of SMDS network components.

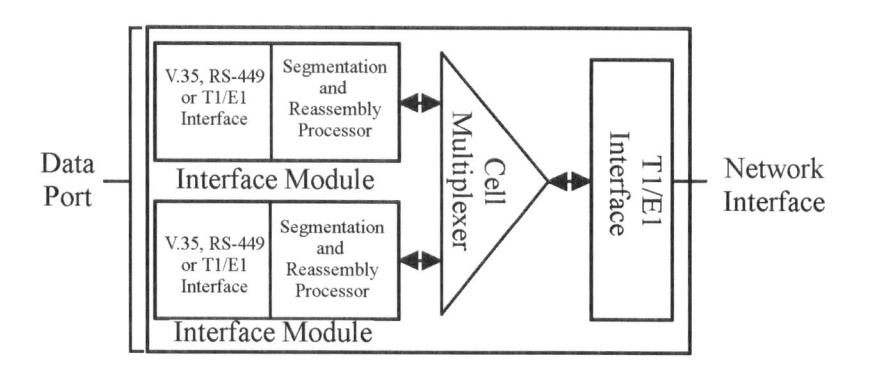

Figure 8-5.
Mapping SIP to the OSI Reference Model.

datagram-based WAN networking technology that allows communication over public data networks. The networking technology operates over fiber-based or copper-based media and supports speeds of 1.544 Mbps over DS-1 transmission facilities and 44.736 Mbps over DS-3 transmission facilities. In addition, SMDS has large data units that can encapsulate entire IEEE 802.3, IEEE 802.5, and FDDI frames.

SMDS networks provide high-speed data service through customer premises equipment, carrier equipment, and the subscriber network interface (SNI). Going back to Chapter 9, CPE consists of terminal equipment typically owned and maintained by the customer and includes end nodes such as terminals and personal computers, and intermediate nodes such as routers, modems, and multiplexers. In addition to finding intermediate nodes at the customer premises, SMDS carriers also provide intermediate nodes.

Carrier equipment consists of high-speed WAN switches that must conform to certain network equipment specifications that define network operations, the interface between a local carrier network and a long-distance carrier network, and the interface between two switches inside a single carrier network.

The SNI provides the interface between CPE and carrier equipment and exists as the point where the customer network ends and the carrier network begins. Along with those functions, SNI also ensures that the carrier SMDS network and operation remain transparent to the customer. *Figure 8-6* shows the relationship between the three components of an SMDS network.

Figure 8-6.
SIP Level 3 PDU fields.

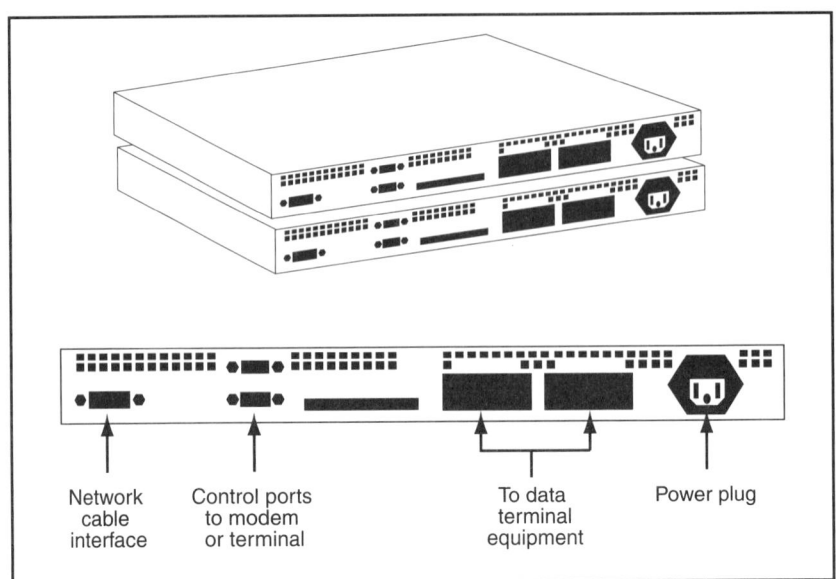

[Diagram showing device with labels: Network cable interface; Control ports to modem or terminal; To data terminal equipment; Power plug]

SMDS networks rely on the SMDS Interface Protocol (SIP) for communications between customer premises equipment and SMDS carrier equipment. The SIP provides connectionless service across the subscriber-network interface—allowing the CPE to access the SMDS network—and has a basis with the IEEE 802.6 Distributed Queue Dual Bus (DQDB) standard for cell relay across metropolitan-area networks.

SMDS Interface Protocol

SIP Level 3 operation begins with the passing of user information in the form of SMDS service data units (SDUs). The SMDS SIP header and trailer encapsulate the SDUs and produce a frame called the Level 3 protocol data unit. SIP Level 3 PDUs pass to SIP Level 2 that operates at the MAC sublayer of the data level layer. The protocol segments the PDUs into the uniformly sized 53-octet ATM cells.

From that point, the cells pass to SIP Level 1 for placement on the physical medium. SIP Level 1 operates at the physical layer and provides the physical-link protocol that operates at DS-1 or DS-3 rates between CPE devices and the network. The SIP Level 1 consists of the transmission system and Physical Layer Convergency Protocol (PLCP) sublayers. The transmission system sublayer defines the characteristics and method of attachment to a DS-1 or DS-3 transmission link. While the PLCP specifies the arrangement of SIP Level 2 cells relative to the DS-1 or DS-3 frame, it also defines other management information. *Figure 8-7* illustrates how SIP maps to the OSI reference model, including the IEEE data link sublayers.

SMDS Protocol Data Units

SMDS protocol data units carry both a source and a destination address. SMDS addresses have 10-digit values resembling conventional telephone numbers and offer group addressing and security features. With SMDS group addresses, a single address refers to multiple CPE stations that specify the group address in the destination address field of the PDU. The network delivers multiple copies of the PDU to all members of the group. Group addresses reduce the amount of network resources required for distributing routing information, resolving addresses, and dynamically discovering network resources. SMDS group addressing provides similar benefits as those seen with multicasting on LANs.

SMDS implements two security features called source address validation and address screening. Source address validation ensures the legitimate assignment of the PDU source address to the SNI from that originated the source address. Moreover, source address validation prevents address spoofing—a situation where illegal traffic assumes the source address of a

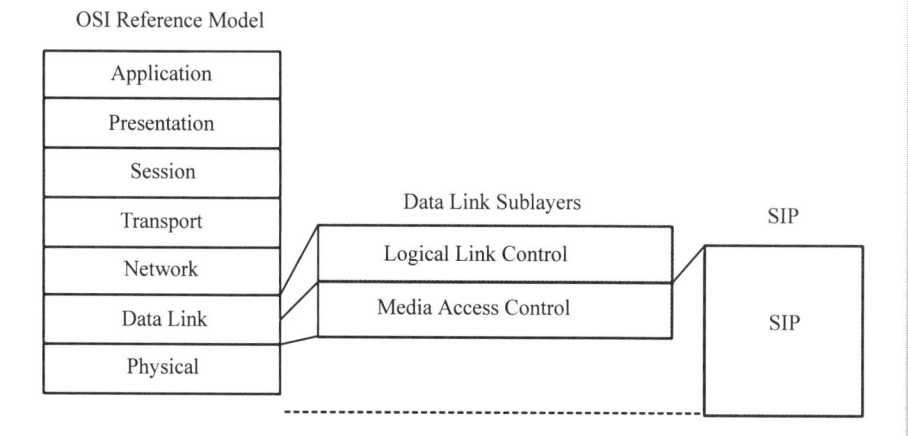

Figure 8-7.
ATM network using ATM switches and endpoints.

legitimate device. Address screening allows a subscriber to establish a private virtual network that excludes unwanted traffic.

The following descriptions define the function of the SIP Level 3 PDU fields illustrated in *Figure 8-8*.

- X+—Ensures that the SIP PDU format aligns with the DQDB protocol format. SMDS doesn't process or change the values in these fields, which may be used by systems connected to the SMDS network.

- RSVD—Consists of zeros.

- BEtag—Forms an association between the first and last segments of a segmented SIP Level 3 PDU. Both fields contain identical values and are used to detect a condition in which the last segment of one PDU and the first segment of the next PDU are both lost, which results in the receipt of an invalid Level 3 PDU.

- BAsize—Contains the buffer allocation size.

- Destination Address (DA)—Consists of two parts:

    1) Address Type: Occupies the four most significant bits of the field. The Address Type can be either 1100 or 1110. The former indicates a 60-bit individual address, while the latter indicates a 60-bit group address.
    2) Address: The individual or group SMDS address for the destination. SMDS address formats are consistent with the North American Numbering Plan (NANP). The four most significant bits of the Destination Address subfield contain the value 0001 (the internationally defined country code for North America). The next 40 bits contain the binary-encoded value of the 10-digit SMDS address. The final 16 (least-significant) bits are populated with ones for padding.

- Source Address (SA)—Consists of two parts:

    1) Address type: Occupies the four most significant bits of the field. The Source Address Type field can indicate only an individual address.

2) Address: Occupies the individual SMDS address of the source. This field follows the same format as the Address subfield of the Destination Address field.

- Higher Layer Protocol Identifier (HLPI)—Indicates the type of protocol encapsulated in the Information field. The value isn't important to SMDS, but can be used by certain systems connected to the network.

- Header Extension Length (HEL)—Indicates the number of 32-bit words in the Header Extension (HE) field. Currently, the field size for SMDS is fixed at 12 bytes. (Thus, the HEL value is always 0011.)

- Header Extension (HE)—Contains the SMDS version number. This field also conveys the carrier-selection value, which is used to select the particular interexchange carrier to carry SMDS traffic from one local carrier network to another.

- Information and Padding (Info + Pad)—Contains an encapsulated SMDS service data unit (SDU) and padding that ensures that the field ends on a 32-bit boundary.

- Cyclic Redundancy Check (CRC)—Contains a value used for error checking.

- Length—Indicates the length of the PDU.

The following descriptions briefly summarize the functions of the SIP Level 2 PDU fields illustrated in Figure 8-8.

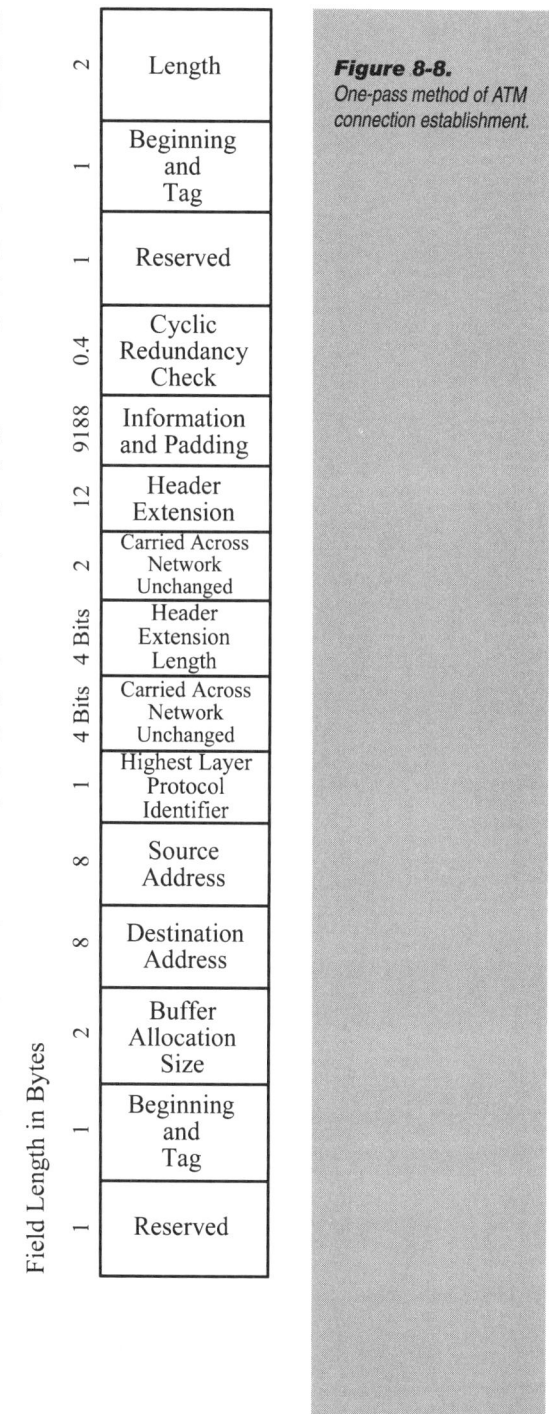

Figure 8-8.
One-pass method of ATM connection establishment.

- Access Control—Contains different values, depending on the direction of information flow. If the cell was sent from a switch to a CPE device, only the indication of whether the Level 3 PDU contains information is important. If the cell was sent from a CPE device to a switch, and if the CPE configuration is multi-CPE, this field can carry request bits that indicate bids for cells on the bus going from the switch to the CPE device.

- Network Control Information—Contains a value indicating whether the PDU contains information.

- Segment Type—Indicates whether the cell is the first, last, or a middle cell from a segmented level 3 PDU.

- Level 3 PDU. Four possible Segment Type values exist:

    1) 00: Continuation of message
    2) 01: End of message
    3) 10: Beginning of message
    4) 11: Single-segment message

- Message ID—Associates Level 2 cells with a Level 3 PDU. The Message ID is the same for all of the segments of a given Level 3 PDU. In a multi-CPE configuration, Level 3 PDUs originating from different CPE devices must have a different Message ID. This allows the SMDS network receiving interleaved cells from different Level 3 PDUs to associate each Level 2 cell with the correct Level 3 PDU.

- Segmentation Unit—Contains the data portion of the cell. If the Level 2 cell is empty, this field is populated with zeros.

- Payload Length—Indicates how many bytes of a Level 3 PDU actually are contained in the Segmentation Unit field. If the Level 2 cell is empty, this field is populated with zeros.

- Payload CRC—Contains a CRC value used to detect errors in the following fields:

    1) Segment Type

2) Message ID
3) Segmentation Unit
4) Payload Length
5) Payload CRC

- The Payload CRC value does not cover the Access Control or the Network Control Information fields.

Distributed Queue Dual Bus

The Distributed Queue Dual Bus provides an open standard that supports all the SMDS service features at the data-link layer level. In addition, DQDB ensures compatibility with current carrier transmission standards, aligns with emerging standards for broadband ISDN, and works with MAN and WAN technologies. As a result, SMDS networks can interoperate with broadband video and voice services.

DQDB ensures compatibility with current carrier transmission standards, aligns with emerging standards for broadband ISDN, and works with MAN and WAN technologies.

The IEEE 802.6 DQDB standard specifies a network topology composed of two unidirectional logical buses that interconnect multiple systems. An access DQDB only describes the operation of the DQDB protocol across a user-network interface and distinguishes this action from the operation of a DQDB protocol between carrier equipment within the SMDS Public Data Network. The access DQDB consists of:

- Carrier equipment—A switch in the SMDS network operates as one station on the bus.
- CPE—One or more CPE devices operate as stations on the bus.
- SNI—The SNI acts as the interface between the CPE and the carrier equipment.

SMDS Access Classes

SMDS networks can accommodate a broad range of traffic requirements and equipment capabilities through the use of access classes that establish a maximum sustained information transfer rate and a maximum allowed degree of traffic. Access classes maintain a sustained or average rate of data-transfer "burstiness"—or sudden increases in bandwidth demand on a network—for CPE devices. Five access classes are supported for DS-3-rate access at data rates of 4, 10, 16, 25, and 34 Mbps.

ATM Adaptation Layer 5

The ATM Adaptation Layer 5 serves as the primary AAL for data and supports both connection-oriented and connectionless data. AAL5 transfers most non-SMDS data, such as IP over ATM and provides LAN Emulation (LANE). Many designers refer to AAL5 as the simple and efficient adaptation layer (SEAL), because the SAR sublayer accepts and segments the CS-PDU into 48-octet SAR-PDUs without adding any additional fields.

AAL5 prepares a cell for transmission with the CS sublayer appending a variable-length pad and an eight-byte trailer to a frame. The pad ensures that the resulting PDU falls on the 48-byte boundary of an ATM cell, while the trailer includes the length of the frame and a 32-bit CRC computed across the entire PDU. With this, the AAL5 receiving process detects bit errors and out-of-sequence cells. From there, the SAR sublayer segments the CS-PDU into 48-byte blocks. Since AAL5 doesn't add either a header or a trailer, the protocol layer cannot interleave messages. At the end of the cell preparation, the ATM layer places each block into the Payload field of an ATM cell. With the exception of the last cell, AAL5 sets a bit in the Payload Type field to zero. With this, AAL5 shows which cell exists as the last cell in a series that represents a single frame. For the last cell, AAL5 sets the bit in the PT field to one.

ATM Devices

An ATM network consists of an ATM switch, media, and ATM endpoints. An ATM switch provides for cell transit through an ATM network. During operation, the ATM switch accepts the incoming cell from an ATM endpoint or another ATM switch, reads and updates the cell-header information, and quickly switches the cell to an output interface toward its destination. An ATM endpoint contains an ATM network interface adapter and may appear as a workstation, router, DSU, LAN switch, or CODEC. *Figure 8-9* illustrates an ATM network made up of ATM switches and ATM endpoints.

ATM Connections

ATM networks support point-to-point and point-to-multipoint connections. Point-to-point connections attach two ATM end systems and may have

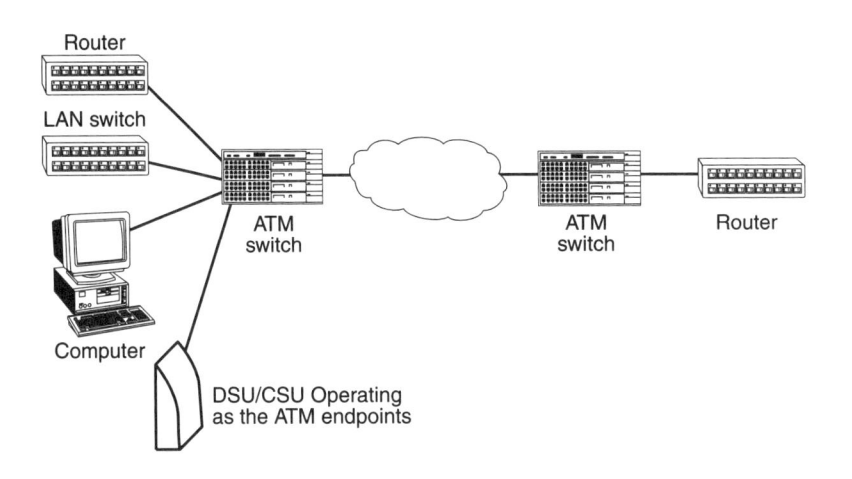

Figure 8-9.
ATM addresses used for private networks.

either unidirectional or bidirectional capabilities. In contrast, point-to-multipoint connections attach a single-source end system to multiple destination end systems and have only unidirectional capabilities. A single-source end system can transmit to multiple destination end systems. The reverse, however, cannot occur. The ATM switches replicate the cell when the connection splits into two or more branches.

Establishing an ATM Network Connection

When an ATM device wants to establish a connection with another ATM device, the originating device sends a signaling-request packet to the direct-connected ATM switch. The signaling-request packet contains the ATM address of the desired ATM endpoint and any quality-of-service parameters required for the connection. ATM signaling protocols occur according to the use of UNI or NNI links.

ATM signaling uses the one-pass method of connection setup seen with all modern telecommunication and all telephone networks. An ATM connection setup begins with the source end system sending a connection-signaling request that propagates through the network.

As a result, the connection-signaling request establishes connections through the network. The final network destination either accepts or rejects the connection request.

An ATM routing protocol oversees the routing of the connection request on destination and source addresses, traffic, and the quality-of-service parameters requested by the source end system. Typically, the negotiation of a connection request rejected by the destination cannot occur, because of call routing based on initial connection parameters. Changing the parameters to fit the connection request could affect the connection routing. *Figure 8-10* shows the one-pass method of ATM connection establishment.

Connection-management messages—such as setup, call proceeding, connect, and release—establish and tear down an ATM connection. The source end system sends a setup message that includes the destination end system address and any traffic Quality-of-Service parameters when it wants to establish a connection. In turn, the first switch in the network sends a call proceeding message back to the source in response to the setup message and invokes an ATM routing protocol. Next, the switch attached to the destination end system sends a connect message when accepting the connection and allows the data transfer to begin. The destination end system sends a release message back to the source end system if rejecting the connection and clears the connection.

ATM Quality-of-Service

In addition to one of the service categories, each ATM connection may request a Quality-of-Service parameter. The use of the parameters varies according to the utilization of a permanent virtual circuit or a switched

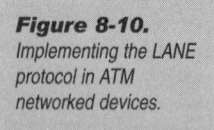

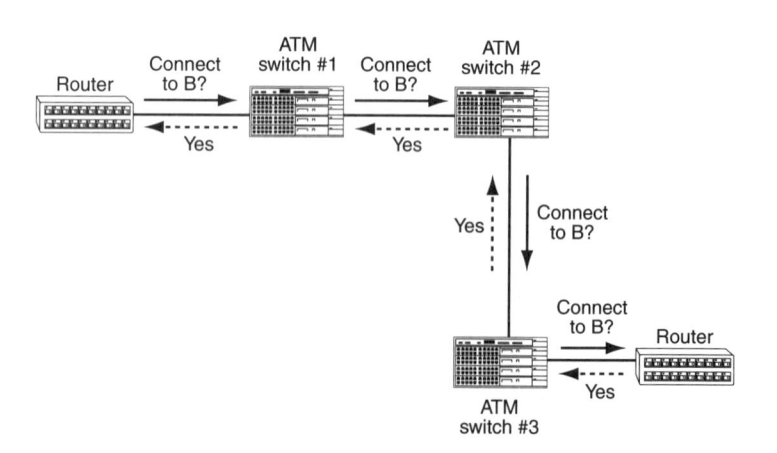

Figure 8-10.
Implementing the LANE protocol in ATM networked devices.

virtual circuit. With a permanent virtual circuit, the ATM connection will request a parameter that corresponds with network functions. With a switched virtual circuit, the ATM connection requests a parameter that corresponds with signaling.

ATM supports Quality-of-Service (QoS) guarantees composed of traffic contract, traffic shaping, and traffic policing. The QoS parameters include Cell Loss Ratio (CLR), Maximum Cell Transfer Delay (maxCTD), and Cell Delay Variation (CDV). A traffic contract specifies an envelope that describes values for peak bandwidth, average sustained bandwidth, and burst size as defined by a Connection Traffic Descriptor and a QoS parameter. As long as the end system continues to send traffic across the UNI as specified by the Connection Traffic Descriptor, the network will enforce the negotiated QoS.

The Connection Traffic Descriptor includes the Cell Delay Variation Tolerance (CDVT) and the Source Traffic Descriptor. In turn, the Source Traffic Descriptor consists of a set of parameters that describes the connection's expected bandwidth utilization. These parameters include the Peak Cell Rate (PC+R), Sustainable Cell Rate (SCR), Maximum Burst Size (MBS), and Minimum Cell Rate (MCR).

While the network delivers traffic under the SCR rate, it discards traffic sent above the SCR rate that also exceeds the MBS. Sustainable Cell Rate establishes the amount of traffic that an end point can burst onto the network. Maximum Burst Size defines the maximum amount of cells accepted over a period of time.

With these parameters, an ATM end system describes average and peak bandwidth requirements and maximum packet size. In addition, the network determines whether it can establish the connection and continue to meet the required QoS level for all connections including the new one. We can refer to this calculation as the Connection Admission Control (CAC).

If sufficient resources exist at each network element and the requested connection won't affect the Quality-of-Service at current connections, the CAC will accept the connection. During operation, the CAC considers the following factors.

- Source Traffic Descriptor and QoS of requested connections
- Traffic contracts of connections currently supported
- Previously allocated bandwidth at port level (logical and physical)
- Shared buffer and output queue occupancy
- A user-specific overbooking parameter

The Usage Parameter Control (UPC) function works as a policing function and determines conformance at the input edge of the network. The policing function ensures that a connection complies with the parameters of the traffic contract. If conformance exists, QoS guarantees to other connections remain in place. The UPC denies admission to the network to noncompliant cells or admits and tags as the cells as noncompliant.

Performance parameters also include the CLR, CTD, and CDV. The cell-to-lost cell ratio (CLR) measures the ratio of lost cells to total transmitted cells. An ATM switch malfunction may cause the loss of cells. In addition, an ATM switch will discard cells due to noncompliance with ATM standards or as a response to network congestion. The importance of proper cell discarding becomes more evident because of the dangers of congestion and the potential loss of an entire data packet.

A higher-layer protocol may require end-to-end error recovery through packet retransmission and cause mild congestion to escalate to severe congestion. To prevent this from occurring, an ATM switch can discard cells on a frame basis by using Early Packet Discard (EPD) and Partial Packet Discard (PPD) mechanisms. EPD discards an entire incoming packet after detecting the potential for congestion. PPD discards all of the remaining cells with the exception of the End-of-Packet cell of a partially discarded packet.

When considering TCP/IP applications, combined EPD and PPD can significantly reduce partial packet loss and improve overall packet throughput. Network administrators can achieve better TCP/IP performance with the addition of per-VC discard. Since EPD/PPD schemes may discard TCP/IP packets traveling through a congested switch at the same time, related end systems react to packet loss in a synchronous fashion. In turn, congestion and reduced throughput occur. Staggering or randomizing EPD/PPD actions on Virtual Channels can avoid global synchronization.

CTD measures the elapsed time between a cell exit event at the source and the corresponding cell entry event at the destination. CTD is the sum of the total inter-ATM node transmission delay and the total ATM node processing delay between source and destination. Sometimes called cell jitter, CDV refers to factors that disrupt evenly spaced cells and cause cell clumping and gaps in the cell stream.

ATM networks require more than just segmenting frames or packets into cells. Traffic shaping covers the use of queues to constrain data bursts, limit peak data rate, and smooth jitters so that traffic fits within the promised envelope. In short, traffic shaping ensures that the interval between cells conforms to the peak cell rate. ATM devices adhere to the contract through traffic shaping, while ATM switches can use traffic policing to enforce the contract.

As an example of network action without traffic shaping, an ATM end system supporting Ethernet LANs may request access to an ATM network. Since any port on the Ethernet LANs has a peak bandwidth of 10 Mbps, the user establishes PVCs with a peak cell rate of 10 Mbps per second. For the short term, everything works fine. As the network adds more customers and traffic grows, the carrier upgrades the network by enabling traffic policing so that the QoS of all virtual connections remains in place.

But the ATM access device fails to limit its transmissions to the peak cell rate. For very short packets, the tolerance built into the policing algorithm allows the cells to enter the network. For longer packets, however, the policing function discards cells because of the peak cell rate. The discarding of longer packets occurs despite the retransmission of cells. As a result, normally functioning applications begin to fail.

To avoid traffic policing problems, a user may renegotiate the traffic contract for a peak cell rate at the full line rate of 155 Mbps for all virtual connections. Yet this action could dramatically increase the cost of the service. Second, the user may implement traffic shaping by installing an ATM switch capable of traffic shaping between the Ethernet product and the network.

Traffic shaping at the outer edge of a network reduces CDV across the UNI to the ATM end system. As a result, the quality degradation that results

> *Sometimes called cell jitter, CDV refers to factors that disrupt evenly spaced cells and cause cell clumping and gaps in the cell stream.*

from buffer overruns or cell loss for real-time ATM services will not occur. In addition to performance benefits, traffic shaping provides economic benefits to ATM WANs.

For example, an ATM end system may specify a Source Traffic Descriptor that results in a lower tariff service while achieving the same throughput as a higher tariff that would have resulted without traffic shaping. Lower network costs also occur through a higher level of trunk utilization while supporting the same level of traffic. Shaping allows interswitch trunks to operate at a higher level of utilization while maintaining a fixed CLR.

The switch can measure actual traffic flow and compare the measurement against the specified traffic envelope. If the switch finds that traffic exceeds or falls below those parameters, it can set the cell-loss priority (CLP) bit of the offending cells. Setting the CLP bit sets the cell discard as eligible and allows any switch handling the cell to drop the cell during periods of congestion.

ATM Services

ATM services exist as permanent virtual circuits, switched virtual circuits, and the connectionless service seen with SMDS.

ATM services exist as permanent virtual circuits, switched virtual circuits, and the connectionless service seen with SMDS. Since PVC allows direct connectivity between sites, the permanent virtual circuit appears similar to a leased line. As you know, a PVC guarantees availability of a connection and doesn't require call setup procedures between switches. Comparable to a telephone call, a switched virtual circuit occurs and releases dynamically while remaining in use only during a data transfer. Dynamic call control requires a signaling protocol between the ATM endpoint and the ATM switch. An SVC offers connection flexibility and call setup that a networking device can handle automatically.

ATM Virtual Connections

Since ATM networks rely on a connection, the network must establish a virtual channel or a virtual path across the ATM network prior to any data transfer. Virtual path identifiers pinpoint virtual paths, while a combination of VPIs and virtual channel identifiers identify virtual channels. All VCIs and VPIs have only local significance across a particular link. For this situation, a virtual channel equals a virtual circuit. A virtual path consists of a bundle of virtual channels that switch transparently across the ATM network on the basis of the common VPI.

ATM Service Classes

An ATM network establishes Virtual Path or Virtual Channel connections with five distinct levels of service called Constant Bit Rate, Variable Bit Rate-real time, Variable Bit Rate-non-real time, Unspecified Bit Rate, and Available Bit Rate. Constant Bit Rate (CBR) resembles a leased-line service and works with network connections that request a static amount of bandwidth that remains continuously available during the duration of the connection. The Peak Cell Rate (PCR) value defines the maximum amount of traffic that the network can accept under CBR conditions. Although CBR wastes bandwidth, typical applications include videoconferencing, telephony, television applications such as distance learning and pay-per-view, video-on-demand, and audio library.

Variable Bit Rate-real time (VBR-rt) works best for real-time applications that require tightly constrained delay and delay variation. Typically, this type of application doesn't require a fixed transmission rate. As with CBR connections, VBR-rt connections cannot exceed specifications set by the Peak Cell Rate. In addition, VBR-rt connections also must maintain levels according to the Sustainable Cell Rate (SCR) and the Maximum Burst Size (MBS).

VBR-rt can operate within native ATM voice applications that apply bandwidth compression and silence suppression. Applications for VBR-rt include some types of multimedia communications along with compressed voice and video. VBR-rt offers a low-latency guarantee, tighter connection-admission control and more congestion control.

End systems using CBR or VBR-rt service indicate their CTD requirement by negotiating with the network maxCTD. For end systems running voice, video, and multimedia applications over CBR and VBR-rt connections, CDV can become an issue. If the network cannot properly control CDV, some real-time services may have distorted communications. An end system using CBR or VBR-rt service indicates its end-to-end cell delay variation requirement by negotiating a peak-to-peak CDV with the network.

As the name implies, VBR-nrt works with non-real-time bursty applications that require service guarantees from the network. Like VBR-rt, VBR-nrt connections adhere to PCR, SCR, and MBS levels. Typical applications for VBR-rt include data transfers for action-processing applications such as airline reservation record maintenance, banking transactions, and process monitoring. In addition, Frame Relay traffic can also use VBR-nrt service.

Unspecified Bit Rate doesn't offer any guaranteed bandwidth or latency and works well for non-real-time, bursty applications that remain tolerant of delay and loss. Since UBR service doesn't specify service guarantees, many refer to the service class as the "best effort" service. Applications for UBR include text/data/image transfers, telecommuting, electronic mail, store-and-forward networks, LAN interconnection, LAN emulation, supercomputer applications, distributed file services, and computer process swapping/paging.

Available Bit Rate specifies a bandwidth setting but doesn't use a maximum delay. With ABR, the network may buffer cells and signal the sender to reduce the data transfer rate. As a result, ABR applications can tolerate a Minimum Cell Rate (MCR), while retaining the ability to adapt to feedback from the network for the purpose of taking advantage of available bandwidth. On the establishment of an ABR connection, the end system specifies both a PCR and an MCR. A flow-control mechanism that supports several types of feedback allocates the available bandwidth among the ABR connections.

ATM Switching Operations

ATM switching receives a cell across a link on a known VCI or VPI value. The switch looks up the connection value in a local translation table to determine the outgoing port or ports of the connection and the new VPI/VCI value of the connection on that link. As the switching process continues, the switch then retransmits the cell on the outgoing link with the appropriate connection identifiers. Since all VCIs and VPIs have only local significance across a particular link, the switch remaps the values as necessary.

ATM Addressing

ATM public and private networks use addressing based on the ITU-T standard and on the use of E.164 addresses that appear similar to telephone numbers. In addition, the use of the subnetwork or overlay model of addressing places responsibility for mapping network-layer addresses to ATM addresses. This subnetwork model works as an alternative to the use of network-layer protocol addresses—such as IP and IPX—and existing routing protocols.

During operation, the subnetwork model of addressing decouples the ATM layer from any existing higher-layer protocols. As a result, the model requires an entirely new addressing scheme and routing protocol. Each ATM system must have an assigned an ATM address along with higher-layer protocol addresses. The assignment of ATM addresses requires an ATM address resolution protocol (ATM ARP) to map higher-layer addresses to corresponding ATM addresses.

ATM private networks can base addressing on the telephone-like E.164 address of the connected public UNI and can take the address prefix from the E.164 number that identifies local nodes by the lower-order bits. All ATM addresses consist of the authority and format identifier (AFI), the initial domain identifier (IDI), and the domain specific part (DSP). The AFI identifies the type and format of the IDI that, in turn, identifies the address allocation and administrative authority. While the DSP contains actual routing information, the IDI may signify an E.164 number, a data country code, or an international code designator.

Figure 8-11 illustrates the three formats of ATM addresses used for private networks and the following list defines the address fields.

- AFI—Identifies the type and format of the address (DCC, ICD, or E.164).

The assignment of ATM addresses requires an ATM address resolution protocol (ATM ARP) to map higher-layer addresses to corresponding ATM addresses.

Figure 8-11.
LANE data connections.

DCC ATM Format

ICD ATM Format

NASP Format E.164

- DCC—Identifies particular countries.

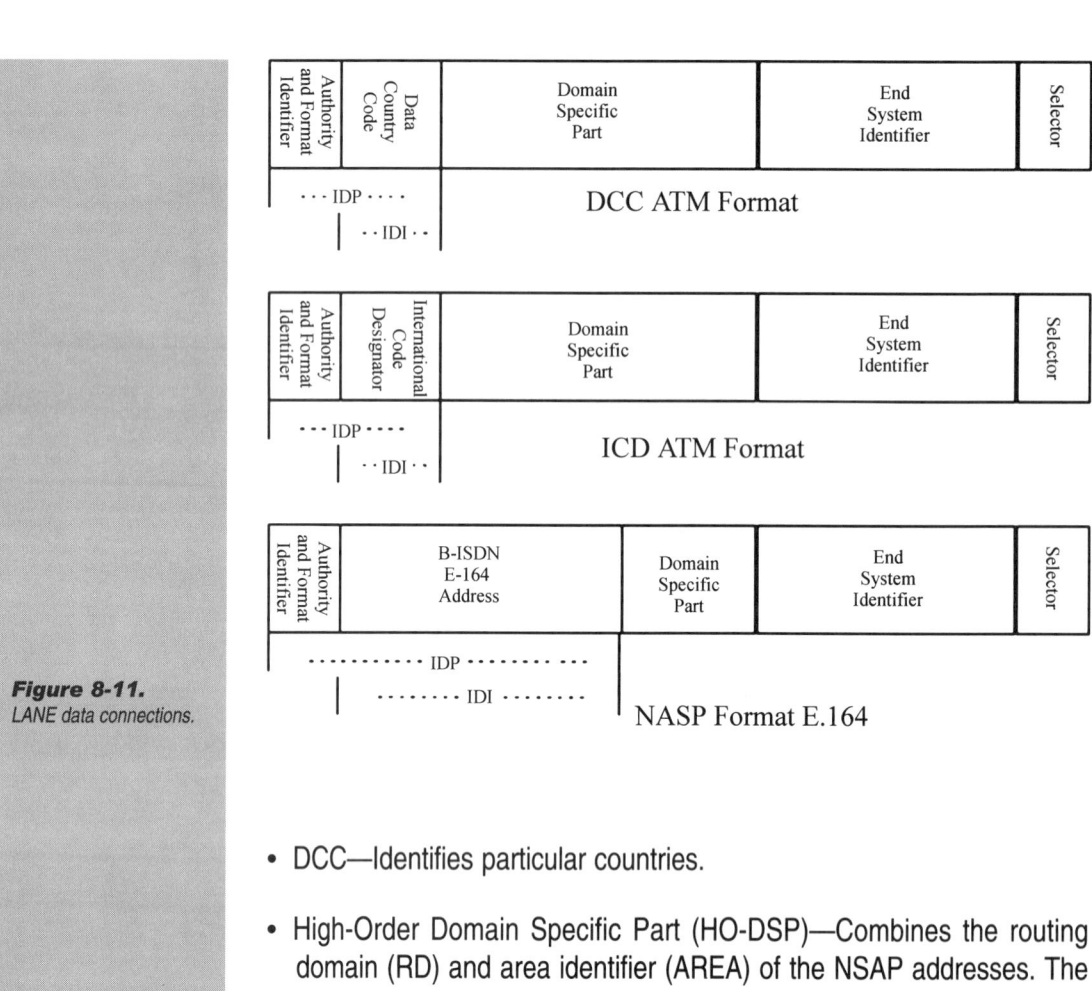

- High-Order Domain Specific Part (HO-DSP)—Combines the routing domain (RD) and area identifier (AREA) of the NSAP addresses. The ATM Forum combined these fields to support a flexible, multilevel addressing hierarchy for prefix-based routing protocols.

- End System Identifier (ESI)—Specifies the 48-bit MAC address, as administered by the Institute of Electrical and Electronic Engineers (IEEE).

- Selector (SEL)—Used for local multiplexing within end stations and has no network significance.

- ICD—Identifies particular international organizations.

- E.164—Indicates the BISDN E.164 address.

ATM LAN Emulation

ATM LAN Emulation (LANE) gives stations attached through an ATM network the same capabilities obtained from Ethernet and Token Ring networks. The LANE protocol emulates a LAN on top of an ATM network by defining mechanisms for emulating either an IEEE 802.3 Ethernet or an 802.5 Token Ring LAN. In addition, the LANE protocol defines a service interface for network layer protocols that allows the service interface to appear identical to service interfaces employed existing LANs.

The appropriate LAN MAC packet format encapsulates data sent across an ATM network. As a result, the LANE protocols cause an ATM network to look and behave like an Ethernet or Token Ring LAN. But the LANE protocols allow the LAN to have much faster data transfer speeds. ATM LAN Emulation doesn't attempt to emulate the actual MAC protocol of the specific LAN such as CSMA/CD for Ethernet or token passing for IEEE 802.5. As a result, LANE requires no modifications to higher-layer protocols to enable their operation over an ATM network. Since the LANE service presents the same service interface of existing MAC protocols to network-layer drivers, the drivers don't require any changes.

LANE protocols cause an ATM network to look and behave like an Ethernet or Token Ring LAN. But the LANE protocols allow the LAN to have much faster data transfer speeds.

ATM network-interface cards, along with ATM-attached LAN switches and routers, deploy the LANE protocol. ATM NICs implement the LANE protocol and interface to the ATM network, while presenting the current LAN service interface to the higher-level protocol drivers within the attached end system. The network-layer protocols on the end system continue to communicate as if connected to a traditional Ethernet or Token Ring LAN.

ATM-attached LAN switches and routers work with directly attached ATM hosts equipped with ATM NICs and provide a virtual LAN (VLAN). The LANE protocol defines the operation of a single VLAN. With the VLAN service, ports on the LAN switches become assigned to a particular VLAN independently of physical location. *Figure 8-12* shows the LANE protocol architecture implemented in ATM network devices. The Emulated LAN consists of:

- LAN emulation client (LEC)—The LEC is an entity in an end system that performs data forwarding, address resolution, and registration of

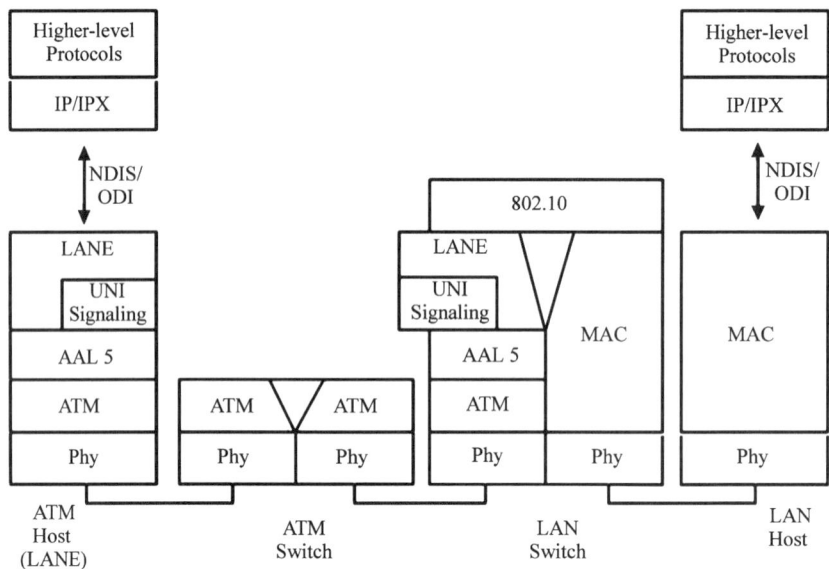

Figure 8-12.
Customer premises connected directly to an ATM network.

MAC addresses with the LAN emulation server (LES). The LEC also provides a standard LAN interface to higher-level protocols on legacy LANs. An ATM end system that connects to multiple Emulated LANs has one LEC per ELAN.

- LES—The LES provides a central control point for LECs to forward registration and control information. Only one LES exists per ELAN.

- Broadcast and unknown server (BUS)—The BUS is a multicast server that is used to flood unknown destination address traffic and to forward multicast and broadcast traffic to clients within a particular ELAN. Each LEC is associated with only one BUS per ELAN.

- LAN emulation configuration server (LECS)—The LECS maintains a database of LECs and the Emulated LANs to which they belong. This server accepts queries from LECs and responds with the appropriate ELAN identifier, namely the ATM address of the LES that serves the appropriate ELAN. One LECS per administrative domain serves all ELANs within that domain.

Figure 8-13 shows the LANE data connections and illustrates the operation of a LANE system and connected components. LAN Emulation Client

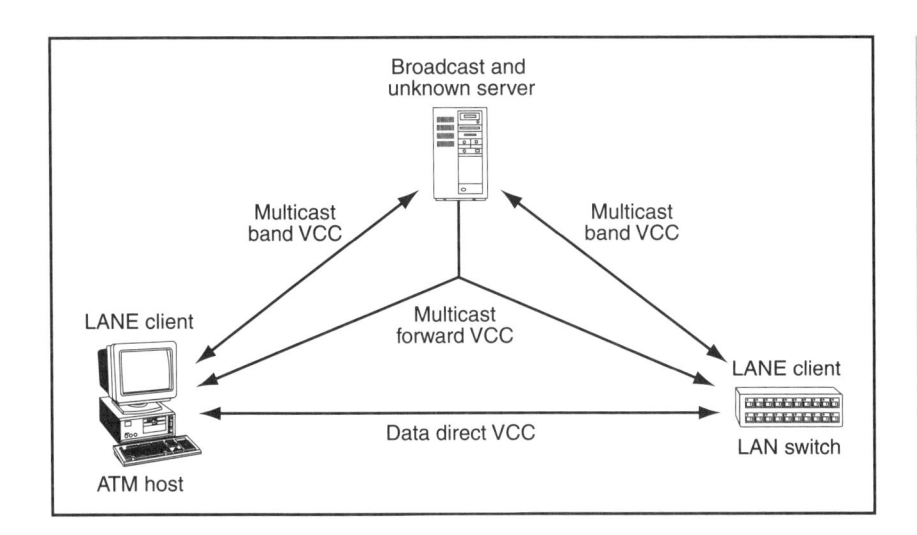

Figure 8-13.
ATM and internetworking.

operation involves initialization and configuration, joining and registering with the LAN Emulation Server, finding and joining the Broadcast and Unknown Server, and data transfer. Upon initialization, a LEC obtains required configuration information. The LEC begins this process by obtaining an ATM address through address registration.

Then the LAN emulation client determines the location of the LAN emulation configuration server. To do this, the LEC determines the LECS address by using a well-known LECS address or by using a well-known permanent connection to the LECS through a VPI and VCI. After locating the LECS, the LEC sends an LE_CONFIGURE_REQUEST. If the LEC finds a matching entry, the LAN emulation configuration server returns an LE_CONFIGURE_RESPONSE to the LEC. The configuration information allows the LEC to connect to its target VLAN and includes the ATM address of the LES, type of LAN for emulation, maximum packet size on the ELAN, and ELAN name.

When an LEC joins the LES and registers ATM and MAC addresses, the LEC optionally clears the connection to the LECS and sends an LE_JOIN_REQUEST through a virtual channel. This process allows the LEC to register its own MAC and ATM addresses with the LES and any other MAC addresses used for proxy services. As a result, no two clients will register the same MAC or ATM address.

After receipt of the LE_JOIN_REQUEST, the LES checks with the LECS through an open connection, verifies the request, and confirms the membership of the client. With verification in place, the LES adds the LEC as a node of the point-to-multipoint control-distribute virtual channel. Then the LES issues the LEC a successful LE_JOIN_RESPONSE that contains a unique LAN Emulation Client ID (LECID). The LEC uses the LECID to filter its own broadcasts from the BUS.

Once the LEC has successfully joined the VLAN, the client must find the BUS/s ATM address to join the broadcast group and become a member of the emulated LAN. The process begins with the LEC creating an LE_ARP_REQUEST packet with the MAC address 0xFFFFFFFF. Then the LEC sends the special LE_ARP packet on the control-direct VCC to the LES. The LES recognizes that the LEC has begun to search for the BUS and responds with the ATM address of the BUS on the control-distribute VCC. When the LEC has the ATM address of the BUS, the LEC joins the BUS by creating a signaling packet with the ATM address of the BUS and setting up a multicast-send VCC with the BUS. Upon receipt of the signaling request, the BUS adds the LEC as a client on its point-to-multipoint multicast forward Virtual-Channel Connection (VCC).

The LEC now operates as part of the ELAN and stands ready for data transfer. The transfer of data across the ELAN involves resolving the ATM address of the destination LEC. When a LEC has a data packet to send to an unknown-destination MAC address, the LEC must discover the ATM address of the destination LEC. To accomplish this, the LEC sends the data frame to the BUS for distribution to all LECs on the ELAN through the multicast forward VCC.

This multicast must occur, because resolving the ATM address may consume time that, in turn, produces network delays. The LEC then sends a LAN Emulation Address Resolution Protocol Request (LE_ARP_Request) control frame to the LES through a control-direct VCC. If the LES can respond to the request, it sends the ATM address of the LEC that owns the required MAC address. If the LES can't respond to the LE_ARP_Request, it floods the request to some or all LECs over control-direct and control-distribute Virtual-Channel Connections. If bridge or switching devices with

LEC software participating in the ELAN exist, the devices translate and forward the ARP on their LAN interfaces.

The LEC sets up a data-direct VCC to the destination node and uses the channel for data transfer rather than the BUS path. Before doing this, the LEC usually uses the LANE flush procedure that ensures the delivery of all packets previously sent to the BUS to the destination prior to the use of the data-direct Virtual-Channel Connection. The LEC then waits until the destination acknowledges receipt of the flush packet before using the second path to send packets.

ATM and Internetworking

With the exponential growth of Internet usage, many Internet service providers begun to look for solutions to saturated DS0 and X.25 links. The introduction of Frame Relay has allowed planned expansions and the growth of the Internet to continue. Frame Relay, with its statistical bandwidth usage, provides an option for replacing leased lines as the interconnection choice of the ISPs.

Despite any associated cost disadvantages, however, leased lines continue to provide fully dedicated bandwidth and can guarantee the delivery of delay-sensitive traffic such as voice and video. The lack of QoS standards for Frame Relay and its status as a statistically multiplexed service rules out Frame Relay for those services. ATM serves as a bridge between the requirements, due to its mature class of service distinctions, QoS operation, and the capability to support leased line traffic with the same guarantees for delay sensitive traffic.

Internetworking allows higher layer network elements to operate in an integrated and comprehensive manner, while fully utilizing the basic capabilities of the related services to move information through the network across multiple service elements. Features of different networks integrate into the internetwork. For example, the functionality of ATM technologies supports Frame Relay and leased lines without loss of capability. Consequently, some carriers have begun to consolidate all service offerings over ATM.

Features of different networks integrate into the internetwork. For example, the functionality of ATM technologies supports Frame Relay and leased lines without loss of capability.

Telecommunication Technologies

The deployment of the new xDSL technologies as the "final mile" solution to the house allows leased-line service providers to offer a wide variety of rates and speeds. Integrating these rates and speeds into a Frame Relay network will provide a seamless evolutionary path that protects investments in legacy equipment, while allowing customers to take advantage of the newly emerging technologies.

ATM, Frame Relay, and Leased Lines

ATM can efficiently provide a universal transport layer for all of these needs in addition to Quality of Service and Traffic Management. An ATM-based model called Circuit Emulation Service (CES) supports both low-speed leased lines and Frame Relay. CES can transport low-speed leased-line traffic while supporting Frame Relay applications with standard internetworking functionality. The introduction of CES permits edge-users to combine DS0 channels into E1/T1 trunks that multiplex into the standard ATM fabric speed, switch to destinations, and demultiplex into the original DS0 channels while preserving their real-time characteristics.

Carriers can establish CES functionality at the Central Office to provide connections for large numbers of low-speed ATM ports overlaid onto a single high-speed ATM switch port. By using inexpensive customer premise access concentrators, the carriers can offer the benefits of dedicated private line services.

ATM and Frame Relay. Relatively slower speeds and lower costs have allowed Frame Relay to begin supplanting legacy-leased lines and to support local access for Internet service providers.

In turn, the long haul carrying capability of ATM meshes with the infrastructure needs of network providers. As a result, the economies of scale of ATM maximized through high bandwidth payloads combine with the less expensive Frame Relay deployed at the local level.

Since Frame Relay data rates range from 56 Kbps to 45 Mbps, Frame Relay can serve as a link between analog modem connectivity and entry-level ATM capabilities. Moreover, additional benefits become apparent as Frame Relay operates over a highly reliable physical transport medium

Relatively slower speeds and lower costs have allowed Frame Relay to begin supplanting legacy-leased lines and to support local access for Internet service providers.

and doesn't need the rigorous error detection schemes employed by X.25. With Frame Relay packets taking the form of variably sized frames, Frame Relay accommodates data packets of different sizes and native protocols with little protocol translation.

Delay-sensitive traffic with an 8K-octet maximum frame size may not flow efficiently over Frame Relay because of significant variations in frame size. The smaller packet size seen with ATM smooths out the traffic flow and prevents the transmission of large pieces of data from blocking the transmission of small real-time packets. In the event of errors, retransmissions can occur quickly.

Several definite benefits exist with the use of Frame Relay. While ATM must use a 48 octet, Frame Relay can function with a packet size of just one octet if necessary. Frame Relay virtual circuits may exist as point-to-point, point-to-multipoint, or multipoint-to-multipoint and arrange into closed user groups for security purposes. Frame Relay handles congestion a little less precisely than ATM by defining a Committed Information Rate (CIR).

The Committed Information Rate is the rate that the network supports for a particular frame connection. If the network transmits data in excess of the CIR, the excess data could become discarded when congestion occurs. ATM observes guarantees of service in a much stricter fashion. Unstructured CES functions with equipment that does not support standard framing formats such as clear channel T1/E1 circuits. As a result, a CES-based solution connects customer premise equipment to a private or public ATM network and integrates voice and video with data.

Frame Relay devices work independently of the ATM backbone connection, because the ATM WAN switches provide the necessary internetworking function. An ATM backbone can support multiple Frame Relay networks and establish a scalable high-speed option for Frame Relay users and network designers that doesn't require changes to customer premises equipment. Thus, the network infrastructure combines the flexible and cost-effective access provided by Frame Relay and the high-speed and throughput capabilities of ATM.

ATM and Leased Lines. Despite the growing popularity of Frame Relay, leased lines remain as a viable option for internetworking. Leased lines provide constant bit rate, uninterrupted traffic flow and support voice and video along with many data applications. As mentioned, ATM provides a universal transport layer for all of these needs in addition to Quality of Service and traffic management. The process of internetworking leased lines and ATM technologies must maintain the straightforward, nonnegotiable characteristics seen with leased lines.

To accomplish this goal, the ATM function must allow transparent, nonblocking, nondelayed traffic flow from end to end for each ATM equivalent virtual circuit. Leased lines of Nx64Kbps or Nx56Kbps services can assimilate into the backbone through the ATM CES function. In this case, the network must support all appropriate modem functions and have the capability to rebuild the off-loaded signal at the distant end so that the signal remains compatible with the receiving CPE.

In addition, the ATM backbone must accommodate digital leased lines that operate up to T1/E1 and higher use CSU/DSU technology. *Figure 8-14* illustrates customer premises connected directly into the ATM network through an on-premises ATM access device. In the figure, the internetworking function works from within the device that connects directly to the ATM cloud. With this, internetworking and ATM based QoS assurance occurs at the customer premises. Furthermore, network efficiency increases, because ATM switches handle only ATM cells.

The benefits of a Frame Relay-ATM network go beyond providing a method for frame-based user equipment to access ATM networks at the user to network interface.

Network Internetworking/Service Internetworking

By carrying Frame Relay and leased-line traffic on ATM, those traffic streams can benefit from the fully provisioned QoS capabilities of ATM. As a result, the user gains low-cost, right-sized access trunks through Frame Relay and leased lines, and high-speed, QoS-managed wide-area service through ATM. In *Figure 8-15*, a Frame Relay-ATM network internetwork connects separate Frame Relay networks using an ATM cloud as the backbone. From a user perspective, though, the connection appears as one network.

Figure 8-14.
ATM and Frame Relay.

Figure 8-15.
Data and voice over Frame Relay.

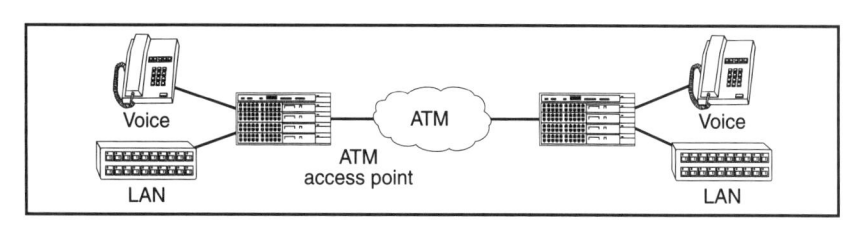

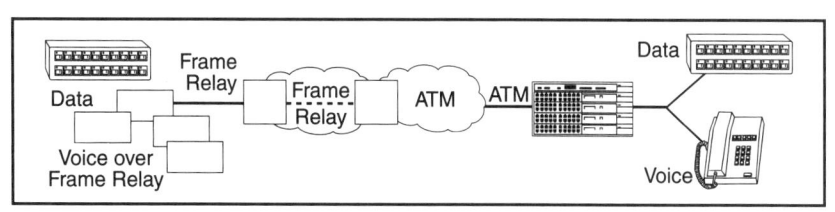

The benefits of a Frame Relay-ATM network go beyond providing a method for frame-based user equipment to access ATM networks at the user to network interface. Interoperability through network and service internetworking may stand as the most important benefit. Interoperability at the network and service levels also allows a network manager to avoid the selection of one network type over another. With both network types working together seamlessly, the network manager can implement individual network components with either technology according to need.

At the Frame Relay-ATM network level, network internetworking provides a transport between two Frame Relay devices or entities. Service internetworking enables an ATM user to transparently exchange data with a Frame Relay user. Internetworking can occur in either the transparent or translation mode. In transparent mode, the customer premises equipment—such as an ATM access concentrator—accepts the Frame Relay traffic with no changes. In the translation mode, internetworking unbundles the Frame Relay frame and maps the frame overhead onto an ATM cell, replaces the frame protocol, and reduces overhead.

The Network Internetworking Function (NIF) facilitates the transparent transport of Frame Relay user data and PVC signaling from one Frame Relay network connected to another by an ATM backbone. During the process, the NIF completely encapsulates the Frame Relay frame and places the frame into a series of ATM cells.

Telecommunication Technologies

The Network Internetworking Function integrates into either the ATM or the Frame Relay switch and maps each Frame Relay PVC directly onto an ATM PVC. The internetworking function can occur inside the Frame Relay network while attaching as an ATM UNI into the backbone. In addition, the ATM connection point can perform the internetworking function within the switch performing the UNI. In this case, the translation occurs in the Service Specific Convergence sublayer of the ATM protocol.

Another option configures the sending node to interoperate with the distant node. The sending node will deliver an intact Frame Relay frame to the distant end of the ATM network.

Internetworking at the service level begins with the conversion of Frame Relay PVC status signaling to ATM OAM cells and OAM cells to Frame Relay status signaling. Under this arrangement, a failure in one network causes a notification of the other network its appropriate protocol. Service level internetworking maps other conditions such as congestion and discard/eligible/cell loss priority appropriately. Specifically, the SIWF maps the Frame Relay DLCI onto an ATM VPC/VCI address, the FECN into the payload field and the DE bit onto the CLP bit. Since the Frame Relay multiprotocol encapsulation procedures don't match ATM protocol encapsulation procedures, service internetworking also converts the multiprotocol data unit headers from the Frame Relay format into ATM format and from ATM to Frame Relay.

Each of the following chapters will continue the discussion of ATM and internetworking.

Summary

Chapter 8 covered ATM, Frame Relay, and SMDS technologies. The chapter used this discussion to view the relationship between the ATM and OSI reference models. Referring back to this section, ATM technology relies on adaptation layers to prepare data contained within cells for transmission. The chapter progressed from considering the process of transferring data over an ATM network to a brief overview of hardware used to provide connectivity for the network. With the coverage of those topics, the chapter progressed to a detailed consideration of ATM and internetworking. Each of the following chapters will continue this discussion.

Chatper 9

Optical Networks

Introduction

Chapter 9 provides a follow-up to our discussions about ISDN and ATM with coverage of optical networks. More specifically, Chapter 9 offers information about Synchronous Optical Network (SONET) technologies and Fiber Distributed Data Interface (FDDI) technologies. Both play a major role in establishing the framework for internetworking across the country.

As an example of the transport potential given by SONET, the standard can move the 650 megabytes of multimedia information from one coast to the other within one second. Given this capability, many businesses and corporations have begun to add SONET rings as interconnections between buildings. Although SONET complements ATM networks, the potential for transporting packets-over-SONET has begun to grow to the extent where packets-over-SONET may replace ATM-based local- and wide-area networks.

Synchronous Optical Network (SONET)

In 1984, the Exchange Carriers Standards Association (ECSA) put together specifications for an optical telecommunications transport network for the American National Standards Institute. ANSI sets industry-wide standards for telecommunications. The combined labor of the two entities produced the SONET standard. Most experts predict that the SONET/synchronous digital hierarchy (SONET/SDH) will become the transport infrastructure for worldwide telecommunications for the next 30 years.

Before the advent of SONET/SDH, fiber-optic systems found in the public telephone network relied on proprietary architectures, equipment, line

Although SONET complements ATM networks, the potential for transporting packets-over-SONET has begun to grow to the extent where packets-over-SONET may replace ATM-based local- and wide-area networks.

codes, multiplexing formats, and maintenance procedures. As a result, the RBOCs and IXCs couldn't mix and match equipment from different vendors because of the lack of standards. SONET/SDH provided a solution for those problems.

As a standard, SONET/SDH defines optical carrier (OC) levels and electrically equivalent synchronous transport signals (STSs) for the fiber optic-based transmission hierarchy. At the physical Optical Carrier level, data travels through either wave-division multiplexing (WDM) or dense-wave-division multiplexing (DWDM). With wave-division multiplexing, data transfers through pulses at a single laser. Given a maximum of OC-48 speeds, faster laser pulses allow the pushing of more data through the fiber.

Higher bandwidths reach beyond the capabilities of WDM and the single laser. Dense-wave-division multiplexing (DWDM) combines multiple OC-48 WDM lasers that individually operate at a different wavelength. As a result, DWDM uses the same fiber conduit but enlarges the bandwidth by transmitting more wavelengths of light.

Even though designers intended SONET as a method for eliminating the electrical transmission of data, a SONET network relies on STS for very short distances that typically only span across the switch cabinet and for electrical switching of the signals.

Any level of STS describes electrical frame generation within a switch. Optical carrier levels describe transmission of the signal from point to point. Since SONET sends 8,000 STS frames per second—or at the same frame rate found with early DS-1—it easily incorporates different data-transfer rates. In addition, SONET/SDH offers advantages such as:

- Reduction in equipment requirements

- Increase in network reliability

- Availability of a set of generic standards that enable connectivity between different vendor products

- Definition of a flexible architecture that can accommodate future applications and transmission rates

- Definition of a synchronous multiplexing format for carrying lower level digital signals

- Definition of a synchronous structure that simplifies the interface to digital switches, digital cross-connect switches, and add-drop multiplexers

- Provision of overhead bytes that permit management of payload bytes on an individual basis and facilitate fault detection

SONET Signals

As the name shows, SONET/SDH relies on the synchronization of digital signals. With this, the digital transitions in the signals occur at exactly the same rate. All clock signals refer back to an extremely accurate primary reference clock (PRC). Any phase difference between the transitions occurs within specified limits. Phase differences may occur because of propagation time delays or jitter introduced into the transmission network.

The use of synchronized digital signals contrasts against the asynchronous transmission systems that we have discussed in which each terminal in the network runs from an individual clock. With digital signals, clocking—or the use of a series of repetitive pulses to keep the bit rate of data constant—exists as one of the most important functions of the system. Clocking also indicates the location of the 1s and 0s in the data stream.

The absence of synchronized clock signals allows large variations in the clock and signal bit rates to occur. Asynchronous data transmission relies on multiple stages, the multiplexing of signals, and the adding of bits to account for variations in the individual bit streams. The adding of bits and multiplexing of the signals must occur to allow the transfer of data between nonsynchronous lines with different bit rates.

Synchronous data transmission makes use of the fact that all clocks in the system will have the same average frequency. As a result, a system can multiplex lines without adding bits to compensate for variations. This allows the accessing of lines with a lower line rate by lines with a higher line rate.

With digital signals, clocking—or the use of a series of repetitive pulses to keep the bit rate of data constant—exists as one of the most important functions of the system.

Table 9-1.
SONET signal hierarchy.

Signal	Bit Rate (Mbps)
STS-1, OC-1	51.840
STS-3, OC-3	155.520
STS-12, OC-12	622.080
STS-48, OC-48	2,488.320
STS-192, OC-192	9,953.280

Note: STS = Synchronous Transport Signal; OC = Optical Carrier.

Multiplexing SONET Signals

SONET relies on a byte-interleaved multiplexing scheme that simplifies multiplexing and offers end-to-end network management. When the SONET multiplexing process begins, the system generates a base signal called the synchronous transport signal/level 1 (STS-1). The base signal operates at 51.84 Mbps. Higher level signals occur as multiples of STS-1 and act as a family of STS-N signals. Each STS-N signal consists of N byte interleaved STS-1 signals. *Table 9-1* shows the SONET signal hierarchy.

Within the SONET, digital switches and cross-connect systems operate as part of the hierarchy. A master/slave relationship exists with clocks of higher-level nodes feeding timing signals to clocks of lower-level nodes. As mentioned, all clocks within the nodes link to a primary reference clock. The internal clock of a SONET may derive its timing signal from a building integrated timing supply (BITS) used by the switching systems and other equipment. Consequently, the terminal could serve as a master for other SONET nodes by providing timing on an outgoing optical carrier signal. The other nodes operate in a loop timing mode and have internal clocks timed by the optical carrier signal.

Virtual Tributary Signals

Multiplexing also allows SONET networks to interleave and transport low-speed virtual tributary (VT) signals at higher rates. As *Table 9-2* shows, the VT signals have data transport rates at sub-STS-1 levels. At low speeds, synchronous VT-1.5 signals transport DS-1 lines at a constant rate of 1.728 Mbps. Single-step multiplexing up to the STS-1 level does not require the bit stuffing seen with asynchronous multiplexing and allows the easy access of the virtual tributaries.

VT Type	Bit Rate (Mbps)	Size
VT 1.5	1.728	9 rows, 3 columns
VT 2	2.304	9 rows, 4 columns
VT 3	3.456	9 rows, 6 columns
VT 6	6.912	9 rows, 12 columns

Table 9-2.
Virtual tributary signal rates and sizes.

Grouping the VTs together accommodates the mixing of different VT types within an STS-1 SPE. Each STS-1 SPE that carries virtual tributary signals divides into seven VT groups. In turn, each VT group uses 12 columns of the STS-1 SPE. Moreover, each VT group can contain only one type of VT, while an STS-1 SPE may contain a mix of different Virtual Tributaries. As a result, an STS-1 SPE could have a mix of all seven groups.

The VT groups have no overhead or pointers and occur for organizational purposes only. As the table shows, each of the VT types consists of columns. The 12 columns that make up a VT group don't have any definite order within the SPE. Instead, the columns interleave column by column with respect to the other VT groups. Column one works as the POH.

VT Superframe

While reading previous sections, we found that VTs fit into VT groups. Yet a larger structure called a VT superframe also exists for each VT. Four consecutive and equal-sized VT-structured STS-1 SPE frames make up the superframe. In addition, the VT superframe covers 500 microseconds and contains the V1 and V2 bytes, or the VT payload pointer, and the VT envelope capacity. While the VT envelope capacity contains the VT SPE, both vary with each the size of the VT. The VT payload pointer sets the flexible and dynamic alignment of the VT SPE within the VT envelope capacity and works independently of other VT SPEs. Each VT SPE contains four bytes of VT POH. The remaining bytes make up the payload capacity.

Pointers

Established as an offset value that points to the byte where an SPE begins, pointers accommodate differences in the reference source frequencies and phase. In addition, the use of pointers prevents frequency differences from appearing during synchronization failures. The pointers also allow the transparent transport of synchronous payload envelopes between nodes with separate network clocks having almost the same timing. As a result, the use of pointers prevents delays and loss of data from occurring when the SONET encounters large slip buffers for synchronization.

During operation of the SONET, pointers provide a method for dynamically aligning the phase of STS and VT payloads. With this, SONET devices can drop and insert data or cross-connect devices. The pointers also allow the network to minimize transmission signal variations.

SONET Frame Format Structure

Referring to Table 9-2, we see that the frame format structure for SONET builds from integer multiples of the base rate. An STS-1 frame structure consists of 810 bytes that include overhead bytes and an envelope capacity for transporting data payloads. We can view the frame as a 90-column by nine-row structure that has a length of 8,000 frames per second. Bytes transmit row by row from top to bottom and from left to right. Given the frame specifics, we can confirm that the data transport rate for STS-1 equals:

(9) rows x (90 bytes/frame) x (8 bits/byte) x (8,000/s) = 51,840,000 bps or 51,840 Mbps

Moving to *Figure 9-1*, we see that the SONET frame divides into the transport overhead and the synchronous payload envelope (SPE). SONET reserves the first three nine-byte columns of the frame for the transport overhead. The overhead information allows simpler multiplexing and greatly expanded operations, administration, maintenance, and provisioning (OAM&P) capabilities. Within the 27 bytes of transport overhead, the section overhead requires nine bytes and the line overhead requires 18 bytes.

Section overhead establishes communications between adjacent network elements, while line overhead applies the signal between STS-N multiplexers. The section overhead supports performance monitoring, data communication channels for OAM&P, and framing. *Table 9-3* provides a more specific listing of section overhead functions.

Figure 9-1.
SONET frame division.

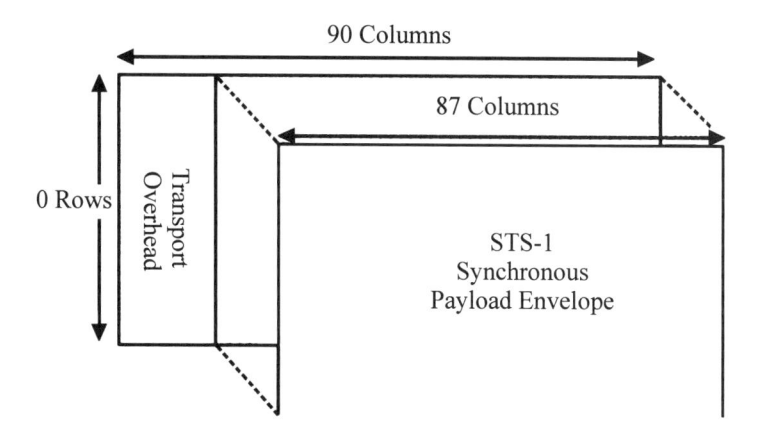

Line overhead contains information accessed, generated, and processed by line terminating equipment and extends from rows 4 to 9 of columns 1 to 9 of the frame structure. In addition, the line overhead supports locating the SPE in the frame, multiplexing signals, performance monitoring, automatic protection switching, and line maintenance. *Table 9-4* provides specific information about the line overhead functions.

Byte	Function
A1 and A2	Framing Bytes—Indicate the beginning of an STS-1 frame.
J0	Section Trace
B1	Section Bit-interleaved Parity Code—Used to check for transmission errors over a regenerator section. Defined only for the STS-1 number 1 of an STS-N signal.
E1	Section Orderwire Byte—Allocated for use as a local orderwire channel for voice communication between regenerators, hubs, and remote terminal locations.
F1	Section User Channel Byte—Allocated for user purposes. It terminates all section-terminating equipment within a line. The section user channel byte can be read and written to at each section-terminating equipment in that line.
D1, D2, and D3	Section Data Communications Channel Bytes—Form a 192 kbps message channel that provides a message-based channel for OAM&P between section-terminating equipment. The channel is used at the central location for alarms, control, monitoring, administration, and other communication needs.
Z0	Section Growth

Table 9-3.
Section overhead functions.

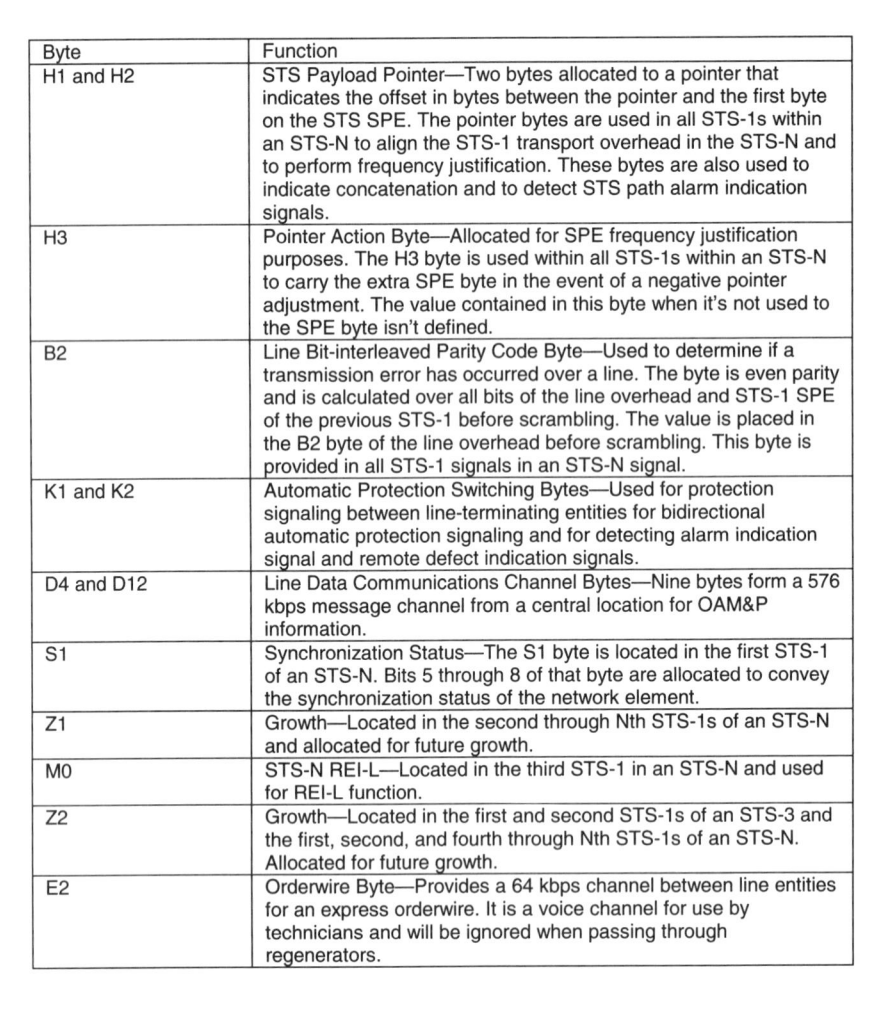

Table 9-4.
Line overhead functions.

Byte	Function
H1 and H2	STS Payload Pointer—Two bytes allocated to a pointer that indicates the offset in bytes between the pointer and the first byte on the STS SPE. The pointer bytes are used in all STS-1s within an STS-N to align the STS-1 transport overhead in the STS-N and to perform frequency justification. These bytes are also used to indicate concatenation and to detect STS path alarm indication signals.
H3	Pointer Action Byte—Allocated for SPE frequency justification purposes. The H3 byte is used within all STS-1s within an STS-N to carry the extra SPE byte in the event of a negative pointer adjustment. The value contained in this byte when it's not used to the SPE byte isn't defined.
B2	Line Bit-interleaved Parity Code Byte—Used to determine if a transmission error has occurred over a line. The byte is even parity and is calculated over all bits of the line overhead and STS-1 SPE of the previous STS-1 before scrambling. The value is placed in the B2 byte of the line overhead before scrambling. This byte is provided in all STS-1 signals in an STS-N signal.
K1 and K2	Automatic Protection Switching Bytes—Used for protection signaling between line-terminating entities for bidirectional automatic protection signaling and for detecting alarm indication signal and remote defect indication signals.
D4 and D12	Line Data Communications Channel Bytes—Nine bytes form a 576 kbps message channel from a central location for OAM&P information.
S1	Synchronization Status—The S1 byte is located in the first STS-1 of an STS-N. Bits 5 through 8 of that byte are allocated to convey the synchronization status of the network element.
Z1	Growth—Located in the second through Nth STS-1s of an STS-N and allocated for future growth.
M0	STS-N REI-L—Located in the third STS-1 in an STS-N and used for REI-L function.
Z2	Growth—Located in the first and second STS-1s of an STS-3 and the first, second, and fourth through Nth STS-1s of an STS-N. Allocated for future growth.
E2	Orderwire Byte—Provides a 64 kbps channel between line entities for an express orderwire. It is a voice channel for use by technicians and will be ignored when passing through regenerators.

The Synchronous Payload Envelope. The remaining 87 columns of the frame make up the synchronous payload envelope. In turn, the SPE separates into the STS path overhead (POH) and the payload. In the most basic sense, the payload consists of revenue-producing traffic transported and routed over the SONET network. After the multiplexing of the payload into the synchronous payload envelope, transporting and switching of the payload through the SONET can occur without any examination or demultiplexing at local or intermediate nodes. An STS-1 payload has the capacity to transport up to:

- 28 DS-1 lines
- One DS-3 line

- 21 2.048 Mbps signals
- A combination of DS-1 and DS-3 lines and 2.048 Mbps signals

STS Path Overhead. The STS path overhead contains nine evenly distributed POH bytes that separate every 125 microseconds after starting at the first byte of the Synchronous Payload Envelope. Path overhead establishes communication between the originating and disassembly points of the SPE. While establishing communication, the POH supports performance monitoring of the SPE, signal labeling of the SPE content, path status, and path trace. *Table 9-5* provides a detailed listing of POH functions.

Virtual Tributary Path Overhead

The virtual tributary path overhead (VT POH) contains four evenly distributed POH bytes within the VT SPE, with the first POH bytes starting at the first byte of the VT SPE. Of the four bytes, the V5 byte occupies the first space of the VT SPE, becomes the first byte pointed to by the VT payload pointer, and provides error checking, signal label, and path status. The J2, Z6, and Z7 bytes occupy the remaining portion of the VT superframe. During SONET

Byte	Function
J1	STS Path Trace Byte—User programmable byte repetitively transmits a 64-byte or 16-byte format string. This allows the receiving terminal in a path to verify its continued connection to the intended transmitting terminal.
B3	STS Path Bit-interleaved Parity Code—Even parity code used to determine if a transmission error has occurred over a path. The value of the Path BIP-8 is calculated over all bits of the previous SPE before scrambling.
C2	STS Path Signal Label Byte—Indicates the content of the STS SPE, including the status of mapped payloads.
G1	Path Status Byte—Conveys the path-terminating status and performance back to the originating path-terminating equipment. With this, the entire duplex path can be monitored from either end or from any point along the path. Bits 1 through 4 are allocated for an STS path REI function. Bits 5, 6, and 7 are allocated for an STS path RDI signal. Bit 8 is undefined.
F2	Path User Channel Byte—Used for user communication between path elements.
H4	VT Multiframe Indicator Byte—Provides a generalized multiframe indicator for payload containers. Used only for tributary unit structured payloads.

Table 9-5.
STS POH functions.

operation, the virtual tributary path overhead establishes communication between the point of creation and point of disassembly of a VT SPE.

Figure 9-2 represents the V5 byte of the VT POH. As the figure shows, bits 1 and 2 monitor error performance, while bit 3 provides the remote error indication (REI) signal. Bit 4 of the V5 byte yields the remote failure indication (RFI) signal in DS-1 mapping. While bits 5 through 7 work as the signal label and indicate the content of the VT SPE, bit 8 provides the remote defect indication (RDI) signal.

SONET Multiplexing

SONET can carry large payloads that range above 50 Mbps. This capability extends from the subdivision of the SPE into the smaller virtual tributaries and allows the transport and switching of payloads smaller than the STS-1 rate. SONET accepts any type of service such as voice, high-speed data, or video with service adapters. Each service adapter maps the signal into the payload envelope of the STS-1 or VT. The addition of new service adapters at the edge of the SONET network enables the transportation of new services and signals.

With the exception of concatenated signals, the system converts all inputs into the base format of the STS-1 signal. Lower-speed inputs become first-byte multiplexed into VTs. Several synchronous STS-1 lines multiplex together in either a single-stage or two-stage process that forms an electrical STS-N signal.

Figure 9-2.
The V5 byte of the VT POH. Bits one and two monitor error performance, while bit three provides the remote error indication (REI) signal.

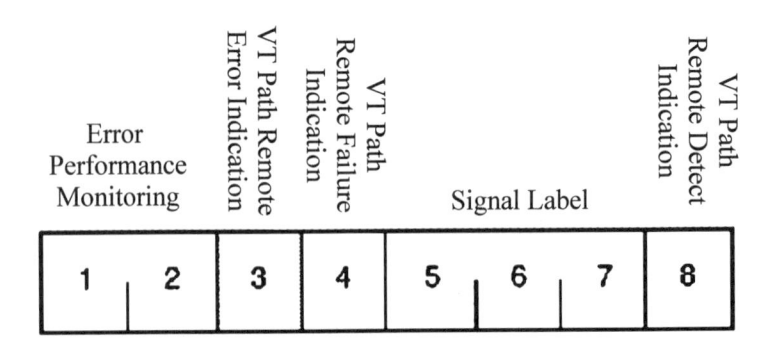

Error Performance Monitoring	VT Path Remote Error Indication	VT Path Remote Failure Indication	Signal Label	VT Path Remote Detect Indication
1 , 2	3	4	5 , 6 , 7	8

STS multiplexing occurs at the byte interleave synchronous multiplexer. During the process, the bytes interleave together in a format that leaves low-speed signals visible. With the exception of the direct conversion from electrical signal to optical signal format, no additional signal processing occurs.

SONET Network Components

Terminal Multiplexer

For SONET networks, a terminal multiplexer operates as a concentrator for tributary signals. Moving to *Figure 9-3*, we see that a simple configuration for a terminal multiplexer involves linking two of the devices with fiber optics.

Regenerator

Regenerators in SONET networks provide the same type of task as a repeater in an Ethernet network. When the signal level on the fiber link between two multiplexers becomes too weak, the regenerator boosts the signal. During operation, the regenerator clocks from the received signal and replaces the section overhead bytes before retransmitting the signal. Regeneration of the signal does not change the line overhead, payload, or POH bytes.

Add/Drop Multiplexer

An add/drop multiplexer (ADM) supports interfaces between different types of signal inputs and the SONET signal. In addition, an ADM inserts or

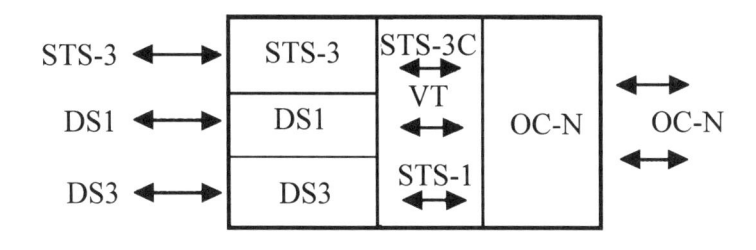

Figure 9-3.
A simple configuration for a terminal multiplexer.

drops only the signals that the network needs to access at a particular time. The capability to insert or drop a signal allows traffic to continue through the network element without requiring special pass-through units or additional signal processing.

With rural-based applications, an ADM can install at a terminal site or intermediate location for the purpose of consolidating traffic from widely dispersed locations. In addition, a network administrator could install several ADMs as a ring architecture. With this, the SONET enables drop and continue where the signal terminates at one node, repeats through the ADM, and then travels through the network to the next node. The use of ADMs in the ring architecture provides alternate routing for traffic passing through interconnecting rings in a matched-nodes configuration. If the node doesn't support the connection, the ADM repeats and passes the signal along an alternate route.

Wideband Digital Cross-Connects

A SONET wideband digital cross-connect operates as a hub as it accepts different optical carrier transport rates, accesses STS-1 signals, and switches at the STS level. In comparison with an ADM, the wideband digital cross-connect can interconnect a larger number of STS-1 lines. With this, the wideband digital cross-connect becomes a valuable tool for broadband traffic management. During operation, the device may segregate high bandwidth traffic such as video information from low-bandwidth traffic such as voice information. After separating the two streams, the device will send the data to appropriate switches.

Wideband digital cross-connects also offer the advantage of less multiplexing and demultiplexing than ADM devices, because of the accessing and switching of only the required tributaries. In addition, the switching occurs at the VT level rather than at the STS level. As a result, the accessing and switching of low-speed tributaries can occur without demultiplexing. Since wideband digital cross-connects can automatically cross-connect VTs and DS-1s, the devices have become valuable for network management.

Broadband Digital Cross-Connect Switch

Operating as a hub, the broadband digital cross-connect switch interconnects SONET signals and DS-3 signals. When operating, the device accesses STS-1 signals and switches at the STS level. A broadband digital cross-connect switch accepts optical signals and maintains overhead for OAM&P. *Figure 9-4A* is a block diagram of an optical cross-connect, while *Figure 9-4B* shows the signals found between port A and D and port A and C.

Digital Loop Carrier

A digital loop carrier consists of a system of multiplexers and switches designed to concentrate signals from the remote terminals to the local central office. As a result, digital loop carriers work as concentrators for low-speed services prior to distribution at the local central office. While a

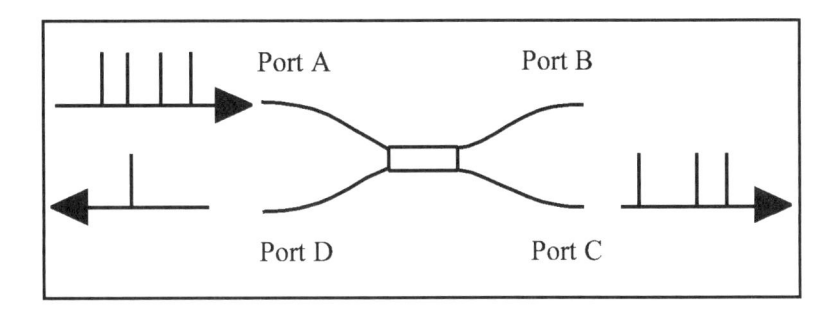

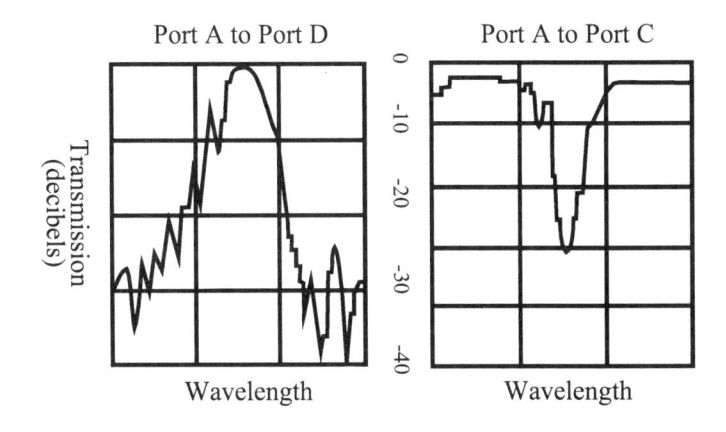

Figure 9-4A.
A block diagram of an optical cross-connect.

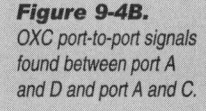

Figure 9-4B.
OXC port-to-port signals found between port A and D and port A and C.

Figure 9-5.
Schematic diagram of a 2.5-Gbps fiber-optic receiver.

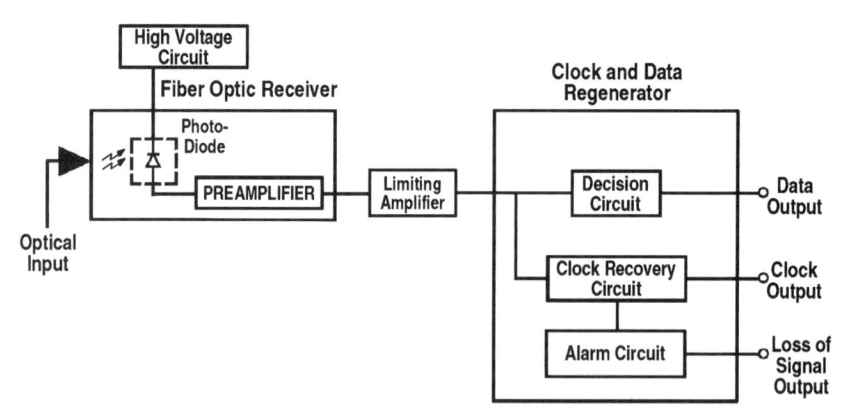

SONET multiplexer installs at the customer premise site, DLCs install at the central office or at a carrier site. Without the use of a digital loop carrier, the network would limit the number of subscribers according to the number of lines served by the local central office.

Optical Receiver

Featured in *Figure 9-5*, the 2.5 Gbps fiber-optic receiver establishes SONET requirements for OC-48 system applications. The complete receiver includes clock recovery and data regeneration. To accomplish those tasks, a photo-detector diode converts an incoming optical signal to an electrical current. The preamplifier stage amplifies the weak electrical current from the photo-detector and converts the signal into a voltage output. From there, a limiting amplifier stage further amplifies the signal but also limits its output amplitude in order to remain within the maximum input limits of the next stage. Along with that circuit operation, a decision circuit decodes the incoming electrical signal into a digital signal that represents the originally transmitted data. The clock-recovery circuit extracts a timing signal synchronized to the data rate.

SONET Network Configurations

Point-to-Point

With the point-to-point configuration shown in *Figure 9-6*, the SONET and end-to-end service path remain identical. The point-to-point configuration

Chapter 9: Optical Networks

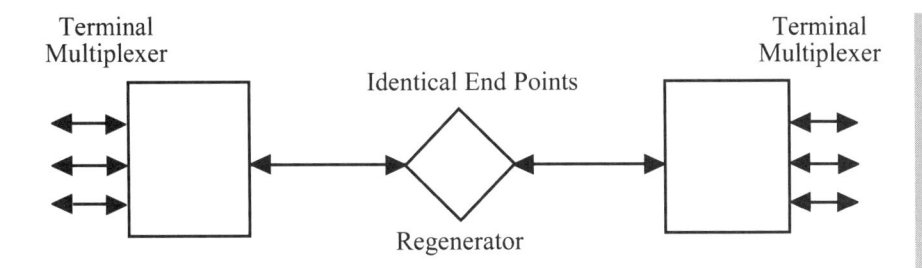

Figure 9-6.
SONET point-to-point configuration. The SONET and end-to-end service path remain identical.

consists of two terminal multiplexers connected through fiber optics with or without a regenerator in the link. Operationally, the synchronous island created by this arrangement can exist within an asynchronous environment.

Looking into the future, point-to-point service path connections will reach across the entire network with the originating and terminating points found at a multiplexer.

Point-to-Multipoint

The point-to-multipoint configuration allows the adding and dropping of circuits through the use of the SONET add/drop multiplexer. Specifically designed for point-to-multipoint configurations, the SONET ADM by-passes the process of demultiplexing, cross-connecting, adding and dropping channels, and multiplexing the data.

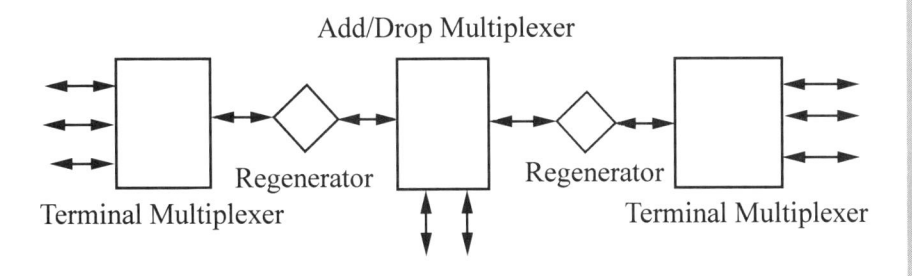

As *Figure 9-7* shows, the point-to-multipoint configuration places the ADM along the SONET link to enable the adding and dropping of tributary channels at intermediate points along the network.

Figure 9-7.
SONET point-to-multipoint configuration places the ADM along the SONET link.

269

Hub Network

Shown in *Figure 9-8*, a SONET hub network allows unexpected growth or changes to occur and offers much more flexibility than that seen with point-to-point networks. The hub concentrates traffic at a central site and allows reprovisioning of the circuits.

Hub networks may consist of two or more ADMs and a wideband cross-connect switch. This configuration allows the cross-connecting of tributary services at the tributary level. A second type of hub network configuration relies on a broadband digital cross-connect switch and allows cross-connecting at both the SONET and tributary levels.

Ring Architecture

As with the point-to-multipoint configuration, the SONET ring configuration relies on the placement of an ADM or multiple ADMs in the circuit. Moving to *Figure 9-9*, we see that the ring architecture allows either bidirectional or unidirectional traffic. The SONET ring architecture network compares favorably with the Token Ring networks in that the cutting of a fiber link in the ring will not reduce survivability. When a cut occurs, multiplexers send the affected services through an alternate path without interruption.

Figure 9-8.
A SONET hub network allows unexpected growth or changes to occur.

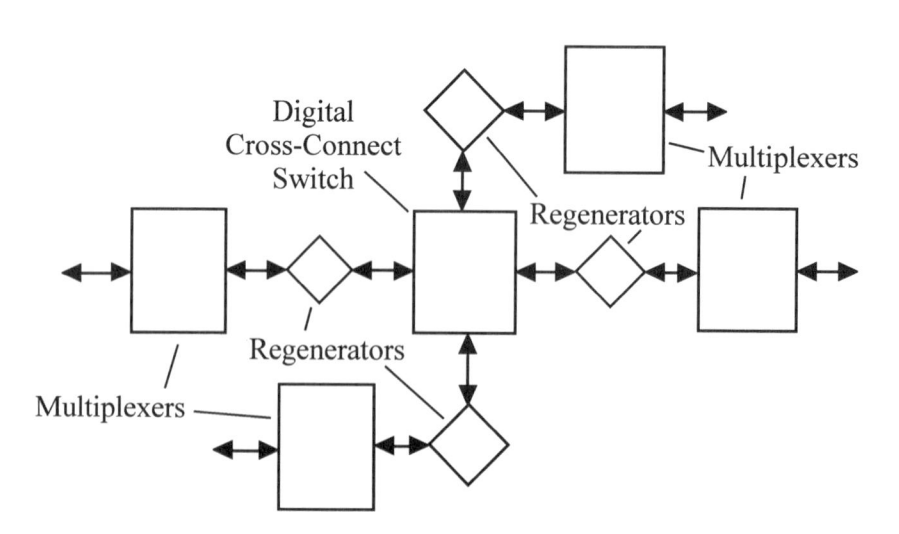

Synchronous Digital Hierarchy

After the development of the SONET standard by ANSI, the CCITT began to define a synchronization standard that allows internetworking to occur between the hierarchies developed by the two entities. The joint development effort produced the synchronous development hierarchy (SDH) standards. Interestingly, SDH serves as a world standard, while the earlier SONET exists as a subset of SDH.

ATM Networks and SONET/SDH

SONET provides the flexibility needed to transport broadband information through ATM networks. With ATM multiplexing payloads into a cell format, the need for the additional bandwidth capacity offered by SONET becomes more apparent. In brief review, ATM networks use a fast-packet switching technology to transport cells that contain 53 octets or bytes. Moreover, ATM networks have transparent bandwidth characteristics that enable the handling of a variety of dynamic services and different data-transfer rates.

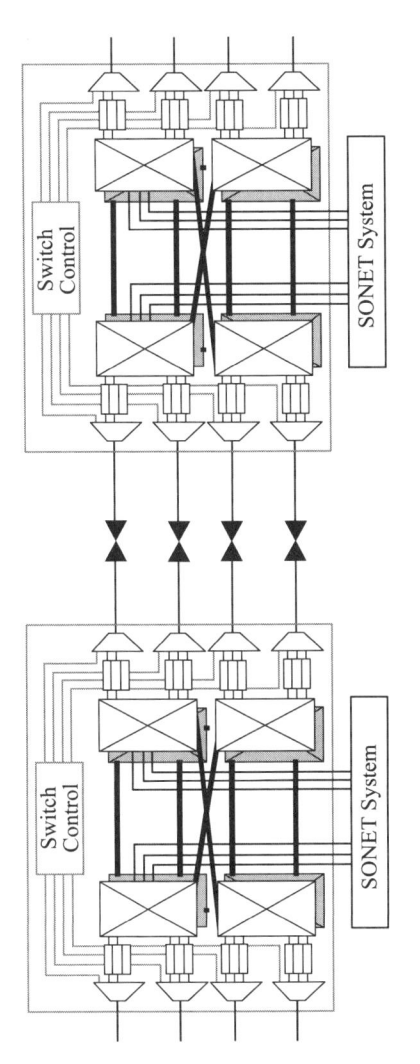

Figure 9-9.
SONET ring architecture allows either bi-directional or unidirectional traffic.

Fiber Distributed Data Interface

Another type of optical network called the Fiber Distributed Data Interface (FDDI) uses a 100-Mbps token-passing, dual-ring LAN architecture and fiber-optic cable. Referring to *Figure 9-10*, we see that the use of a dual-

Figure 9-10.
SONET dual-ring architecture allows counter-rotating traffic to flow on each ring.

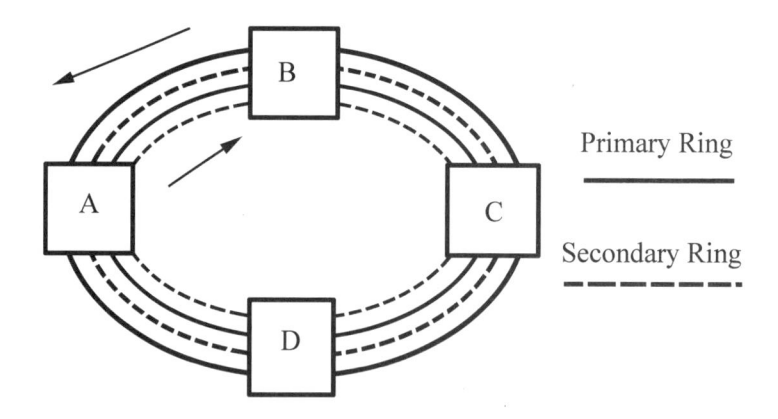

ring architecture allows counter-rotating traffic to flow on each ring. With the traffic moving in opposite directions, the dual-rings consist of a primary and a secondary ring. Due to its support for high-bandwidth applications and its capability to carry data more than two kilometers between stations with multimode fiber and longer distances with single-mode fiber, FDDI has become a high-speed backbone technology.

FDDI and the OSI Model

FDDI operates as a link-layer protocol and allows upper-layer protocols to operate independently of the FDDI protocol. As we have seen with Ethernet or Token Ring networks, applications pass packet-level data using higher-layer protocols down to the logical link control layer. Since FDDI uses a different physical layer protocol than Ethernet and Token Ring, however, it requires a different method for bridging or routing traffic on and off an FDDI ring. In addition, FDDI allows larger packet sizes than the lower-speed networking standards. For this reason, connections between FDDI and Ethernet or Token Ring LANs require the disassembly and reassembly of frames.

Developed in the mid-1980s by the ANSI standards committee, FDDI relies on the physical and media-access portions of the OSI reference model. Rather than existing as a single specification, FDDI relies on Media Access Control (MAC), the Physical Layer Protocol (PHY), the Physical-Medium Dependent (PMD), and the Station Management (SMT) specifications. As a combination, the four specifications have the capability to provide high-speed connectivity between upper-layer protocols, such as TCP/IP and IPX, and the Physical Layer.

The MAC specification defines the frame format, token handling, addressing, algorithms for calculating cyclic redundancy check (CRC), value, and error-recovery mechanisms. In addition, the MAC layer communicates with higher-layer protocols, such as TCP/IP, SNA, IPX, DECnet, DEC LAT, and AppleTalk. The FDDI MAC layer accepts Protocol Data Units (PDUs) of up to 9,000 symbols from the upper-layer protocols, adds the MAC header, and then passes packets of up to 4,500 bytes to the PHY layer.

Along with clock synchronization, the PHY layer handles the encoding and decoding of packet data into symbol streams, while the PMD specification defines fiber-optic links, power levels, bit-error rates, optical components, and connectors. With this, the PMD layer handles the analog baseband transmission between nodes on the physical media. PMD standards include TP-PMD for twisted-pair copper wires and Fiber-PMD for fiber-optic cable. With TP-PMD in place, network managers can implement FDDI over UTP cable over maximum distances of 100 meters and cut cabling costs by approximately one-third.

The SMT specification defines FDDI station configuration, ring configuration, and ring control features, including station insertion and removal, initialization, fault isolation and recovery, scheduling, and statistics collection. SMT operates as an overlay function that handles the management of the FDDI ring. Functions handled by SMT include neighbor identification and the monitoring of traffic statistics.

Figure 9-11 illustrates the relationships between the four FDDI specifications and the IEEE-defined Logical-Link Control sublayer of the OSI model. Given these relationships, the FDDI frame format remains similar to IEEE 802.3 Ethernet and IEEE 802.5 Token Ring with the exception of providing a larger frame size at 4,500 bytes. FDDI provides connectivity between upper OSI layers of common protocols and the media used to connect network devices.

Figure 9-12 shows the frame format of an FDDI data frame and token fields, while the following list provides additional detail.

Telecommunication Technologies

Figure 9-11.
The relationships between the four FDDI specifications and the IEEE-defined Logical-Link Control sublayer of the OSI model.

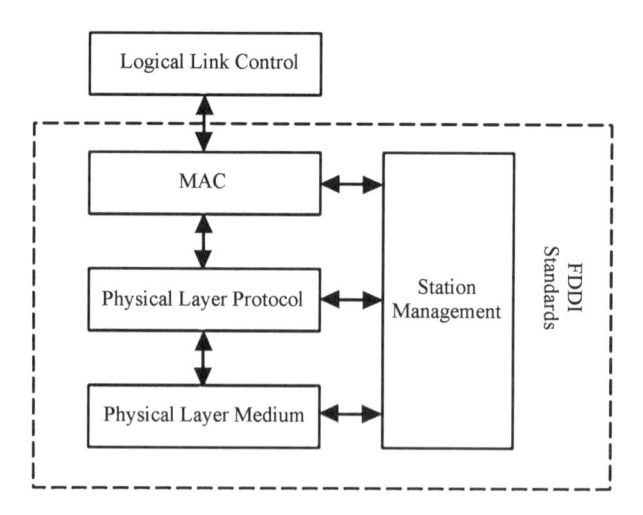

- Preamble—A unique sequence that prepares each station for an upcoming frame.

- Start Delimiter—Indicates the beginning of a frame by employing a signaling pattern that differentiates it from the rest of the frame.

- Frame Control—Indicates the size of the address fields and whether the frame contains asynchronous or synchronous data, among other control information.

- Destination Address—Contains a unicast (singular), multicast (group), or broadcast (every station) address. As with Ethernet and Token Ring addresses, FDDI destination addresses are six bytes long.

Figure 9-12.
The frame format of an FDDI data frame and token fields.

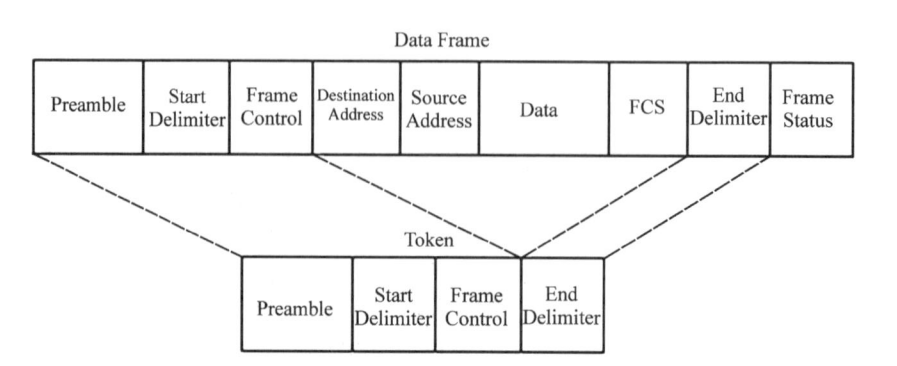

- Source Address—Identifies the single station that sent the frame. As with Ethernet and Token Ring addresses, FDDI source addresses are six bytes long.

- Data—Contains either information destined for an upper-layer protocol or control information.

- Frame Check Sequence (FCS)—Filed by the source station with a calculated cyclic redundancy check value dependent on frame contents (as with Token Ring and Ethernet). The destination address recalculates the value to determine whether the frame was damaged in transit. If so, the frame is discarded.

- End Delimiter—Contains unique symbols, which cannot be data symbols, that indicate the end of the frame.

FDDI Devices

When compared to other networking schemes, FDDI has unique qualities, because a variety of methods exist for attaching devices to the network. FDDI defines three types of devices called the single-attachment station (SAS), dual-attachment station (DAS), and a concentrator. An SAS attaches to the primary ring through a concentrator and provides the advantage of not having any effect on the FDDI ring if a device disconnects or powers down.

An FDDI concentrator stands as the foundation piece of an FDDI network. Also called a dual-attachment concentrator (DAC), the device attaches directly to both the primary and secondary rings. During operation, the concentrator ensures that a failure or power-down of any SAS won't damage the performance of the ring. *Figure 9-13* shows the ring attachments of an FDDI SAS, DAS, and concentrator.

Each FDDI DAS has two ports designated as A and B that connect the device to the dual FDDI ring. The use of a dual-attachment station establishes a connection for both the primary and the secondary ring. As opposed to connections utilizing an SAS, the disconnection or powering down of DAS devices will affect the ring. *Figure 9-14* shows FDDI DAS A and B ports with attachments to the primary and secondary rings.

Figure 9-13.
The ring attachments of an FDDI SAS, DAS, and concentrator.

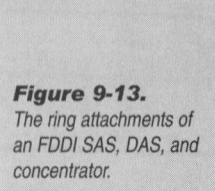

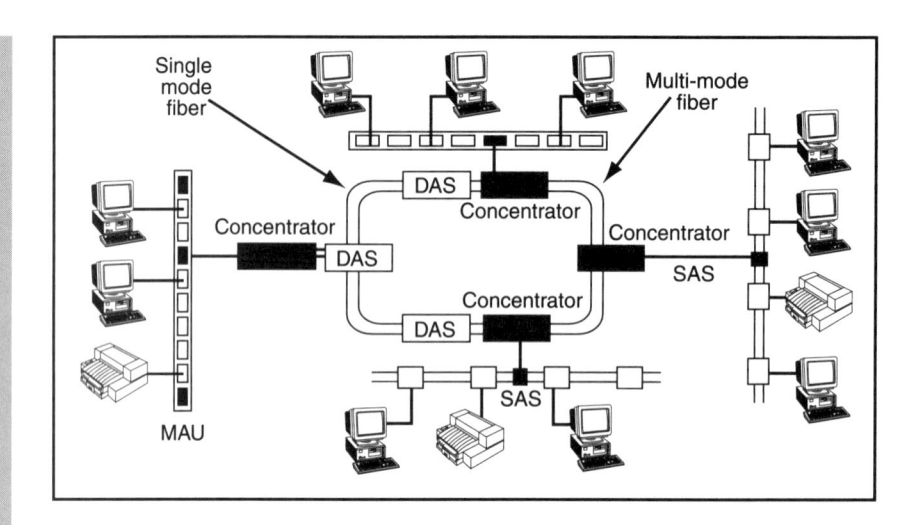

Figure 9-14.
FDDI DAS A and B ports with attachments to the primary and secondary rings.

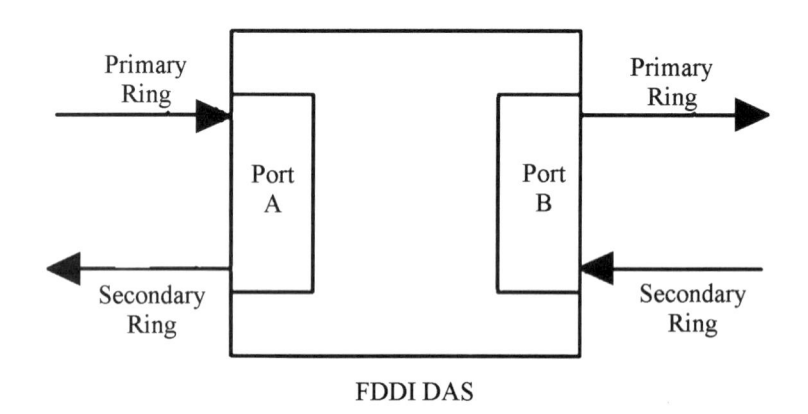

Optical Bypass Switches

An optical bypass switch protects the continuous dual-ring operation from ring segmentation and eliminates failed stations from a ring during the failure of a DAS device. To accomplish this task, the optical bypass switch uses internal mirrors to pass light from the ring directly to the DAS device during normal operation. The failure or powering down of the DAS device causes the light through optical bypass switch. As a result, the optical bypass switch maintains the integrity of the ring. *Figure 9-15* illustrates the operation of an optical bypass switch in an FDDI network.

Chapter 9: Optical Networks

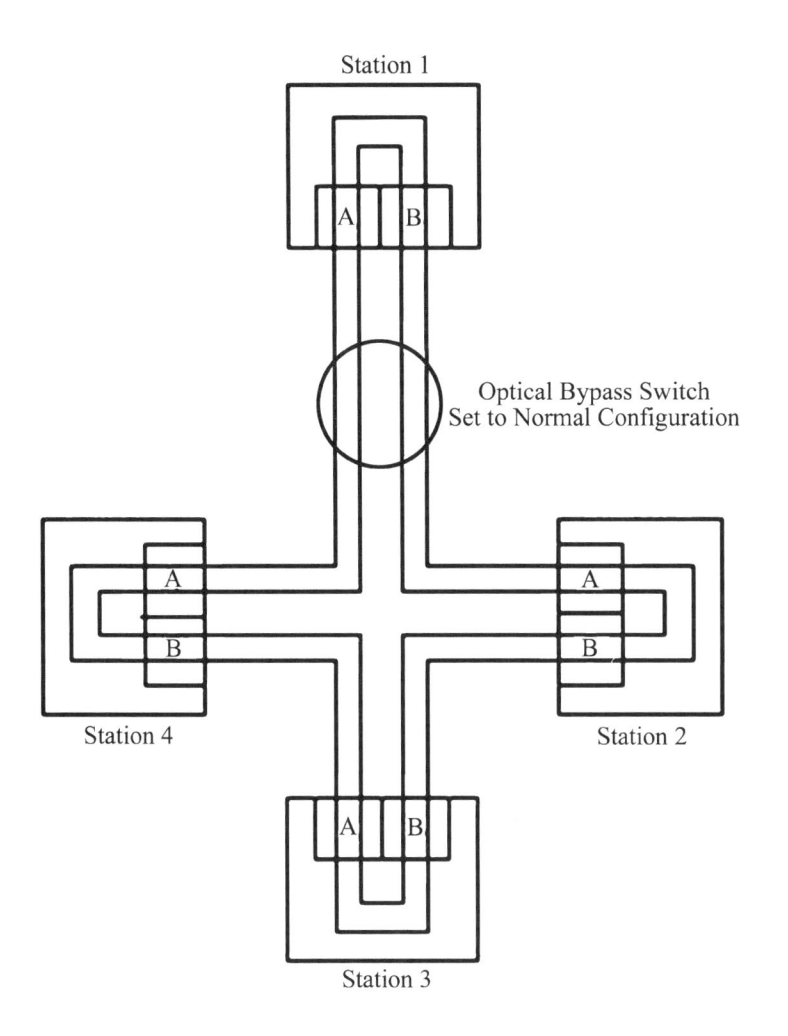

Figure 9-15.
The operation of an optical bypass switch in an FDDI network.

FDDI Fault Tolerance

Along with the use of the optical bypass switch, FDDI features fault tolerance through the dual-ring environment and dual-homing support. Going back to the implementation of FDDI, stations connect directly together. The dual counter-rotating ring design wraps around the failed node when a failure occurs and becomes a single-ring topology. Data continues to transfer from one station to another. *Figure 9-16* and *Figure 9-17* illustrate the effect of ring wrapping in FDDI. But the dual counter-rotating ring design has limitations when two nodes fail. With two nodes down, the ring separates

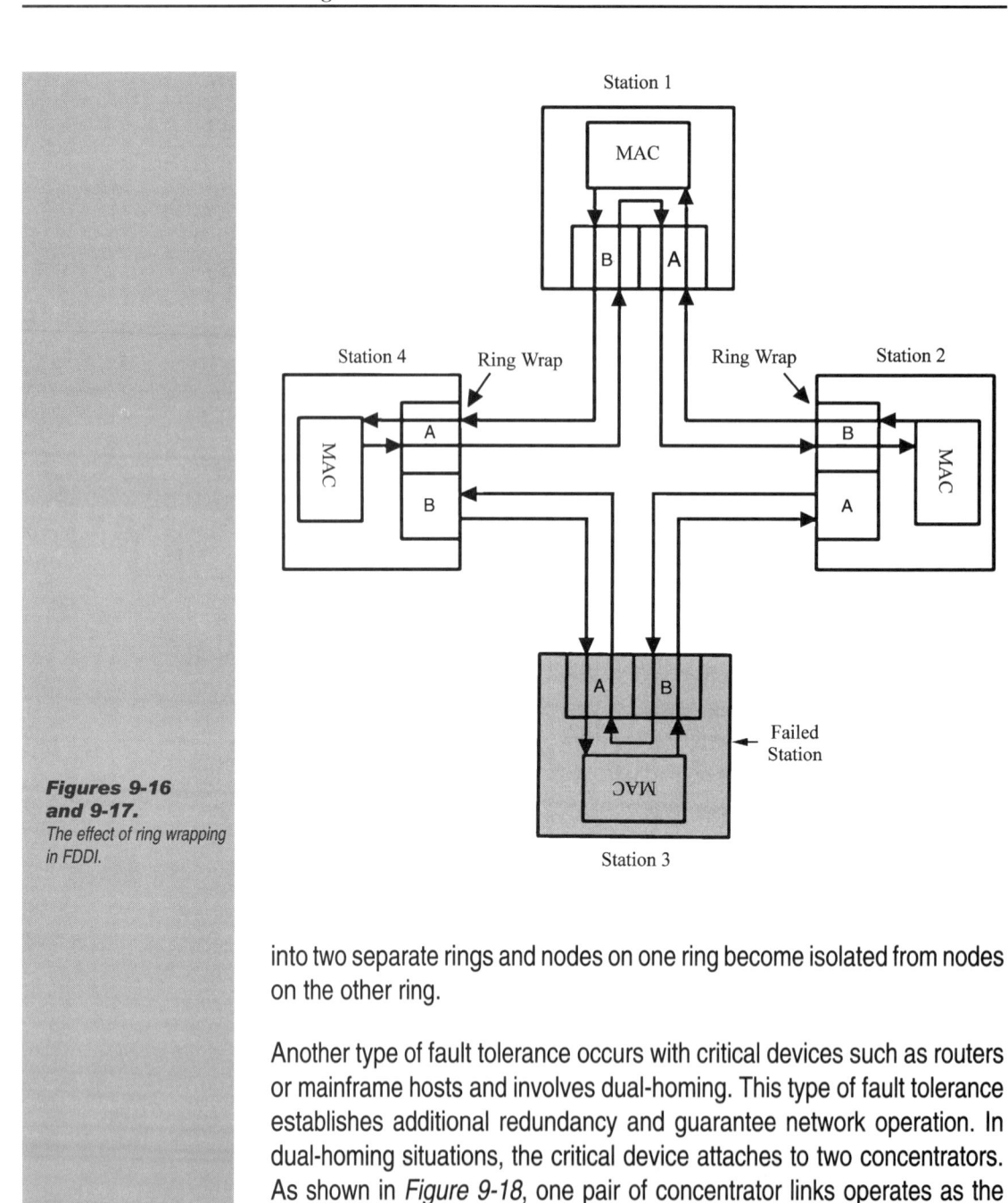

Figures 9-16 and 9-17.
The effect of ring wrapping in FDDI.

into two separate rings and nodes on one ring become isolated from nodes on the other ring.

Another type of fault tolerance occurs with critical devices such as routers or mainframe hosts and involves dual-homing. This type of fault tolerance establishes additional redundancy and guarantee network operation. In dual-homing situations, the critical device attaches to two concentrators. As shown in *Figure 9-18*, one pair of concentrator links operates as the active link, while the other pair operates as the passive link. The passive link remains in backup mode until the primary link or the concentrator attached to the primary link fails. When this type of failure occurs, the passive link automatically activates.

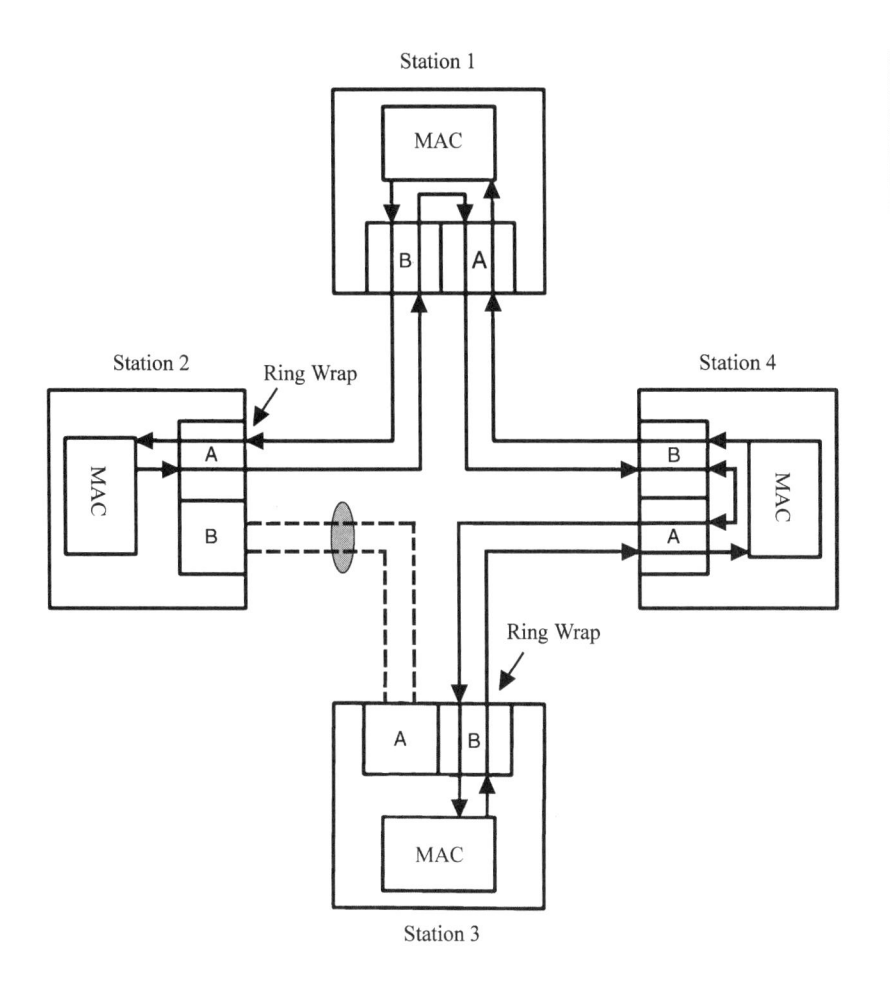

Figure 9-17.

Summary

Chapter 9 provided a detailed look at optical networks such as SONET and FDDI solutions. The chapter built from that discussion an overview of SONET signals and techniques for multiplexing those signals. Chapter 9 also defined a number of devices used to carry SONET signals such as optical receivers and cross-connect switches. The description of those relationships also led to an overview of FDDI solutions. As with the discussion of SONET/SDH, the chapter began by comparing the FDDI reference model with the OSI reference model and then progressed to the operation of the FDDI networking scheme.

Telecommunication Technologies

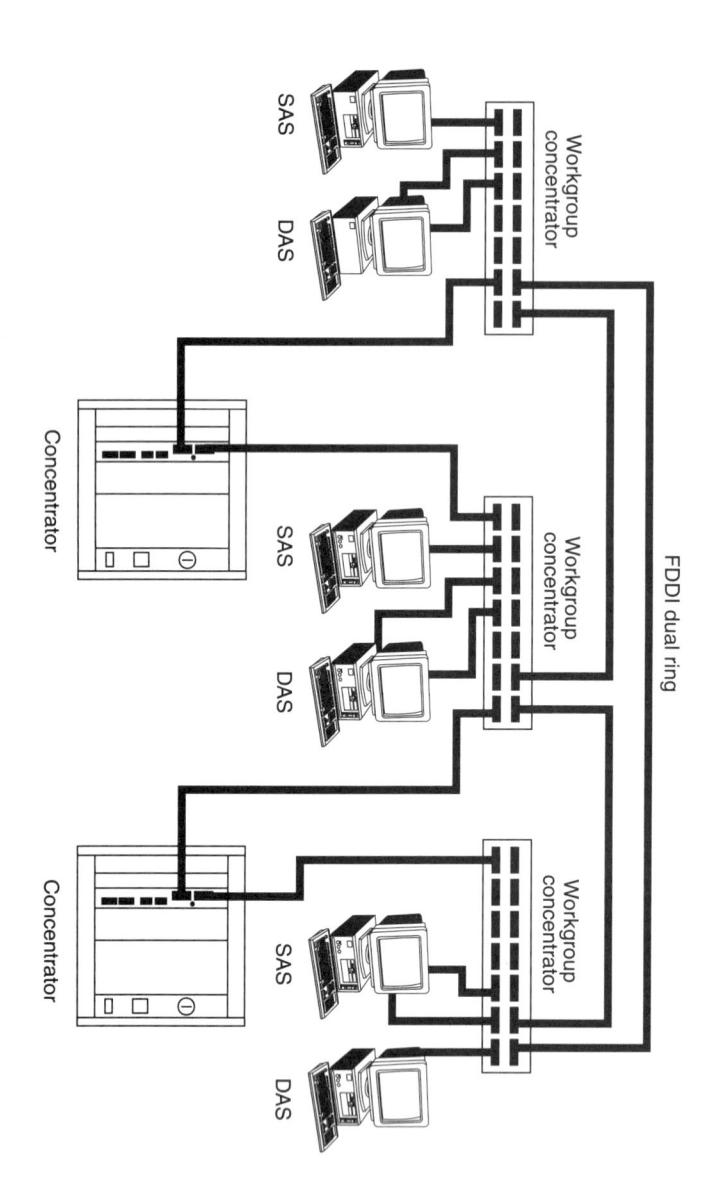

Figure 9-18.
Concentrator links in an FDDI. One pair of links operates as the active link, while the other pair operates as the passive link.

Chapter 10

Cable Modems and Hybrid Fiber/Coax Networks

Introduction

Cable modems allow high-speed access to the Internet through traditional cable television networks. Although cable modems and analog voiceband modems operate from the same principle, cable modems offer more power and can deliver data at a rate 500 times faster than that seen with the voiceband modems. Much of the difference occurs because of the physical media. While voiceband modems and ISDN operate over telephone lines, cable modems transfer data over the coaxial cable networks established by cable television companies. As a result, cable modems work through a physical media that has a theoretical bandwidth of 36 Mbps.

Cable Modem Operation

As with voiceband modems, cable modems modulate and demodulate data signals. The similarities between voiceband and cable modems end, however, with the modulation and demodulation of signals. Since a cable modem operates as a 64/256 QAM RF receiver, it has the capability to transfer approximately 30 to 40 Mbps of data over a 6-MHz-wide cable channel. Data travels from the user to the network under control of the network headend. Diagrammed in *Figure 10-1*, the cable modem uses a QPSK/16 QAM transmitter with data rates ranging from 320 kbps to 10 Mbps to modulate the signal.

Cable modems offer more power and can deliver data at a rate 500 times faster than that seen with the voiceband modems.

Figure 10-1.
A block diagram of a cable modem, which uses a QPSK/16 QAM transmitter with data rates ranging from 320 kbps to 10 Mbps.

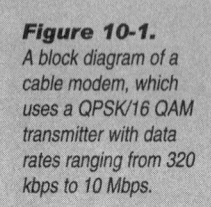

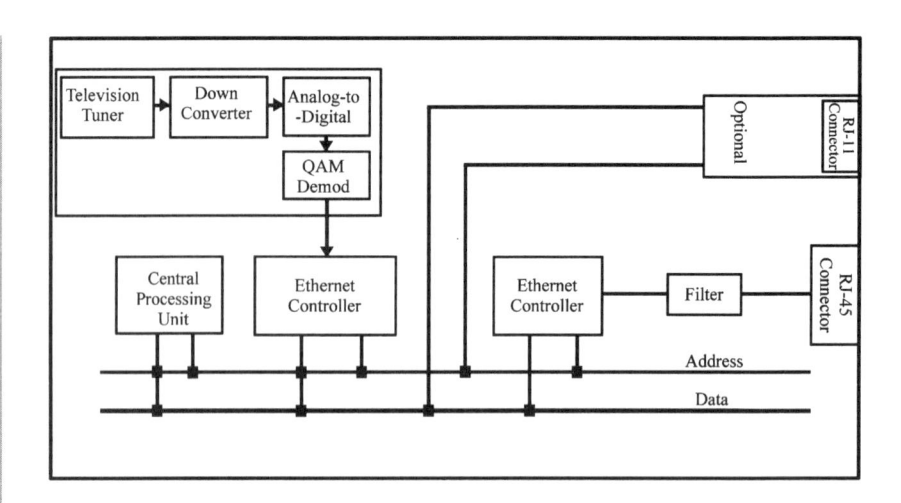

Quadrature Phase Shift Keying Modulation

With digital signals, a different modulation method called Quadrature Phase Shift Keying (QPSK) modulates the digital information onto the carrier. This occurs through the modulating of the phase of the carrier signal. During operation, the modulating circuit senses the type of data and forces the carrier into one of four different phase states called a symbol. Each symbol contains two data bits. As a result, QPSK doubles the potential amount of data transmitted through amplitude or frequency modulation.

Quadrature phase shift key-modulated data rates occur as symbol rates rather than as a bit rates. For example, a symbol rate shown as 20 MS/s or 20 mega-symbols equals 40 Mb/s or 40 megabits bits per second. As *Figure 10-2* depicts, each possible pair of data bits is represented by a different phase angle, while *Figure 10-3* shows a QPSK waveform.

Quadrature Amplitude Modulation

Quadrature Amplitude Modulation (QAM) combines amplitude modulation and phase shift keying. Data traveling to the cable modem and data moving downstream becomes modulated using Quadrature Amplitude Modulation.

Chapter 10: Cable Modems and Hybrid Fiber/Coax Networks

Figure 10-2.
Quadrature Phase Shift Key (QPSK) modulated data bits.

Figure 10-3.
A QPSK waveform.

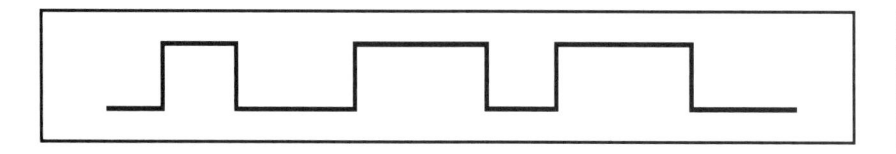

A QAM modulation scheme works well for cable modem applications, because of the rate of data transfer and because of the capability to place information in 6-MHz television carrier without disturbing the video signal.

QAM-modulated data can achieve transfer rates of up to 36 Mbps on a standard coaxial or hybrid fiber/coaxial television system. Quadrature Amplitude Modulation places the modulated, downstream, digital data in the television carrier between the frequencies of 42 MHz and 750 MHz. As a result, the data remains immune to interference from high-band radio transmission and impulse noise. 64 QAM modulation carries data at a transfer rate of 27 Mbps over a 6-MHz channel, while 256 QAM modulation supports a data transfer rate of 36 Mbps.

As an example of QAM operation, a signal transmitting at 3,600 bps, or three bits per baud, represents eight binary combinations. Combining two measures of amplitude and four possible phase shifts produces eight possible waves for transmission. *Table 10-1* shows which waves correspond to which binary combination.

Bit Value	Amplitude	Phase Shift
000	1	None
001	2	None
010	1	1/4
011	2	1/4
100	1	1/2
101	2	1/2
110	1	3/4
111	2	3/4

Table 10-1.
QAM bit stream manipulation.

Note: Encoding the bit stream produces 001010100011101000011110
Breaking the bit stream into three-bit triads shows as 001-010-100-011-101-000-011-110

Cable System Architecture

The architecture of a cable-data network resembles the architecture of a local-area network. Indeed, a combination of hardware devices found at the headend of the cable network establishes an Ethernet network over a wide-area network. Other network configurations may involve the interconnection of cable headends by fiber-optic links. In some instances, a super hub or collection of hubs may control typical network tasks and transactions. Finally, management of growth of a cable data network can occur with the addition of data channels for more bandwidth and with the same methods seen with Ethernet networks.

The Cable System Headend, Super Hub, and Regional Data Center

A cable headend combines downstream data channels with the video, pay-per-view, audio, and other programming usually received by cable subscribers. Transmitters found at the headend transmit the combined signals throughout the cable distribution network. The signals travel to splitters at the customer location with the video and audio signals splitting to a set-top box or HDTV receiver and the data signals splitting to a cable modem and personal computer.

Operating as a cable headend, a super hub provides temperature-controlled facilities for the housing of servers. The super hub also supports the dynamic host configuration protocol, domain name server, and log control servers necessary for administration of the cable-data network. Super hubs may operate as part of a regional data center that supports dial-up modem services and business-to-business Internet services. Network management at the super hub level covers television and data network operations.

Shown in *Figure 10-4*, the regional data center also contains network switches, routers, and servers. Each regional data center connects to other regional data centers through a national backbone along with the Internet and World Wide Web services. The servers contained within the regional data center support electronic mail, web hosting, news, chat, proxy serving,

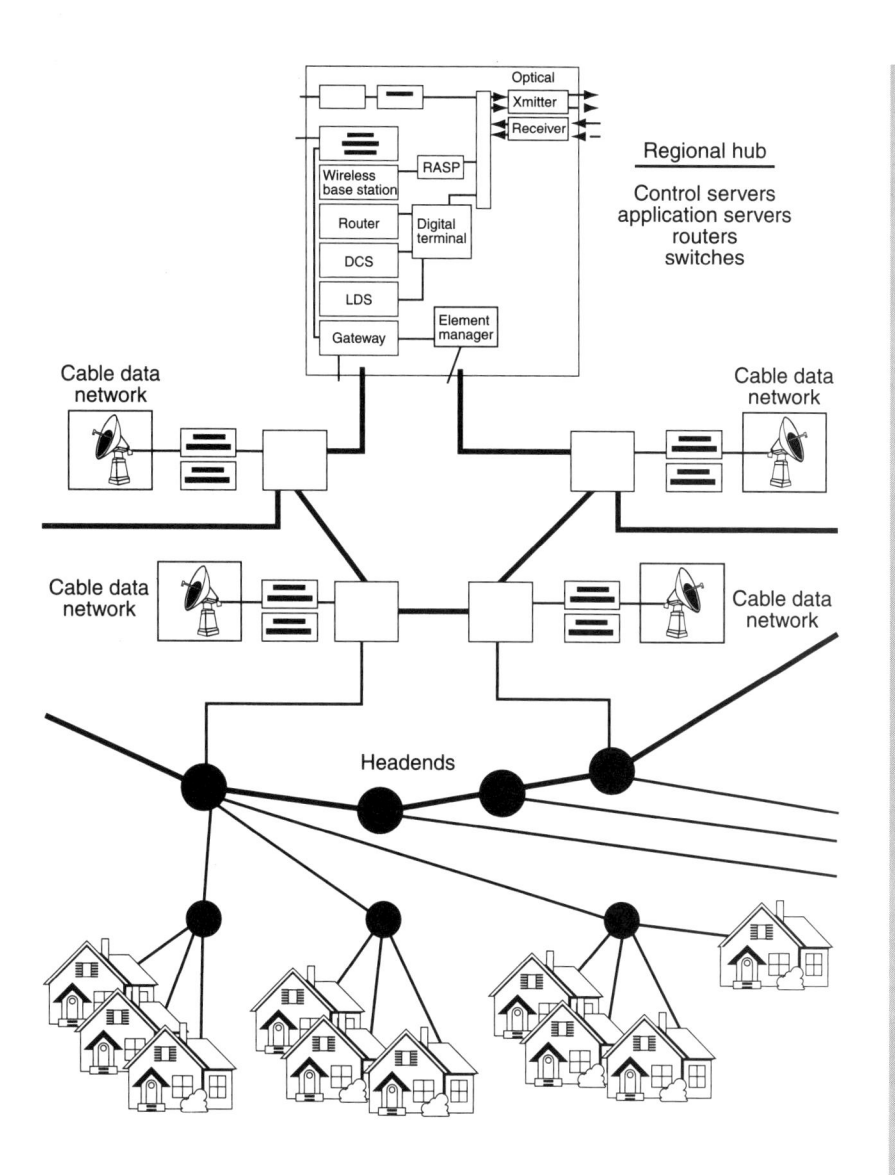

Figure 10-4.
The regional data center, which contains network switches, routers, and servers.

caching, streaming media, and other services. *Table 10-2* lists the services possibly provided by cable data network.

Node Data Controller

At the headend of the cable system, the Node Data Controller (NDC) accomplishes the same function as the cable modem on a much larger scale and operates as the Cable Modem Termination System (CMTS). Multiple

Table 10-2.
Available cable data network services.

Service	Characteristics	Example
Data Symmetrical	Asynchronous Available Bit Rate (ABR)	File Sharing LAN Traffic
Data Asymmetrical	Asynchronous Available Bit Rate (ABR)	Game Downloading
Interactive Video	Store and Forward Constant Bit Rate (CBR) Variable Bit Rate (VBR)	Video on Demand
Messaging	Store and Forward Available Bit Rate (ABR)	E-mail Voice mail Billing
Multimedia	Bi-directional Variable Bit Rate (VBR) Constant Bit Rate (CBR)	Real Time Videoconferencing Document Conferencing Interactive Games
Multimedia	Unidirectional Store and Forward Available Bit Rate (ABR)	Image Retrieval
Telephony	Real Time w/ Fixed Delay Constant Bit Rate (CBR)	POTS ISDN

NDCs can attach to multiple segments of a bridge for the purpose of breaking one cable system into several collision domains. With this, the NDC either reduces or prevents data congestion on the cable system.

During operation, upstream demodulators filter data from individual users for additional processing by the NDC. The Node Data Controller routes data sent by cable-modem users over a multiplexed network interface. In addition, the device receives data from the Internet and provides the switching necessary to route the data to the cable users. Data sent from a users group travels to a 64/256 QAM modulator and becomes modulated into a 6-MHz channel.

Element Management System

The cable headend also includes an Element Management System (EMS) that configures and manages the Node Data Controller as an integrated or separate device. EMS tasks involve day-to-day administration, monitoring, alarms, and the testing of various components. If located at a Network Operations Center, a single EMS can support many Cable Modem Termination System devices for a geographic region.

Distribution Hubs

The minimal data network configuration of a cable-data network consists of basic distribution hubs. While the Node Data Controller provides upstream and downstream data transport and an IP router, the basic hubs distribute data received at the regional data center throughout the network. In addition, the distribution hubs support the aggregation of the data.

Routers, Switches, and Servers

The servers found within a super hub or regional data center run the cable-data networks. Tasks associated with the servers include file transfer, user authorization, accounting, system security, and IP address assignment. From a network management perspective, the servers also provide DHCP, DNS, and data-over-cable-service-interface-specifications (DOCSIS) services.

Cable Modems at the Home

Cable modems manufactured for home use feature cable-in and cable-out jacks on the back for interfacing with the cable system and a 10BaseT connector for attaching a computer or small network. Once connected to the cable system, the modem acts as a tuner or receiver and operates on a preprogrammed pair of frequencies. The cable modem uses one frequency to receive data from the cable system and forward the data to the user's Ethernet network. A second frequency provides a path for transmitting data from the Ethernet network to cable system.

Only four nodes can connect to a small network enabled by a cable modem. Yet the modem also acts as a bridge to other networks. The computer or computers connected to the cable modem belong to the same broadcast domain as all computers connected to cable modems by the same provider.

Cable-Data Systems and the OSI Model

As expected, a cable-data network incorporates many features necessary for broadband communications over a wide-area network. From the perspective of the OSI model, a cable-data network uses the Network Layer along with the Data Link Layer and three sublayers including the MAC layer. Cable data systems rely on the Ethernet frame format for data transmission over upstream and downstream channels.

Cable modems manufactured for home use feature cable-in and cable-out jacks on the back for interfacing with the cable system and a 10BaseT connector for attaching a computer or small network.

The Physical Layer

The Physical Layer for cable data networks establishes downstream and upstream data channels. Along with utilizing the 64 and 256 QAM modulation/demodulation scheme, the downstream data channel uses a concatenation of the Reed-Solomon block code and Trellis code. In addition, the downstream data channel takes advantage of variable interleaving supports and a contiguous serial bit stream with no implied framing. The downstream data channel for cable data networks occupies a 6-MHz wide spectrum that coexists with other signals at the cable plant.

Compared with the downstream data channel, the upstream data channel uses the QPSK and 16 QAM modulation/demodulation formats as well as multiple symbol rates. Along with support for TDMA, the upstream data channel also supports fixed-frame and variable-length protocol data units. The upstream data channel relies on programmable Reed-Solomon block coding, CTMS-provided flexible and programmable cable modem control, frequency agility, and programmable preambles.

The Data Link Layer

The Data Link Layer for cable-data networks consists of the logical link control sublayer, the link security sublayer, and the media access control sublayer. Link security sublayer requirements divide into the baseline privacy interface (BPI), security system interface (SSI), and removable security module interface (RSMI). The baseline privacy interface provides data privacy for cable modem customers by encrypting data traffic between the home cable modem and the superhub. In addition to customer data security, the BPI and RSMI requirements combine with the Element Management System and Cable Modem Termination System to map a cable modem identity to paying subscribers. With this mapping, the system also authorizes access to data network services.

The MAC Layer

The Media Access Control sublayer establishes the collision detection and retransmission functions needed by cable-data network subscribers as they share a single upstream data channel for transmission to the network. In addition, the MAC Layer supports timing and synchronization, allocates bandwidth to cable modems under control of the CTMS, provides error detection and error recovery, and establishes procedures for registering new cable modems on the network. To compensate for cable losses and delays caused by the large geographic range of the network, the MAC Layer allows the cable modem to range, or assess the time delay in transmitting to the headend.

The Network, Transport, and Application Layers

At the Network Layer, cable data networks use the Internet Protocol for communication from the cable modem to the network. In addition, network address translation (NAT) exists at the Network Layer and allows the mapping of multiple computers that have a single high-speed access through the cable modem. The Transport Layer of the cable data network supports both TCP and UDP. Cable data networks use the Application Layer to support all Internet-related applications such as electronic mail, chat, FTP, HTTP, and SNMP.

With the downstream communication using TCP/IP, the Transport Control Protocol uses a form of data flow control that allows very fast downstream data-transfer bursts. The data-flow control, or sliding window protocol, matches the bursts to the high bit rate of the downstream channel. In addition, the sliding window protocol optimizes the burst communication with the delay time in receiving acknowledgment signals sent upstream from the customer's computer. The acknowledgment (ACK) signals acknowledge the receipt of the data packets at the customer's computer.

Cable Data Network Topology

The topology of a typical cable system resembles the tree and branch or a collection of trees and branches shown in *Figure 10-5*. At the root, the headend collects and distributes broadcast signals on the forward channel to a large installed base. The trunk, feeder, and drop cables consist of coaxial cable. While the maximum distance from the headend to any end-user node ranges from 10 to 15 kilometers, the network will support a

Figure 10-5.
The topology of a typical cable system resembles a tree and branch or a collection of trees and branches.

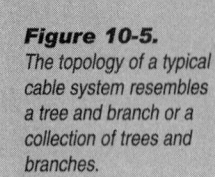

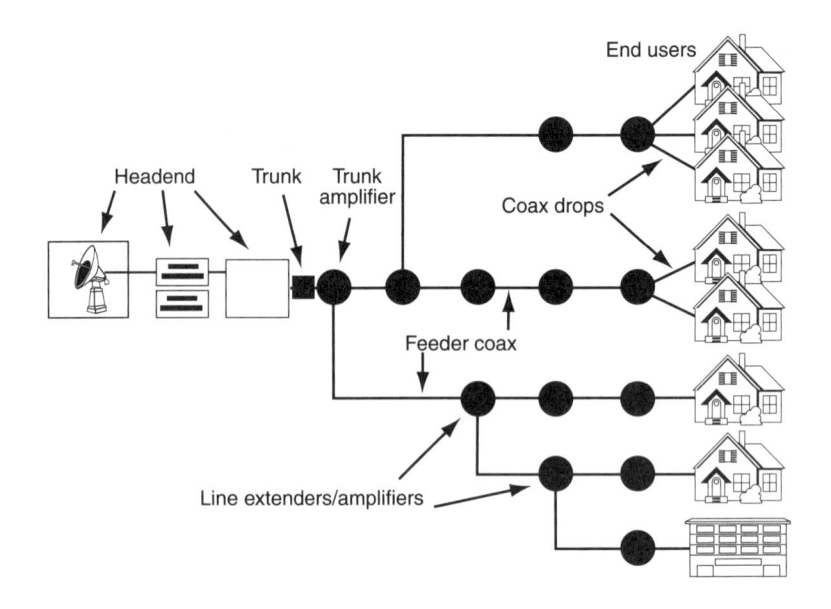

maximum of 125,000 connections per trunk and a maximum of 35 cascaded amplifiers. Newer cable systems that follow the same model use fiber-optic cable for the trunk and coaxial cable for the feeder and drop cables. This approach lengthens the service reach of the system to 80 kilometers. The system supports 500 to 3,000 connections per fiber node and a maximum of four to 10 cascaded amplifiers.

As cable systems have evolved into television and data networks, the tree and branch architecture has also evolved into the ring and bus architecture shown in Figure 10-4. We refer to this architecture as the Regional Hub/Passive Coaxial Network Architecture. The topology for this architecture includes a centralized regional hub to share cost of expensive equipment among multiple cable operators as well as secondary fiber hubs. While connections between central headend and fiber hubs provide ring capability and route diversity, passive coaxial distribution to the home improves electrical characteristics of the network.

Regional Hub/Passive Coaxial Networks employ fiber-optic cables for the supertrunk or ring and the trunk, while using coaxial cable for the feeder and drop cables. The maximum distance from a fiber hub found on the network to a user node is 80 kilometers. While the network supports 200 to 500 connections per fiber node, the architecture does not support cascaded amplifiers.

While the individual fibers or coaxial cables exit the headend as part of branching buses, the topology forms a star of trees emanating from the headend. Since data originating on one tree may travel to another tree, the headend must provide a mechanism to move data from one to another. Multidrop, bidirectional traffic establishes both the forward and reverse channels as operational. With this, the network allows many nodes to transmit to the headend over a reverse channel and the headend to transmit to many nodes over the forward channel. Because of the topology employed with cable networks, the network also uses contention for the available digital bandwidth.

Hybrid Fiber/Coax Networks

The improvements seen with Regional Hub/Passive Coaxial Networks allow upstream and downstream signals to travel over a mixed path of fiber-optic and coaxial cables. Systems utilizing hybrid fiber/coax (HFC) connectivity combine the best features of each transmission media while providing flexible network access. Hybrid fiber/coax supports nearly all transmission technologies such as ATM, Frame Relay, SONET, and SMDS. In addition, hybrid fiber/coax has become the top choice for many cable television and telephony providers.

In addition to using fiber-optic transmissions, a new topology, and other techniques to increase the amount of bandwidth per household, HFC relies on digital technology and QAM modulation to increase downstream and upstream capacity. HFC networks also introduce a node architecture that divides all homes served by the cable network into small areas. With all this, an HFC operates as an asymmetric shared medium-type network.

Hybrid Fiber/Coax Architecture

The fiber portion of hybrid fiber/coax architecture consists of fiber-optic cable carrying video and telephony signals from the headend to the optical node serving a specific neighborhood. At the optical node, circuitry converts the downstream optical signaling found at the fiber trunk to an electrical signal. Then the coax portion of hybrid fiber/coax architecture carries the electrical signal to drops at individual customer locations.

As you know, fiber-optic cable has a much larger data carrying capacity than coaxial cable. Consequently, a single optical node connects to several coaxial distribution feeds that, in turn, connect to Home-Termination Unit within a house. In most instances, one optical node will support four multidrop coaxial cables. While the distance between the headend and fiber node extends to approximately 25 kilometers, the fiber node is less than two kilometers from the home.

Home-Termination Units

A Home-Termination Unit (HTU) terminates the HFC for the processing of the Physical and Mac layers. Depending on the type of service provided through the HFC network, the Home-Termination Unit may take several forms. For example, a set-top box works as a Home-Termination Unit for interactive television services. A cable-data network running through HFC uses a cable modem as the HTU and connects to a personal computer for application processing and display.

Taking Advantage of the HFC Bandwidth

One of the benefits of hybrid fiber/coax technologies comes from the capability to support integrated broadband services that may require a 750-MHz spectrum. While the spectrum serves as an expansion of the standard analog video broadcast spectrum, it also has an additional 250 MHz of space reserved for digital video services, upstream signaling for interactive services, as well as upstream and downstream telephony services.

The HFC bandwidth divides among a number of different services. When transporting video-based technologies, the HFC bandwidth divides into 6-MHz channels. In turn, those channels separate into a maximum of 240 DS-0 lines that support telephone services. While the number of drops on the cable doesn't affect the quality of broadcast video signals, the quality of telephony signals depends on a specific amount of bandwidth for each active connection.

One of the benefits of hybrid fiber/coax technologies comes from the capability to support integrated broadband services that may require a 750-MHz spectrum.

Hybrid Fiber/Coax Topology

Figure 10-6 diagrams the topology of a hybrid fiber/coax network. As the figure shows, the topology combines a star configuration with a mini tree-and-branch configuration. The headend forms the center of the star, while fiber-optic cables extend to each neighborhood terminated at the fiber node. The coaxial cables extending from each of the fiber nodes forms the tree-and-branch configuration and covers 500 homes.

As mentioned, the HFC network operates as an asymmetric network. The homes supported by a node share the 5-42 MHz upstream transmission bandwidth of the HFC network to the headend terminal. From the downstream perspective, however, information travels from the headend to the homes connected the network. With this, downstream communication becomes a matter of a one too many broadcasts with signals above 50 MHz.

The node architecture seen with HFC networks enhances reliability and upstream transmission capacity. Reliability improves because of the division of the service area into small areas that remain independent of other areas. As a result, a failure at a fiber trunk or feeder cable affects only a particular area, rather than the entire service area. Reliability also improves through the use of fiber-optic cables and the reduced usage of amplifiers throughout the network.

Upstream capacity increases because of the capability of the node architecture to share downstream and upstream bandwidth among fewer homes. As a result, the bandwidth allocated per individual subscriber increases substantially. The network employs a MAC protocol to coordinate transmissions, provide a method for avoiding contention between subscribers, optimize the bandwidth, and support multicast capability. In addition, a lesser number of subscribers sharing bandwidth decreases the amount of combined noise. Less noise translates into greater capacity for desired information.

Along with sharing bandwidth among fewer homes and reducing noise, HFC networks also use several methods to maintain the upstream capacity. Return amplifiers operate in the return spectrum on the coaxial return path of the cable plant. Diplex filters ensure that only signals between the 5- to 42-MHz

Telecommunication Technologies

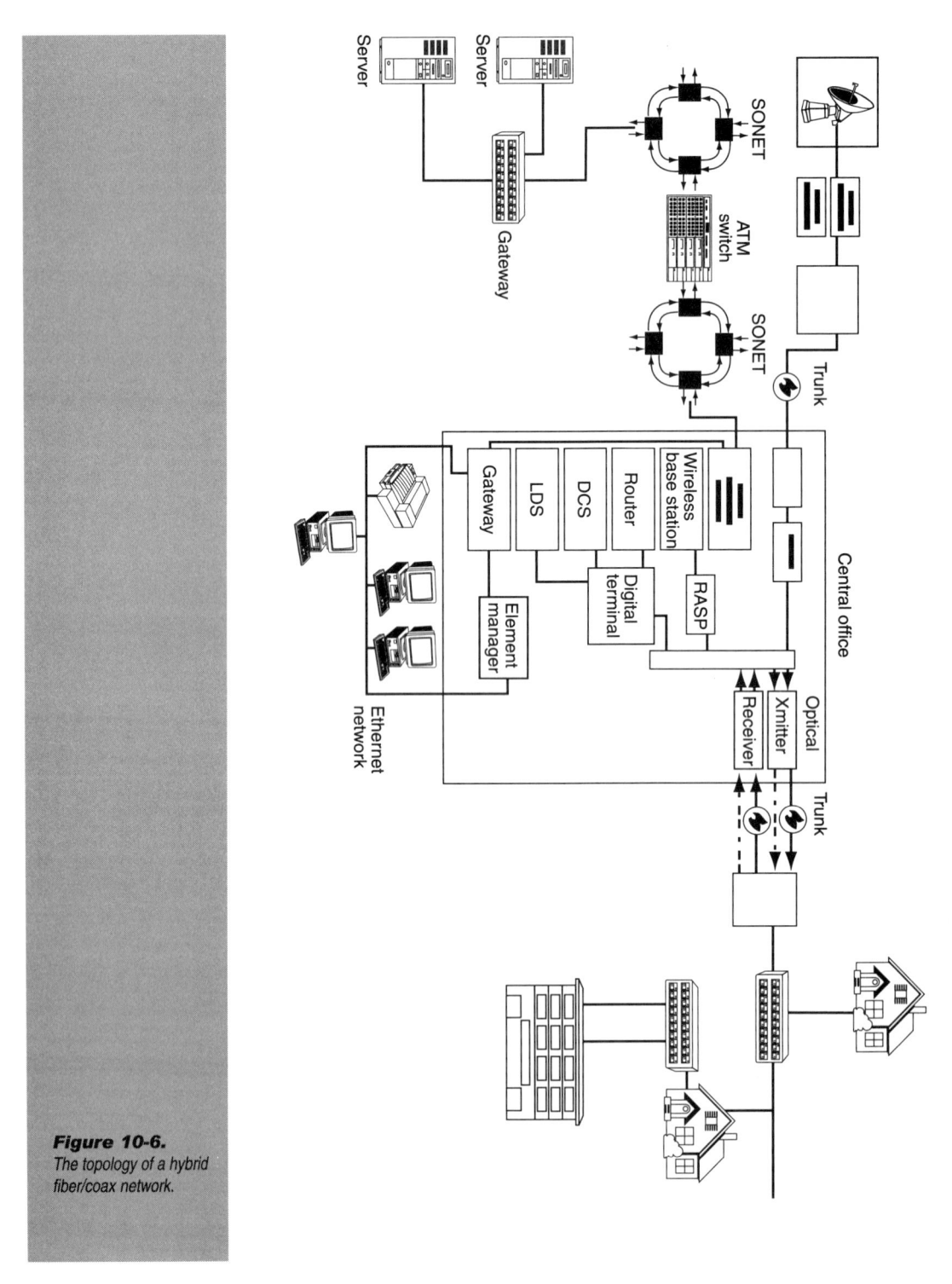

Figure 10-6.
The topology of a hybrid fiber/coax network.

spectrum pass upstream, while the network transmits signals above 50 MHz downstream. Moreover, the use of lasers that operate in the 5- to 42-MHz spectrum in the return direction of fiber trunks improves the capability of the network to move data upstream. Finally, coaxial distribution plants have improved measures to achieve noise immunity throughout the system.

Switched Hybrid Fiber/Coax Networks

The prior discussion focused on the use of hybrid fiber/coax networks as a broadcast architecture. But HFC also functions as a switched architecture (with the addition of a switch), or as an ATM network in the headend and adds two-way interactive capabilities to the network. The switching device or network switches the individual information streams to different homes. In addition, switching allows upstream information to travel from personal computers or set-top boxes residing in the home to external networks, servers at the headend, or the Internet.

Summary

Chapter 10 strengthened the coverage of broadband technologies while providing information on cable modem and hybrid fiber/coax networks. As the chapter described those technologies, it provided a detailed overview of cable modem operation and different modulation techniques. With that as a basis, the chapter also described cable system architecture and defined the operation of headends, distribution hubs, superhubs, and cable data centers.

Chapter 10 also discussed Physical Layer and Network Layer devices used in the cable system network such as routers, switches, and servers. All this led to a comparison of cable system implementation methods and the layers that make up the OSI reference model. Along with that discussion, the chapter also defined topologies for cable modem and hybrid fiber/coax networks.

The prior discussion focused on the use of hybrid fiber/coax networks as a broadcast architecture. But HFC also functions as a switched architecture.

Chapter 11

DSL Technologies

Introduction

Developed in the early 1990s as a method for delivering video on demand and interactive TV over copper wires, digital subscriber line (DSL) technologies use sophisticated modems to send and receive high-bandwidth digital signals over existing twisted-pair telephone lines at multimegabit speeds. Since new signal-processing techniques allow voice and data to travel simultaneously over the same analog line, DSL technologies can transfer data at high rates to highly separated locations.

The acronym "xDSL" serves as an overall name for similar yet competing derivatives of DSL, including asymmetric digital subscriber line (ADSL), symmetric digital subscriber line (SDSL), high-bit-rate digital subscriber line (HDSL), rate-adaptive asymmetric digital subscriber line (RADSL), and very-high-data-rate digital subscriber line (VDSL). Despite the differences between the different types, all DSL services operate through dedicated, point-to-point, public network access.

Several factors have pushed the implementation of ADSL and VDSL technologies and architectures. The Internet and similar data-intensive, high-bandwidth applications have driven the demand for DSL service. With DSL technologies providing the rapid transmission of data over the public switched network, the need for the expensive construction of new fiber networks has begun to decrease.

Current estimates show that the creation of an extensive fiber-to-the-curb (FTTC) network will cost approximately $1,500 per current telephone customer. If we extend those numbers across the nation, the cost of replacing the entire copper network range between one-half and three-quarters of a trillion dollars. Since installation must occur

Since new signal-processing techniques allow voice and data to travel simultaneously over the same analog line, digital subscriber line technologies can transfer data at high rates to highly separated locations.

before customers can receive benefits and pay subscription fees, providers of the FTTC would gamble that customers will subscribe when the technology becomes available.

Advantages of xDSL

An xDSL-based network offers several advantages for residential broadband services that include improved usage of available bandwidth, decreased signal distortion, the use of Digital Signal Processor technologies, point-to-point transmission, always connected status, simultaneous POTS support, and rate adaptive functionality. The improved usage of available bandwidth occurs because xDSL technologies take advantage of a significantly wider frequency spectrum. Compared to the 4-kHz bandwidth seen with analog modems, DSL transfers data along a bandwidth that spans more than 1 MHz.

The contrast between analog and DSL bandwidth exists because of the different types of implementation. Analog modems support end-to-end communications across a telephone network and adhere to the frequency bandwidth limitations set for each voice channel. xDSL technologies operate within only the last mile of the network.

Attenuation and interference decrease with the use of xDSL technologies because of operation over the shorter distance. Due to the use of copper wiring for DSL, however, attenuation and interference increase with longer line length, higher signal frequencies, and decreasing wire diameters. As a result, DSL service remains limited to the local loop, or the last mile between a network service provider's Central Office and the customer site and local loops created either across buildings or across a campus.

Much of the recent success associated with xDSL technologies can be attributed to the use of Digital Signal Processors (DSPs). A DSP application converts analog signals to the digital domain for processing and then back for playback or display. With a DSP, all input signals arrive from the real world. Due to this direct connection to real-world signals, the DSP must

> *An xDSL-based network offers several advantages for residential broadband services that include improved usage of available bandwidth, decreased signal distortion, the use of Digital Signal Processor technologies, and point-to-point transmission.*

react in real time and must measure and convert signals—such as analog voice—to digital numbers.

In addition to those advantages, xDSL technologies also provide point-to-point transmissions. All xDSL technologies can deploy on a per-subscriber basis that, in turn, reduces the cost of deployment. The per-subscriber basis allows a network operator to selectively deploy the xDSL technologies to targeted subscribers in each neighborhood, rather than attempting to cover an area that includes nonsubscribers and subscribers. In addition, point-to-point transmission offers dedicated bandwidth to each subscriber.

An xDSL-based residential broadband network appears as "always-connected" without tying up network resources. This advantage occurs because the xDSL modems operate at the Physical Layer, terminate at the Central Office, and represent dedicated bandwidth for each subscriber. Most importantly, however, xDSL-based traffic flowing beyond the Central Office goes through a packet-switched network rather than a PSTN. With this, xDSL technologies can "always connect" through layer 2 connectionless architecture (such as a local-area network) and layer 2 connected architecture (such as ATM).

xDSL technologies can support the simultaneous transmission of voice telephone and high-speed data services over the same twisted-pair copper line. While the voice telephone signals remain on a limited 4-kHz bandwidth, a separate 1-MHz-wide spectrum carries the high-speed data services. As a result, the installation of DSL modems in a home can occur without the installation of additional telephone lines.

An xDSL modem pair can adapt to specified copper loop characteristics and provide maximum bandwidth while maintaining a set bit error rate objective. Rate adaptation occurs through the loop testing sequence found at the beginning of modem pair operation. The loop testing sequence finds the maximum data-transfer rate. Due to the application of the "always connected" mode, xDSL modems perform the test sequence periodically to find and establish the maximum possible rate for a given time.

Differences Between xDSL Technologies

Four factors separate the xDSL family into the different categories shown in *Table 11-1*, xDSL technologies may provide the same or different upstream and downstream bandwidth capabilities. The maximum bandwidth provided through xDSL depends on the loop length and on the upstream/downstream characteristics. In turn, different xDSL technologies support different loop lengths. Finally, xDSL technologies may or may not support the power and voice communication needs of a lifeline POTS. Due to DSL subscriber configuration, an xDSL technology may use the 0-4 kHz spectrum dedicated to the POTS services and not support the simultaneous transmission of voice information and high-speed data.

All DSL technologies run on existing copper phone lines and use modulation to increase data-transfer rates. As the table shows, however, the different xDSL approaches have characteristics appropriate for different applications. When considering each DSL option, we should remember the key points: the trade-offs between signal distance and speed and the differences in symmetry of upstream and downstream traffic.

As these issues become less problematic, xDSL technology could become the foundation for delivering ATM to branch offices and homes over a combination of optical fiber and copper lines. ANSI, ETSI, the ADSL Forum, the ATM Forum, and the Digital Audio-Visual Council (DAVIC) have become heavily involved in setting the standards for the future. The setting of those

Table 11-1.
xDSL technology specifications.

xDSL Technology	Downstream/ Upstream Symmetry	Downstream Bandwidth	Upstream Bandwidth	Maximum Loop Distance	Lifeline POTS Support
ADSL	Asymmetric	1.5 Mbps	64 Kbps	15-18,000 ft	Yes
ADSL	Asymmetric	6 Mbps to 8 Mbps	640 Kbps to 1 Mbps	9-12,000 ft	Yes
DSL (ISDN)	Symmetric	160 kbps	160 kbps	15-18,000 ft	No
HDSL	Symmetric	768 Kbps	768 Kbps	9-12,000 ft	No
HDSL[4]	Symmetric	1.5 Mbps	1.5 Mbps	9-12,000 ft	No
SDSL	Symmetric	384 Kbps	384 Kbps	18,000 ft	Maybe
SDSL	Symmetric	1.5 Mbps	1.5 Mbps	10,000 ft	Maybe
VDSL	Symmetric	12.96 Mbps	12.96 Mbps	3,000 ft	Yes
VDSL	Symmetric	25 Mbps	25 Mbps	1,000 ft	Yes
VDSL	Asymmetric	25 Mbps	3 Mbps	3,000 ft	Yes
VDSL	Asymmetric	52 Mbps	6 Mbps	1,000 ft	Yes

standards will affect the emergence of companies as leading vendors, the transition from testing to the availability of low-cost services, and the demand for DSL services.

Implementing DSL Technologies

DSL service requires the installation of specially designed modems at the customer site and at the service provider's network. Successful transfer of the digital data depends on digital signal processing algorithms that exploit unused bandwidth at the physical connection of the copper wiring. Digital Signal Line technologies use the extra bandwidth to send data to the Central Office and a connection to the high-capacity fiber-optic network owned by the company.

In most instances, service providers place the modem in the Central Office of the Incumbent Local Exchange Carrier (ILEC). But ILECs have displayed some slowness along the path to deploying DSL services. When compared, DSL services have a much lower cost than T1 services already offered by the ILEC. In addition to relying on an ILEC, a service provider may locate modems within the buildings served by the provider. For example, the main cross-connect room of a building may house the DSL modem.

As a result, the arrangement facilitates the provision of DSL service by Competitive Local Exchange Carriers (CLECs). In contrast, CLECs can offer T1 speeds through less expensive DSL technology. Recent regulatory changes have allowed CLECs to place DSL access multiplexers in the Central Office of the RBOCs. With this, CLECs can order unbundled copper pairs and offer broadband connections to other service providers and businesses.

Asymmetric Digital Subscriber Line

Asymmetric digital subscriber line (ADSL) facilitates the simultaneous use of normal telephone services, the transmission of data at a rate of 6 Mbit/s in the downstream, and Basic-rate Access (BRA). As an asymmetric technology, ADSL allows more bandwidth downstream—from the Central Office to the

DSL service requires the installation of specially designed modems at the customer site and at the service provider's network.

customer site—than upstream from the subscriber to the Central Office. The additional downstream bandwidth combines with always-connected access to make ADSL ideal for Internet/intranet surfing, video-on-demand, and remote LAN access, while eliminating call setup.

Video-on-demand services stand as one of the more intriguing possibilities for ADSL. The use of MPEG-coded video allows the delivery of video-quality movies over existing copper-loops to customers. Since high-quality video can occur through a 1.5-Mbps data rate, ADSL could deliver a combination of video-on-demand and services that allow commercial and noncommercial information providers and advertisers to deliver their information. In addition, ADSL gives telephone companies access to next-generation multimedia communication and allows those companies to compete with cable television providers that deliver services through hybrid-fiber/coax technology.

The asymmetric properties of ADSL fit customers' needs in that most users typically download much more information than they send. Moreover, the data-transfer rates seen with ADSL expand existing access capacity by a factor of 50 or more without new cabling. As a result, ADSL can transform the existing public information network from one limited to voice, text, and low-resolution graphics to a powerful, far-ranging system that could deliver full-motion multimedia to every home.

ADSL equipment installed at the Central Office offloads overburdened voice switches by moving data traffic off the public switched telephone network and onto data networks. This problem results from Internet traffic tying up voice networks. In addition, ADSL offers an advantage over ISDN connectivity, because carriers send the power for ADSL over the copper wire. As with conventional phone service, the ADSL line continues to operate when local power fails. ISDN requires a local power supply and a separate phone line for comparable service guarantees. Again compared to ISDN connectivity, ADSL provides two channels for data and one channel for voice information. As a result, voice calls do not affect the data performance of ADSL-based networks.

Since ADSL is a physical-layer transmission protocol for UTP media, work has begun on the development of ADSL system architectures, protocols, and interfaces for major ADSL applications. Technical designs for those architectures, protocols, and interfaces involve:

- ATM over ADSL
- Packet over ADSL
- Customer premises equipment/Central Office configurations and interfaces
- Network management
- Interoperability

An ADSL Network

Figure 11-1 sketches a possible ADSL network and shows that the use of ATM switches in the network creates a flexible method for connecting various servers to device applications. As the diagram shows, the local ATM-switch connects to an access module in a telephone Central Office that connects the ATM network to phone lines. Within the access module, the ATM data stream obtained from the server routes to the corresponding phone lines.

ADSL Network Devices

ADSL requirements vary according to applications and configurations. For data-only traffic, one ADSL ATU-R per LAN suffices. Wide-area networking and internetworking applications require one modem for each subscriber

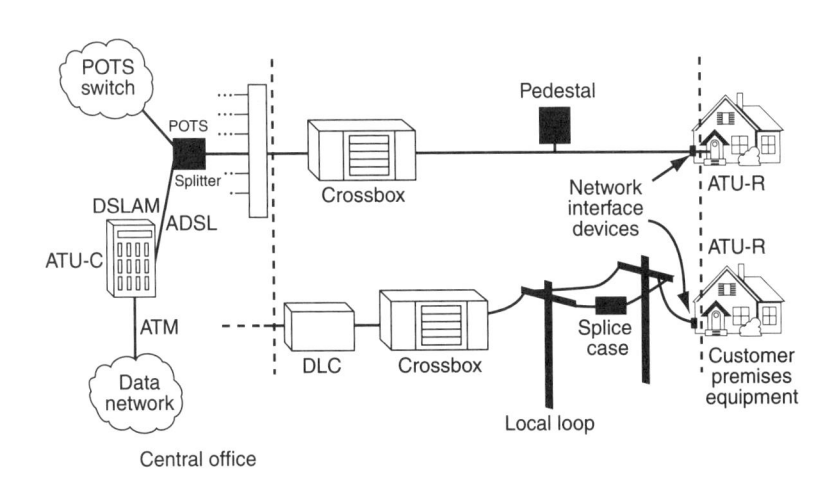

Figure 11-1.
ADSL architecture. The use of ATM switches in the network creates a flexible method for connecting various servers to device applications.

line. With the smaller applications, deploying ADSL involves adding an ADSL interface to the network. Operating as ADSL modems, ATU-R devices sit on the carrier side of the access termination at the subscriber's premises. Routers operate on the enterprise side of the connection and see an ADSL modem as another router interface. Consequently, no need exists for reconfiguring routing tables when a remote user switches from leased lines to ADSL access.

Wide-area networking and internetworking with ADSL becomes more complex because of the asymmetric bandwidth. For example, flow control becomes critical when high-speed LAN traffic hits a lower-speed ADSL link. Since ADSL modems and line splitters slot into digital subscriber line access multiplexers (DSLAMs) that terminate and aggregate incoming ADSL lines for transmission onto voice and data networks, carriers need the ADSL modems and splitters at their end of the network. The splitters divide the ADSL stream and shunt voice onto the public switched network.

Considering *Figures 11-1 and 11-2*, we see that routed data from the access module connects to an ADSL Transceiver Unit-Central Office (ATU-C) that converts the digital data into analog signals. From there, the analog signals travel with POTS signals to remote ends of the network. While the ATU-C also receives and decodes data sent by an ATU-R, or remote transceiver, the splitter either combines or separates the signals depending on the direction of the transmission. In addition, the splitter protects MTS from voice-band interference generated by both ATU's and protects the ATU's from MTS-related signals.

ADSL Modem Operation

ADSL modems provide data rates consistent with the North American T1 and European E1 digital hierarchies and offer various speed ranges and capabilities. While low-end ADSL modems provide a 1.5- or 2.0-Mbps downstream and a 16-kbps duplex channel, high-end ADSL modems have downstream rates up to 8 Mbps and duplex rates up to 640 kbps. With the high-end modems, ADSL can support ATM transport with variable rates. Downstream data rates attained with ADSL modems depend on the length and gauge of the copper line, the presence of bridged taps, and cross-coupled interference.

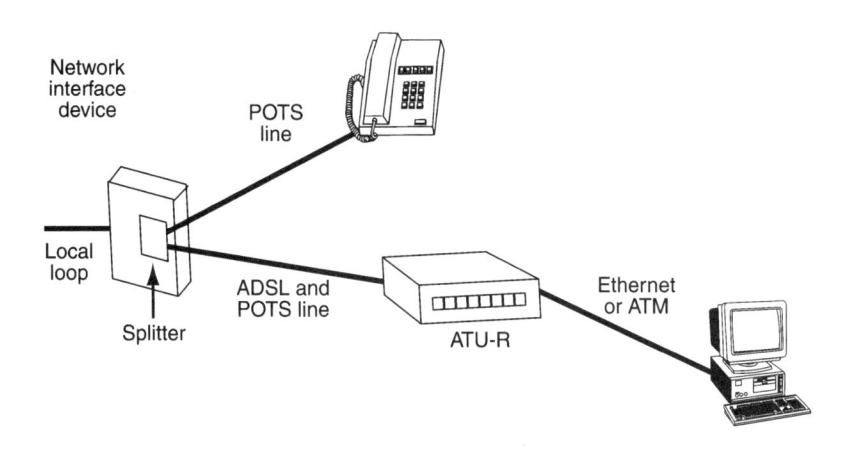

Figure 11-2.
Customer premises equipment. Filters split off the basic telephone service channel from the digital modem and guarantee uninterrupted basic telephone service.

ADSL modems use either frequency-division multiplexing or echo cancellation to divide the available bandwidth of a telephone line and create multiple channels. Frequency-division multiplexing (FDM) assigns one band for upstream data and another band for downstream data. Since ADSL uses FDM, the available bandwidth of a single copper-loop divides into a high-speed downstream channel, a medium-speed duplex channel, and a basic telephone service channel.

As shown in *Figure 11-2*, filters split off the basic telephone service channel from the digital modem and guarantee uninterrupted basic telephone service even if the ADSL service fails. Time-division multiplexing divides the downstream path into one or more high-speed channels and one or more low-speed channels with one low-speed channel reserved for voice. The high-speed channel data-transfer rate typically ranges from 1.5 to 6.1 Mbps, and duplex rates range from 16 to 640 kbps. ADSL modems increase the amount of information that conventional phone lines can carry by using discrete multitone technology (DMT) to divide the bandwidth into many independent subbands and then transmit data on all subbands simultaneously.

DMT operates as a multicarrier system that uses discrete Fourier transforms to create and demodulate individual carriers. For passive network termination configurations, DMT uses FDM for upstream multiplexing. Another type of multitone technology called discrete wavelet multitone (DWMT) also operates as a multicarrier system but uses wavelet transforms to create and demodulate individual carriers. As with DMT, DWMT uses FDM for upstream multiplexing.

Echo cancellation assigns the upstream band to overlap the downstream band. Separation of the two bands occurs through local error cancellation, the same technique used by V.32 and V.34 modems. In practice, echo cancellation balances higher cost and complexity against the more efficient use of bandwidth. With both FDM and echo cancellation, a splitter preceding the ADSL modem allows the allocation of a 4-kHz channel for voice. As a result, customers gain both conventional phone service and digital data over the same wire pair.

After the ADSL modem organizes the aggregate data stream into blocks, it attaches a forward-error correction code to each block. The aggregate data stream exists through the multiplexing of downstream channels, duplex channels, and maintenance channels. Error correction on a symbol-by-symbol basis reduces errors caused by continuous impulse noise coupled into a line. The receiver corrects errors that occur during transmission up to the limits implied by the code and the block length. In addition, the subscriber can use the option to create superblocks by interleaving data within subblocks. With this, the receiver can correct any combination of errors within a specific span of bits and allow for effective transmission of both data and video signals.

ADSL Data Transport Capacity

Three different transport classes for 2.048-Mbps bearers include the 2M-1, 2M-2, and 2M-3 classes, with the 2M-1 class corresponding to the highest rate and shortest range. ADSL downstream transport capacity ranges from 2.048 Mbps to 6.144 Mbps, while the 6.144 Mbps class allows a range of about three kilometers. Lower transmission rates yield longer ranges; the longest distance ranges up to nine kilometers. At distances limited to one mile and a quarter mile, higher data rates of 52 Mbps and 155 Mbps have become possible over fiber optics.

The asymmetric portion of ADSL allows more bandwidth downstream from the Central Office of the carrier to the customer's site than upstream from the customer to the carrier. Both downstream and upstream rates depend

For passive network termination configurations, DMT uses FDM for upstream multiplexing. Another type of multitone technology called discrete wavelet multitone (DWMT) also operates as a multicarrier system but uses wavelet transforms to create and demodulate individual carriers.

on line quality, distance, and wire gauge. In contrast to the downstream rates, the ADSL upstream transport capacity ranges between 0 and 640 kbit/s and remains dependent on the transport class. For up to 18,000 feet, ADSL can move data downstream at 1.544 Mbps, or T1 speeds, using standard 24-gauge wire. At distances of 12,000 feet or less, ADSL establishes a maximum upstream speed of 8 Mbps.

Typical ADSL Chip Set Operation

Semiconductor companies have introduced ADSL transceiver chip sets that combine off-the-shelf components, programmable digital signal processors, and custom Application-Specific Integrated Circuits (ASICs). A typical ADSL transceiver chip set consists of two digital ASICs, a digital signal processor (DSP), an analog front-end (AFE), a daughter board, and a discrete hybrid. The digital interface ASIC (DIA) performs CRC, data scrambling and descrambling, Reed-Solomon encoding and decoding, interleaving and de-interleaving, ADC data insertion and extraction, and indicator-bit insertion and extraction.

Figures 11-3A and 11-3B are block diagrams of the chip sets used within the ATU devices. In Figure 11-3A, we see that a digital signal processor and two ADSL line interface devices combine to produce downstream data rates at 100 times the speed of a standard 56K analog modem. The chip set supports both POTS and ADSL calls and operates at the Central Office and DLC. In addition, the chip set can achieve interactive rates as high as 500 kbps. Moving to Figure 11-3B, we see that the digital signal processor integrates analog modem and ADSL modem functions onto a single integrated circuit. The ATU-R offers plug-and-play operation while remaining independent of any one

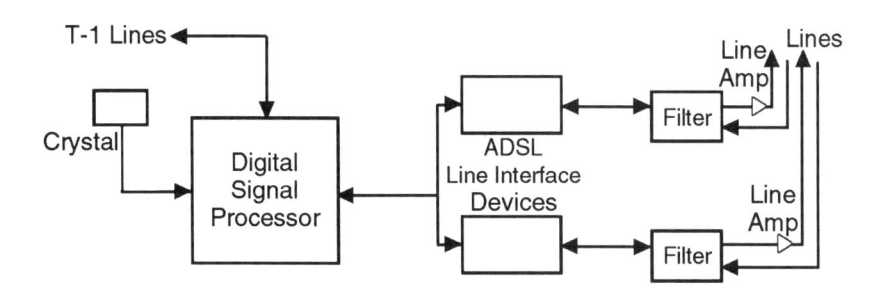

Figure 11-3A.
A block diagram of the ATU-C. A digital signal processor and two ADSL line interface devices combine to produce downstream data rates at 100 times the speed of a standard 56K analog modem.

Figure 11-3B.
A block diagram of the ATU-R. The digital signal processor integrates analog modem and ADSL modem functions onto a single integrated circuit.

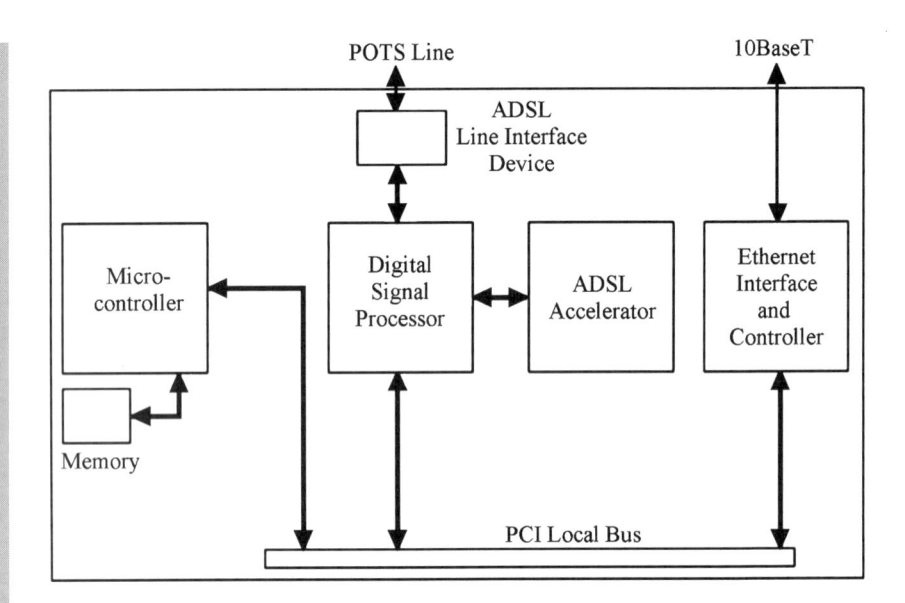

platform or operating system. With the implementation of the chip sets found in both devices, ADSL can support several digital video channels while providing an interactive rate that can handle two-way teleconferencing, high-speed Internet access, and LAN access.

ADSL Framing

DSL chip sets synchronize the downstream and upstream data channels to the 4-kHz ADSL discrete multitone symbol rate and multiplex the channels into either the fast or the interleaved data buffers during initialization. Each superframe consists of 68 ADSL data frames that encode and modulate into DMT symbols. In ADSL DMT-systems the downstream channels divide into 256 4-kHz-wide tones, while the upstream channels divide into 32 subchannels.

Although ADSL uses a standardized DMT, the values differ for the ATU-C and ATU-R devices. From the bit-level and user-data perspective, the DMT symbol rate is 4,000 baud. Because of the sync symbol inserted at the end of each superframe, the transmitted DMT symbol rate is 69/68 * 4,000 baud. The "fast" byte of the fast data buffer carries either CRC, EOC, or synchronization bits.

Symmetric Digital Subscriber Line

Symmetric digital subscriber line (SDSL) transmits both downstream and upstream traffic at a top speed of 768 Kbps over a single twisted-pair copper wire. Although similar to HDSL, SDSL uses only a single wire pair and has a maximum operating range of 10,000 feet. Given these characteristics, SDSL has become suitable for applications such as videoconferencing or collaborative computing. Standards for SDSL remain under development.

Very-High-Data-Rate Digital Subscriber Line

Very-high-data-rate digital subscriber line (VDSL) provides the fastest data-transfer rates of any of the DSL technologies, with downstream rates of 13 to 52 Mbps and upstream rates of 1.5 to 2.3 Mbps over a single wire pair. Yet the range of data-transfer speeds depends on line length and establishes the smallest loop length of all xDSL technologies. VDSL requires the installation of fiber-optic cabling from the Central Office to within a minimum of 1,000 feet and a maximum of 4,500 feet to the home.

With the minimum loop length, VDSL becomes a component of fiber-to-the-curb (FTTC). A typical FTTC network deploys fiber-optic cables to the curb and then terminates the cable with an optical network unit (ONU) that serves eight to 32 homes. VDSL technology can operate over either copper lines or coaxial cable while supporting a bandwidth of 52 Mbps. FTTC supports multiple modems in a home sharing the same copper pair or coaxial cable.

Asymmetric VDSL

With the very fast downstream rates, VDSL has a limited maximum operating distance of only 1,000 feet. Lower speed downstream rates in the range of 13 Mbps increase the maximum operating distance to 4,500 feet. Downstream rates derive from submultiples of the SONET at 155.52 Mbps and SDH at 12.96 Mbps. *Table 11-2* lists the downstream and upstream data-transfer rates for asymmetric VDSL services.

Table 11-2.
Data-transfer rates for asymmetric VDSL services.

Downstream Loop Range (feet)	Downstream Bit Rate (Mbps)	Upstream Loop Range (feet)	Upstream Bit Rate (Mbps)
1000	51.84	1000	6.48
1000	38.88	1000	4.86
1000	29.16	1000	3.24
1000	25.92		
3000	25.92	3000	3.24
3000	22.68	3000	2.43
3000	19.44	3000	1.62
3000	19.44		
3000	16.20		
3000	14.58		
3000	12.96		
4500	12.96	4500	3.24
4500	9.72	4500	2.43
4500	3.24	4500	1.62

Even though ADSL has additional complexity because of operation with large dynamic ranges, VDSL technology has much in common with ADSL technology. We can consider ADSL and VDSL as a set of transmission tools that deliver the maximum amount of data over varying distances of existing telephony infrastructure. While VDSL provides a technology suitable for the transmission of voice, video, and data, ADSL offers the advantage of delivering services over lines that exist today. Since many asymmetric services such as videoconferencing, digital television, virtual CD-ROM access, Internet access, video-on-demand, and remote LAN access can occur at or below T1/E1 rates, the ADSL/VDSL combination seems ideal for the network evolution.

Delivery of each of the services requires a downstream channel that has a higher bandwidth than the upstream channel. For example, high-definition television (HDTV) requires a downstream bandwidth of 18 Mbps for the transmission of video content. The upstream bandwidth requirements for HDTV range into the kbps area, however, due to the need to return only signaling information such as channel change or program selection.

Symmetric VDSL

Asymmetric VDSL has received much attention because of the possibilities for internetworking and wide-area networking. In additional to asymmetric transmission capabilities, VDSL also offers symmetric services for small and midsized networking applications. Symmetric VDSL can replace T1 transmissions on a short-haul basis and symmetrical service between the standard T1 and T3. *Table 11-3* lists the data-transfer rates for symmetric VDSL services.

Loop Range (Feet)	Bit Rate (Mbps)
1000	25.92
1000	19.44
3000	12.96
3000	9.72
3000	6.48

Table 11-3.
Symmetric VDSL service line rates

Looking Inside VDSL

As an example of the simpler operation and resulting lower cost and reduced power, premises VDSL units may need to implement a physical-layer MAC for multiplexing upstream data. But VDSL cannot operate independently of upper-layer protocols. This condition becomes more evident when viewing VDSL from the upstream direction where multiplexing data from more than one CPE may require knowledge of link-layer formats. To achieve error rates compatible with those of compressed video, VDSL incorporates forward-error correction with sufficient interleaving to correct all errors created by impulsive noise events of a specified duration. Interleaving at this level, however, introduces delays that extend to 40 times the maximum length of the correctable impulse.

Symmetric VDSL may also require the use of echo cancellation. Implementing error cancellation requires the maintenance of a large frequency separation between the lowest data channel and basic telephone service. As a result, echo cancellation offers the simple and cost-effective use of basic telephone service splitters.

VDSL Multiplexing and Modulation

Referring to the block diagram shown in *Figure 11-4*, we see that early versions of VDSL use frequency division multiplexing (FDM) to separate downstream from upstream channels and both of them from basic telephone service and ISDN. FDM allocates a channel for each CPE and negates the need for a MAC-level protocol. But the use of FDM either limits data rates available to any one CPE or requires dynamic allocation of bandwidth and inverse multiplexing at each CPE.

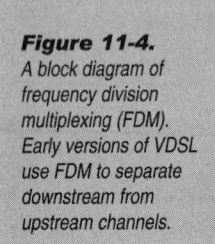

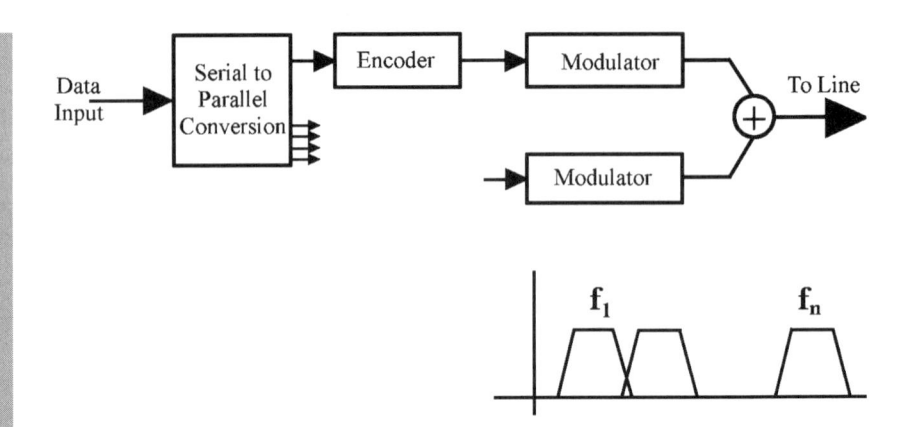

Figure 11-4.
A block diagram of frequency division multiplexing (FDM). Early versions of VDSL use FDM to separate downstream from upstream channels.

All versions of ADSL and later versions of VDSL use discrete multitone technology. Since it offers many advantages for transmission on twisted-pair lines, DMT has been standardized worldwide for ADSL. With DMT operating as a multicarrier modulation method, a DMT transmitter partitions the bandwidth of a channel into a large number of subchannels. Each subchannel has a separate signal-to-noise ratio measured and monitored at the establishment of a connection.

The source bit stream encodes into a set of QAM subsymbols with each subsymbol representing a number of bits determined by the SNR at the midpoint of the respective associated subchannel, the desired overall error probability, and the target bit rate. The set of subsymbols inputs as a block to a complex-to-real inverse discrete Fourier transform (IDFT).

Following the IDFT, a cyclic prefix becomes part of the output samples to lessen intersymbol interference. Conversion circuitry changes the resulting time-domain samples from digital to analog format and applies the samples to the channel. At the receiver, after analog-to-digital conversion, additional circuitry strips the cyclic prefix and transforms the noisy samples back to the frequency domain by a DFT. Each output value scales by a single complex number to compensate for the magnitude and phase of its subchannel's frequency response.

The set of complex numbers, one per subchannel, is called the frequency-domain equalizer (FEQ). After the FEQ, a memoryless, or symbol-by-symbol, detector decodes the resulting subsymbols. In contrast to CAP/

QAM systems, DMT systems don't suffer from error propagation. Each subsymbol decodes independently of all other previous, current, and future subsymbols. *Figure 11-5* uses a block diagram of a DMT transmitter and receiver pair to illustrate the DMT operation.

VDSL and Network Terminations

VDSL offers a much lower cost target than ADSL, because VDSL may connect directly from a wiring center or cable modems. Each provides lower common equipment costs per user. More expense may accompany the implementation of VDSL for passive network terminations than the implementation of VDSL for active NTs. More efficient system management, reliability, and migration favor an active network termination similar to that seen with ADSL and ISDN. Yet, the elimination of any other premises network electronics may establish VDSL for passive NTs as the most cost-effective and desired solution. Passive network interfaces offer hot insertion where a new VDSL premises unit can go on-line without interfering with other modems.

An active network termination configuration operates as a hub and provides either point-to-point or shared-media distribution to multiple CPE on-premises wiring. The distribution remains independent and physically isolated from network wiring. If the active network termination consists

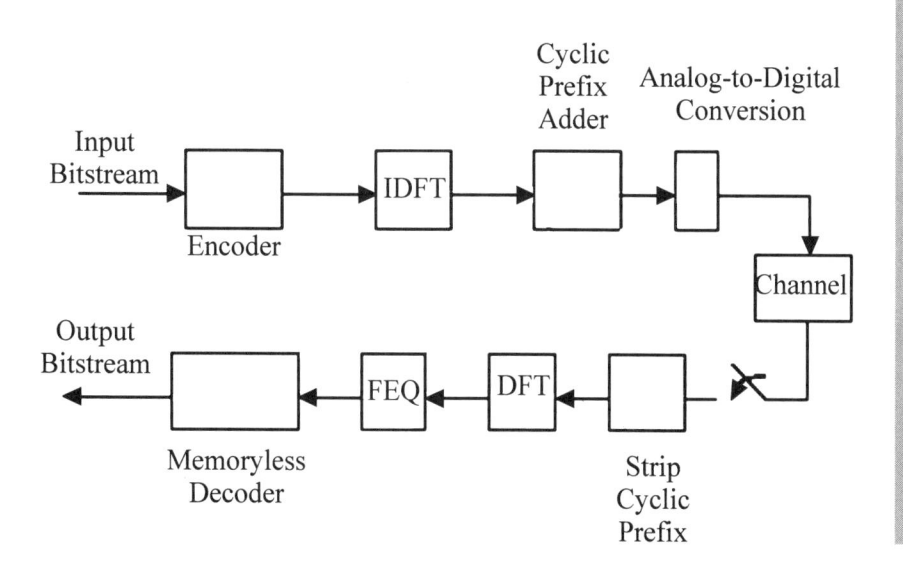

Figure 11-5.
A block diagram of DMT operation. DMT systems don't suffer from error propagation.

of the premises VDSL unit, the method for multiplexing upstream cells or data channels from more than one CPE into a single upstream becomes the responsibility of the premises network. With this, the VDSL unit simply presents raw data streams in both directions. As illustrated in *Figure 11-6*, one type of premises network involves a star connecting each CPE to a switching or multiplexing hub. The hub usually integrates into the premises VDSL unit.

In a passive network termination configuration, each Customer Premises Equipment set combines with an associated VDSL unit. With this configuration, the upstream channels for each CPE must share a common wire. Although a collision-detection system could provide a solution, other solutions for guaranteed bandwidth exist. One solution divides the upstream channel into frequency bands and assigns one band to each CPE. While this method avoids any MAC with its associated overhead, it also restricts the data rate available to any one CPE or imposes a dynamic inverse multiplexing scheme that allows one CPE to send more than its allocation for a period.

The second solution involves starting a cell-grant protocol in which downstream frames contain a few bits that grant access to specific CPE during a specified period subsequent to receiving a frame. A granted CPE can send one upstream cell during this period. The transmitter in the CPE must turn on, send a preamble to condition the receiver, send the cell, and then turn itself off. In turn, the protocol must insert enough silence to allow line ringing clear. One construction of this protocol uses 77 octet intervals to transmit a single 53-octet cell.

Rate-Adaptive Digital Subscriber Line

Rate-adaptive digital subscriber line (RADSL) operates as an asymmetrical service and allows the transfer of data at speeds from 64 Kbps to 8 Mbps on the downstream channels and from 16 Kbps to 1 Mbps on the upstream channels. RADSL also allows simultaneous voice service over the same single twisted-pair copper wire. Although RADSL has the same transmission format as ADSL, it also adjusts transmission speed according to the length and quality of the local line. RADSL establishes the connection speed when

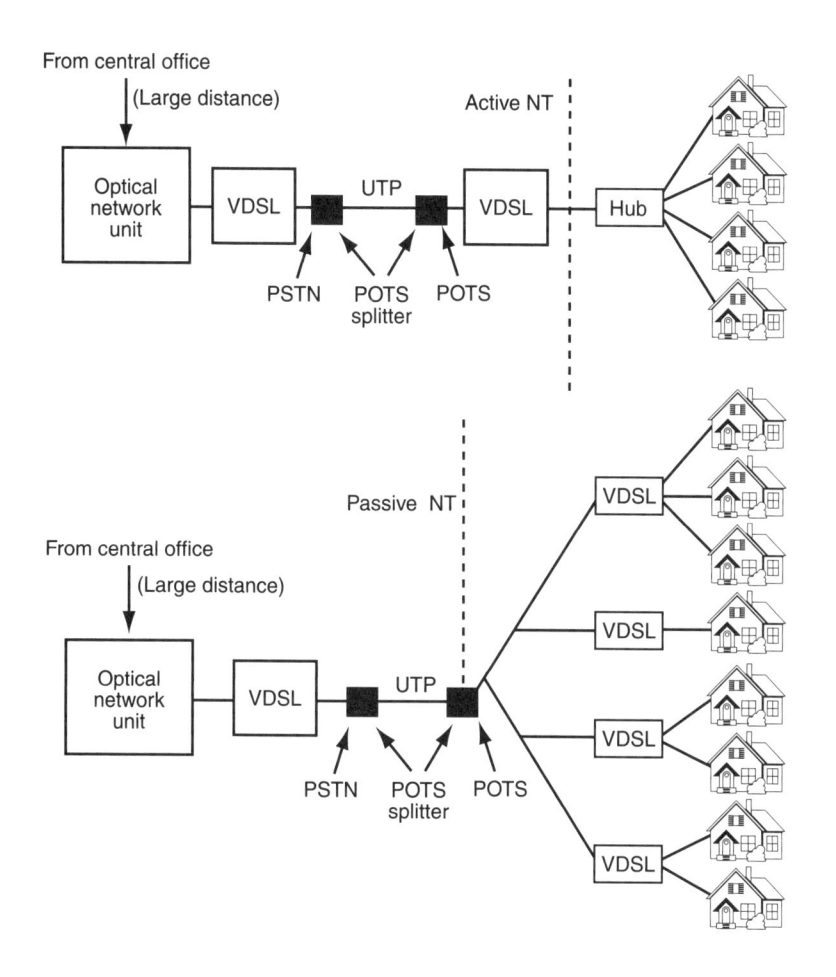

Figure 11-6.
Passive and active VDSL network termination. One type of premises network involves a star connecting each CPE to a switching or multiplexing hub.

line synchronization begins or a signal from the Central Office sets the signal. RADSL applications include Internet/intranet access, video-on-demand, database access, remote LAN access, and lifeline phone service.

High-Bit-Rate Digital Subscriber Line

High-bit-rate digital subscriber line (HDSL) is a symmetrical DSL service that delivers T1 speeds in both directions. To attain those speeds, HDSL divides the data stream in half and simultaneously transmits each half over two twisted-pair copper wires at 784 Kbps. The network recombines the data at the receiving end.

Telecommunication Technologies

HDSL has operated at the feeder plant where the lines extend from Central Offices to remote nodes and also within campus environments. Because of the data-transfer speed obtained over copper lines, telephony providers commonly deploy HDSL as an alternative to T1/E1 lines. With an operating distance of 15,000 feet, HDSL falls short of the operating distance offered through ADSL. But carriers can install signal repeaters to extend the HDSL useful range by 3,000 to 4,000 feet.

In addition, the use of two or three wire-pairs for HDSL transmissions establishes the DSL standard as ideal for connecting PBXs, interexchange carrier points of presence, Internet servers, and campus networks. Given the symmetry and data-transfer speed found with HDSL, carriers have offered HDSL for carrying digital traffic in the local loop, between Telco Central Offices and customer premises.

Even with those benefits, HDSL has become less desirable for providing high-speed services to the home. Aside from requiring an additional wire pair, HDSL has a limited loop length of 9,000 to 12,000 feet. Near-end crosstalk limits the loop length because of the symmetric properties of HDSL. In addition, HDSL uses the same frequency spectrum used by POTS and cannot provide simultaneous POTS service without the installation of additional cabling.

HDSL2

Advanced high-bit-rate digital subscriber line (HDSL2) technology delivers full T1 or E1 capabilities over one copper pair. As a result, HDSL2 offers several benefits not seen with ADSL or HDSL.

Advanced high-bit-rate digital subscriber line (HDSL2) technology delivers full T1 or E1 capabilities over one copper pair. As a result, HDSL2 offers several benefits not seen with ADSL or HDSL. Implementation of HDSL2 allows carriers to meet rapidly increasing demands for high-speed transmission services in areas where copper pair shortages exist. In addition, HDSL2 relies on only one transceiver when operating at full T1 rates, while HDSL requires the use of two transceivers.

More efficient line code used in the HDSL2 standard also allows better performance at many different data rates and over a wide range of loops. Moreover, HDSL2 modems have better noise characteristics than other xDSL technologies. The performance advantage provides either a 30% range improvement for the same data rate or double the data rate over the same range.

Achieving single pair operation at T1 and E1 rates resulted in an increase of the HDSL2 bandwidth with respect to HDSL. Since increased bandwidth and power may result in vulnerability to NEXT, the HDSL2 design incorporates Overlapped Pulse Amplitude Modulated Transmission with Interlocked Spectra (OPTIS). The Interlocked Spectra adopts the ADSL concept of using low frequencies for both upstream and downstream with echo cancellation. While the symbol rates for the downstream and upstream remain the same, higher excess bandwidth used for the downstream path improves performance.

The HDSL2 system transports T1 signals from the provider's Central Office to a subscriber location. At the subscriber end, the H2TU-R modem communicates with the H2TU-C device located at the Central Office. The Central Office has a concentration of many H2TU-C modems, while the subscriber modems are widely distributed over the wire center serving area. Due to this distribution, the affects of crosstalk can disrupt communications at the Central Office, because the combined downstream signals from the H2TU-C design can interfere with the upstream signals from the H2TU-R modems.

The OPTIS solution shapes the spectrum to minimize crosstalk. Both the downstream signals and upstream signals transmit equal Power Spectrum Density (PSD), or the total power transmitted over a frequency range. Increasing the PSD improves signal-to-noise ratios for the H2TU-C devices. In addition, the OPTIS solution increases downstream signal bandwidth. In turn, this reduces the sensitivity to noise and improves its performance under these conditions.

Summary

Chapter 11 discussed another up-and-coming broadband service with its overview of DSL technologies, showing the technical differences between ADSL, RADSL, VDSL, HDSL, HDSL2, and SDL. Throughout the chapter, emphasis remained on the devices used to build xDSL networks. The chapter described ADSL modems, as well as the multiplexing schemes used within the devices. In addition, the chapter provided information about the chip sets used within the ADSL line interfaces.

Chapter 12

Wireless Networks and Satellite Distribution

Introduction

Depending on which frequency band is used, the broadband fixed wireless system provides benefits in that it can quickly provide Internet access to a 10-mile, 20-mile, or 35-mile radius. As a result, the capabilities of the wireless system may allow the telecommunications service provider to team or compete with more traditional providers to serve small-sized and medium-sized business and high-end users. In some instances, typical network solutions may not serve some customers. In others, restrictions may prohibit the installation of cabling required for network access.

Chapter 12 also considers the use of satellite technologies for telecommunications. In recent years, the role of satellite technologies has grown to include not only traditional telephony and TV broadcast services, but also data services. Because of this, new generation smart satellites will incorporate functions such as switching, buffering, and beam switching in addition to signal reproduction. To meet the increasing demand for real-time traffic, networks driven by satellite technologies will feature enhanced channel access and link-layer protocols that will ensure smooth operation over the satellite channel.

To meet the increasing demand for real-time traffic, networks driven by satellite technologies will feature enhanced channel access and link-layer protocols that will ensure smooth operation over the satellite channel.

Planning for the Wireless System

The installation of a wireless network requires much the same basic planning as any wired network. Yet the implementation of wireless access requires additional planning that includes RF path planning, site preparation, and installation of outdoor components such as outdoor units, antennas, lightning protection devices, and cabling suitable for outdoor conditions. In addition, planning for wireless access also calls for an investigation of zoning laws, FCC regulations, and FAA regulations.

A Typical Broadband Wireless System

A broadband wireless system includes several necessary components such as a transmitter, a transmitting antenna, line-of-sight transmission between network access points, and receiving antennas. *Figure 12-1* is a diagram of a typical broadband wireless system. As the diagram shows, the location of the wireless transmitter may differ from the Internet headend location. Due to the line-of-sight requirements for wireless transmission, the location of the receiving antennas may vary from roof-mounted, side-mounted, or window-mounted antennas.

Moving the downstream signal from the Internet point-of-presence to the transmitter stands as a challenge. The uplink path from Internet point of presence to the transmitter has tremendous importance and requires proper design. When we break the path down into media and transmission technologies, we find that media involves copper wiring, fiber optics, or wireless transmission. Transmission technologies include amplitude modulation and frequency modulation.

Since the system relies on the wireless transmission of data, the installation of the system becomes complicated by several factors such as the placement of the transmitter, the distance of the transmitter from the Internet headend location, and the placement of the receiving antennas. In addition, several other factors can affect system performance. Signal power decreases with distance. License availability affects the use of different fixed transmission frequencies. Finally, multipath distortion may harm overall system performance.

Multipath Distortion

When transmitted signals follow several paths between the transmitter and the receiver, a condition called multipath distortion occurs. This condition may stem from the reflection of signals off buildings, water, and other objects.

When transmitted signals follow several paths between the transmitter and the receiver, a condition called multipath distortion occurs. This condition may stem from the reflection of signals off buildings, water, and other objects.

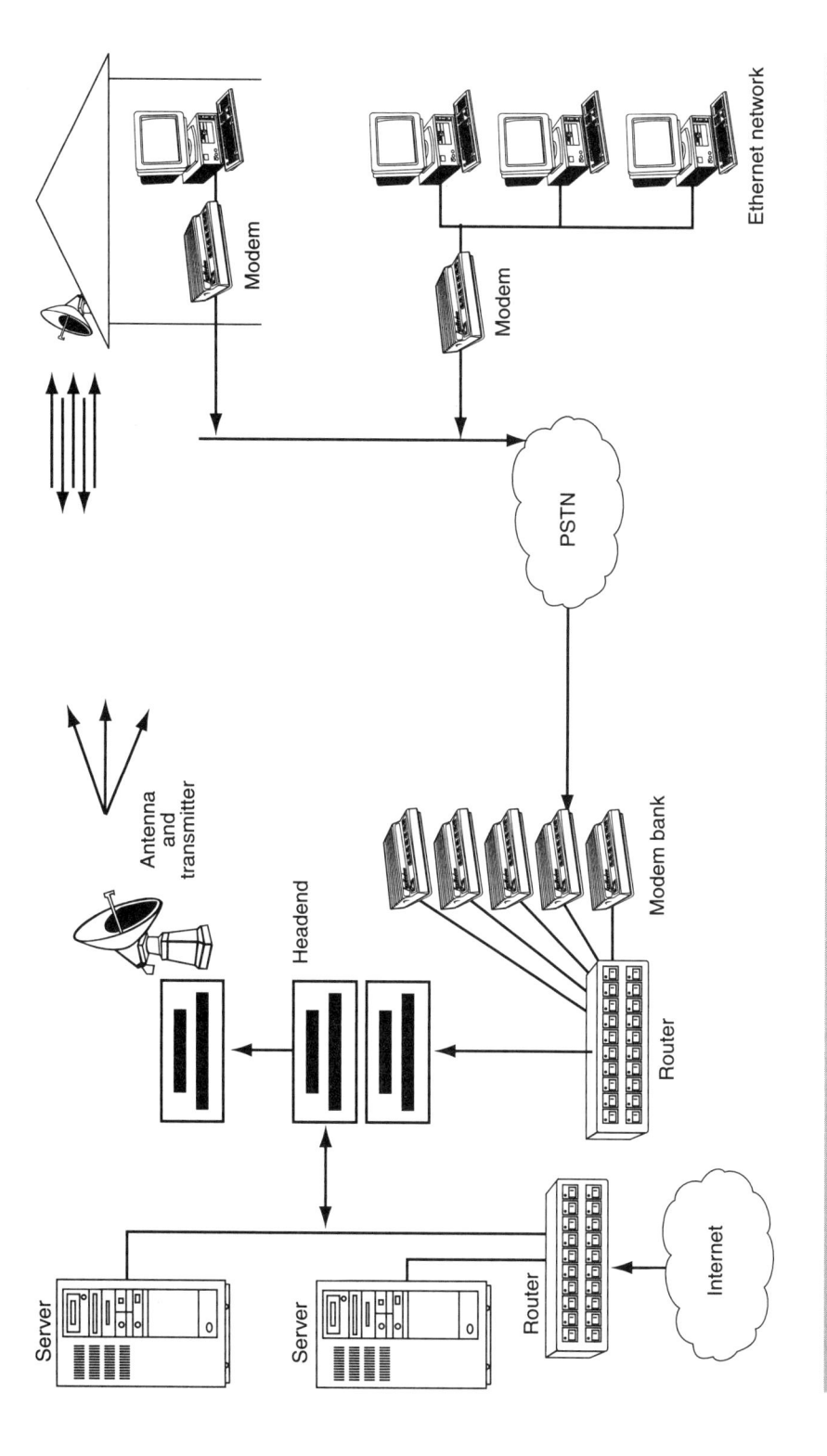

Chapter 12: Wireless Networks and Satellite Distribution

Figure 12-1.
A diagram of a typical wireless broadband system. The location of the wireless transmitter may differ from the Internet headend location.

321

Planning for the physical location of the wireless system components begins with a basic consideration of the sites at each end of the link.

In addition, stratification of the atmosphere can create multiple paths by refracting the signals on long point-to-point radio links. Due to longer path lengths, these reflected or refracted signals take longer to arrive at the receiver. When the signals arrive at the signal through the multiple paths and at different times, interference with the main signal occurs.

Site Selection

Planning for the physical location of the wireless system components begins with a basic consideration of the sites at each end of the link. Since microwave signals travel in a straight line, most plans work toward a clear line of sight between antennas.

Yet the location at one or both ends of the desired link may have been established long before the availability of wireless networking. As a result, planning must consider options that provide the optimal conditions for transmission from antenna to antenna.

Regardless of the conditions, however, site selection should involve the collection of information and the making of decisions. When planning for the installation of the system components, determining the critical level of the information aids in the decision-making process. Site selection issues include:

- Ability to install one or more antennas
    1) Is the roof adequate to support the antenna(s), or will it require structural reinforcement?
    2) Will a tower have to be constructed?
    3) Are permits required?
- Possibility of future obstructions
    1) Will trees grow high enough to interfere with the signal?
    2) Are there plans to erect buildings between the sites that may obstruct the path?
- Availability of grounding—Good grounding is important in all areas of the world, but in areas prone to lightning, it's especially critical.
- Availability of electrical power—Are redundant power systems available if the area is prone to outages?

In addition, the preliminary decision-making about the location of the components should consider weather conditions that may occur at the site. These conditions that may affect the integrity of the system include excessive amounts of rain, the average wind velocity, temperature extremes, and lightning. With the exception of lengthy downpour conditions, rain does not cause signal attenuation of frequencies up to the 6- to 8-GHz range. When the signal frequencies increase to 12 GHz, attenuation becomes a much larger problem with less than downpour conditions. Shorter transmission paths provide the only solution for these conditions.

Antennas and antenna support structures must have sufficient stability to remain unaffected by high wind conditions. Wind loading, or the area presented to wind, has different effects on different antenna designs. Most antenna manufacturers will specify wind loading for each type of antenna manufactured.

Wind Loading

Wind loading stands as another important issue for a satellite receiving system installation. In short, none of us wants to see a dish sailing off a mount or, worse yet, off a roof during a high wind. Therefore, as we select the perfect location for the reflector antenna, we also need to consider the best methods for minimizing wind loading.

For that reason, let's take a longer look at wind loading and methods for prevention. We can calculate the wind load of an object through the following equation.

$$D = 1/2 \, C_d P A V^2$$

where
D = drag force in Newtons
C_d = drag coefficient
A = the frontal characteristics of an object in ft^2
P = the density of fluid stream (1 lbm/ft^3)
V = velocity of fluid stream ft/sec

Since this equation seems rather complex, let's take a closer look at nature and the variables.

None of us wants to see a dish sailing off a mount or, worse yet, off a roof during a high wind. Therefore, as we select the perfect location for the reflector antenna, we also need to consider the best methods for minimizing wind loading.

In the equation, P equals the density of a fluid stream. For our purposes, we can consider the fluid stream as standard air. At an altitude of zero feet with dry conditions and a temperature of 15° centigrade, standard air has a density of 1.225 kg/m3. At an altitude of 5,000 feet, or the same altitude as the city of Denver, the density reduces by approximately 15%. If we lower the temperature of the air, the density increases. For example, lowering the temperature from 15° centigrade to 0° centigrade increases the density by 5%. Water vapor also affects the density of standard air. If we increase the amount of water vapor in the air, the density of the air decreases.

The drag coefficient (C_d) of an object changes according to the amount of surface area directly facing the wind. For example, a 10-foot dish facing directly into the wind can have a drag coefficient as high as 1.5. Adjusting the dish so that it has a 47° look angle into the wind reduces the drag coefficient to 0.9. In addition, yaw, or side-to-side, movement places additional stress and wind resistance on a dish due to asymmetry and lift.

With this knowledge in mind, we can begin to see whether a dish mounted on a pole will have stability in a windstorm. To accomplish this, we can replace variables in the equation with specific amounts. If we replace P with the worst-case scenario where 0% humidity exists at 0° centigrade, then we have:

$$D = 1/2 C_d P A V^2$$
and
$$D = 1/2 C_d^{(1.3)} A V^2$$

From there, we can convert A, or the frontal characteristic of the object, to:

$$A = \pi d^2/4$$

where (d) equals the diameter of the dish. By making these changes, we have converted the original equation into:

$$D = 1/2 C d^{(1.3)} (\pi d^2/4) V^2$$

Reducing the equation again, we have:

$$D = (0.002133) * C_d * d^2(ft) * V^2(mph)$$

Given this equation, we can replace the drag coefficient, the diameter of the reflector, and the velocity with actual numbers. For example, the drag force for a 10-foot diameter reflector with a drag coefficient of 1.5 during 100-mph winds equals:

$D = (0.002133) (1.5) 10^2 * 100^2$ or
$D = (0.002133) * 100 * 10,000$ or
$D = (0.00320) * 1,000,000$ or
$D = 3,200$ lbs.

Planning for a wireless link should always consider the potential for lightning damage to radio equipment. Manufacturers provide a wide variety of lightning protection and grounding devices for use on buildings, towers, antennas, cables, and equipment that could be damaged by a lightning strike. Lightning protection requirements cover exposure at the site, the cost of link downtime, and local building and electrical codes.

To provide effective lightning protection, install antennas at locations unlikely to receive direct lightning strikes or install lightning rods to protect antennas from direct strikes. Install ground cables and equipment to provide low-impedance paths for lightning currents, while installing surge suppressors on telephone lines and power lines. Wireless system manufacturers recommend the placement of lightning protection for both coaxial and control cables at points close to where the cable passes through the bulkhead into the building as well as near the transmitter and receiver. Since the coaxial line carries a dc current to supply power to the transverter, always install gas-discharge surge arrestors rather than quarter-wave stub or solid-state type surge arrestors. When not using steel conduit to encase either the coaxial cable or the control cable, install one surge arrestor within two feet of the building entrance and another within 10 feet of the transmitter or receiver on each cable.

The Wireless Tower and Transmitter

The placement of an antenna in an optimal location sometimes requires the installation of a freestanding tower. In addition, most wireless transmitters require high towers to establish the necessary coverage area for the antenna. The following FCC and FAA regulations and limitations cover the height and location of towers with respect to airports, runways, and airplane approach paths. In most cases, zoning and other property regulations also place the antennas and towers away from populated areas.

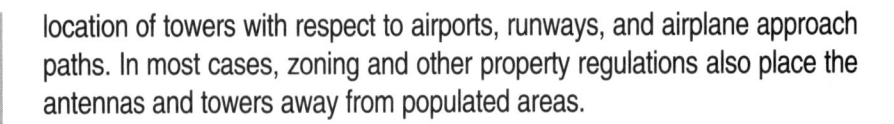

Government Antenna Tower Regulations

"Most antenna structures that are higher than 60.96 meters (200 feet) above ground level or that may interfere with the flight path of a nearby airport must be cleared by the Federal Aviation Administration (FAA) and registered with the FCC. Unless specifically exempted, FAA notification and FCC registration are required for:

1) "Any construction or alteration of more than 60.96 meters (200 feet) in height above ground level at its site.
2) "Any construction or alteration of greater height than an imaginary surface extending outward and upward at one of the following slopes:
   - "100 to 1 for a horizontal distance of 6.10 kilometers (20,000 feet) from the nearest point of the nearest runway of each SPECIFIED AIRPORT with at least one runway more than 0.98 kilometers (3,200 feet) in actual length, excluding heliports;
   - "50 to 1 for a horizontal distance of 3.05 kilometers (10,000 feet) from the nearest point of the nearest runway of each SPECIFIED AIRPORT with its longest runway no more than 0.98 kilometers (3,200 feet) in actual length, excluding heliports; and,
   - "25 to 1 for a horizontal distance of 1.52 kilometers (5,000 feet) from the nearest point of the nearest landing and takeoff area of each heliport at a SPECIFIED AIRPORT.
3) "Any construction or alteration that would be in an instrument approach area and available information indicates it might exceed an obstruction standard of the FAA. In this case, the FAA would specifically ask you to file a notification—you would then be required to register the structure.
4) "Any construction or alteration on a SPECIFIED AIRPORT such as a public use airport listed in the Airport Directory of the current Aeronautical Information Manual or in either the Alaska or Pacific Airman's Guide and Chart Supplement; an airport under construction, that is the subject of a notice or proposal on file with the FAA, and

except for military airports, it is clearly indicated that the airport will be available for public use; or an airport that is operated by an armed force of the United States."

Getting the most value from a wireless system depends on careful path planning. Along with considering the dissipation of microwave radio signals over distance, or free space path loss, we should also consider any other factors that could affect the strength of the signal as it travels through space. Any obstructions in the path will attenuate the signal. Free space path loss is easily calculated for miles or kilometers using one of the following equations.

$Lp = 96.6 + 20 \log 10\, F + 20 \log 10\, D$
where
Lp = free space path loss between antennas (in dB)
F = frequency in GHz
D = path length in miles
H 43.3 D
4F - D 2

$Lp = 92.4 + 20 \log 10\, F + 20 \log 10\, D$
where
Lp = free space path loss between antennas (in dB)
F = frequency in GHz
D = path length in kilometers

The characteristics of a radio signal cause it to occupy a broad cross-section of space—called the Fresnel Zone—between the antennas. *Figure 12-2* shows the area occupied by the First Fresnel Zone. For smooth terrain without obstructions, the total antenna height at each end of the link for paths longer than seven miles equals the height of the First Fresnel Zone plus any additional height required to clear the horizon. Mathematically, the equation for antenna height appears as:

H = Height of the antenna (in feet)
D = Distance between antennas (in miles)
F = Frequency in GHz

Getting the most value from a wireless system depends on careful path planning.

Figure 12-2.
The First Fresnel Zone, which is a broad cross-section of space between antennas.

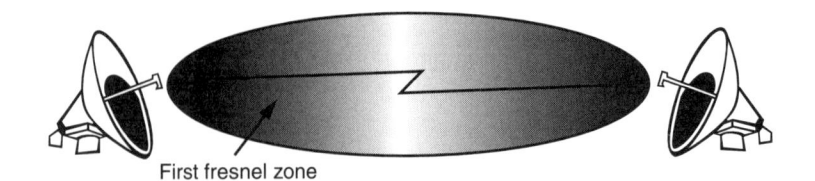

First fresnel zone

Antennas

By most basic definition, antennas focus the microwave radio signal in a specific direction and in a narrow beam while remaining tuned to a specific group of frequencies. Manufacturers provide antennas according to exact measurements such as gain and beamwidth. In addition, antennas respond to polarization. Effectively, the orientation of the antenna will change the orientation of the signal. As a result, the transmitting and receiving antennas in a wireless system must have either horizontal or vertical polarization. Cross-polarization of adjacent antennas can reduce interference between the two antennas.

Polarization

All radio frequency waves—including satellite television signals—have polarization. In brief review, designers use the property of microwave signals to improve the spectrum efficiency in frequency bands. The use of different polarities allows the transmission of more signals from the transmitter without the risk of interference from one channel adjacent to another. As with electrical wiring, electromagnetic wave polarization exists through the direction of electrical and magnetic fields. When the electrical and magnetic fields of a received signal remain within the same plane of the electrical and magnetic fields of the transmitted signal, linear polarization occurs. Horizontally polarized waves travel along the horizontal plane, while vertically polarized waves travel along the vertical plane. *Figure 12-3* uses a diagram to depict linear polarization.

Dish Antenna Gain

The performance of any dish antenna depends on its gain, or the amount of amplification that a dish provides for the received signal. When selecting

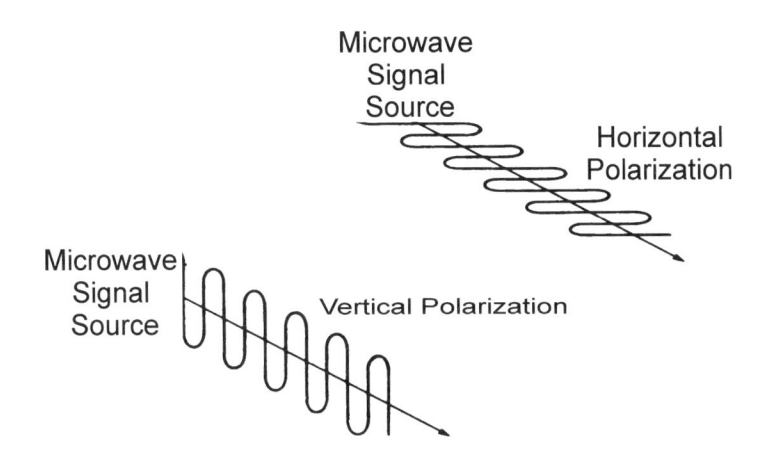

Figure 12-3.
A diagram depicting linear polarization, which occurs when the electrical and magnetic fields of a received signal remain within the same plane of the electrical and magnetic fields of the transmitted signal.

an antenna, always consider the antenna specifications and your applications. Any increase in the signal power with the antenna focused in the desired direction constitutes gain.

By definition, "gain" expresses the ratio of input signal voltage, current, or power to the output signal voltage, current, or power of an amplifier. Any type of gain measurement occurs in decibels (dBs). As an example of how gain measurements work, an amplifier that provides 3 dBs of gain doubles the signal power, while an amplifier providing 10 dBs of gain magnifies the power 10 times. Thirty decibels of gain amplify the signal 1000 times, while 50 dBs of gain multiplies the signal by 100,000.

Antenna gain results from the size and curvature of the dish antenna; occurs by maximizing the signal at the focal point; and depends on the attenuation of any adjacent antenna signals or noise to a minimum level. A parabolic antenna focuses RF signals to the focal point in the same way that an optical parabolic lens in an optical telescope focuses light. As a result, the theoretical size of the gain of a parabola has the relationship with wavelength shown in the following equation.

Gain = $\eta(\pi D/\lambda)^2$
where
η = efficiency
D = the diameter of the reflector in meters
λ = the wavelength of the received signal in meters

Telecommunication Technologies

In practice, we always express gain in decibels and replace the wavelength with the frequency of the signal. By doing so, we arrive at the following equation:

Gain (dB) = 10 logη + 20log D + 20log f + 20.4
where
η = efficiency
D = the diameter of the reflector in meters
f = the operating frequency in gigahertz

When we examine those factors, we find that the usable size and accuracy of the reflective surface affect the dish antenna gain. Dish antenna gain also increases or decreases according to the type of signal received. As signal frequency increases, gain increases with the square of the signal frequency.

Antenna gain increases with antenna size and the efficiency of the antenna. Mathematically, antenna gain appears as:

G_a = 10 log [$(\pi d)^2 p / 100 \lambda^2$] (dB)
where
π = 3.14159
d = the diameter of the antenna
p = the antenna efficiency as a percentage
λ = wavelength

The gain of the antenna combines with the gain provided by the amplifier circuitry found at the antenna feed to produce a usable signal at the receiver. During the early years of home satellite systems, when the amplifiers provided little gain, consumers opted for larger dish antennas. Today, high-gain amplifier electronics couple with the ability to receive higher frequency signals and allow the use of smaller dish sizes.

Engineers use a standard called the isotropic antenna as a reference for measuring the gain of dish antennas. The isotropic standard applies an imaginary aerial antenna design that receives signals equally from all directions and has a gain of zero decibels. Companies demonstrate the capability of antennas used in real installations by showing how much more

---

Engineers use a standard called the isotropic antenna as a reference for measuring the gain of dish antennas. The isotropic standard applies an imaginary aerial antenna design that receives signals equally from all directions .

gain the antenna provides than that seen with the isotropic standard. Each of the three factors that affect the gain of the dish antenna—reflector size, accuracy of the reflective surface, and type of signal received—affect the design. *Table 12-1* provides a listing of typical gain figures for dish antenna sizes.

Dish Size	Gain
6 ft (1.8 meters)	35 dB
8 ft (2.4 meters)	37 dB
10 ft (3.0 meters)	39 dB
12 ft (3.7 meters)	40.5 dB

Table 12-1.
Dish antenna size and gain.

In general, the larger antennas produce higher gain and, unfortunately, require larger masts and support structures. Most manufacturers recommend using the smallest possible antenna while maintaining sufficient protection from interference and enough signal at the far end of the link.

Other considerations include antenna beamwidth, front-to-side ratios, front-to-back ratios, and cross-polarization rejection. Where interference from other licensees on the same channel or adjacent channels becomes an issue, good system performance may dictate the use of an antenna with a narrow beamwidth, high front-to-back and front-to-side ratio, and high cross-polarization rejection.

Interference

Avoiding interference may have the same importance as actually moving data from one location to another in a broadband wireless system. Although external or internal interference can occur, good planning for frequencies and antennas can overcome most interference challenges. Cochannel interference results when two RF links use the same channel frequency. Adjacent-channel interference results when another RF link uses an adjacent channel frequency.

When we consider the use of signals to communicate, we can define any type of disturbance other than the desired signal as interference or noise. These extraneous signals may take a variety of forms and can affect both the transmission and reception of broadcast signals. The ability of a system to reject noise is defined in terms of signal-to-noise ratio. A system with a high signal-to-noise ratio has a greater ability to reject noise. *Table 12-2* defines common types of noise.

Table 12-2.
Types of noise.

Atmospheric Noise: Radio-wave disturbances, such as lightning, that originate in the atmosphere.
Common-mode Interference: Noise caused by the voltage drops across wiring.
Conducted Interference: Interference caused by the direct coupling of noise through wiring or components.
Cosmic Noise: Radio waves generated by extraterrestrial sources.
Crosstalk: Electrical disturbances in one circuit caused by the coupling of that circuit with another circuit.
Electromagnetic Interference (emi): Noise ranging between the subaudio and microwave frequencies.
Electrostatic Induction: Noise signals coupled to a circuit through stray capacitance.
Hum: Electrical disturbance caused by a power supply frequency or harmonics of a power supply frequency.
Impulse Noise: Noise generated by a dc motor or generator. Impulse noise takes the form of a discrete, constant energy burst.
Magnetic Induction: Noise caused by magnetic fields.
Radiated Interference: Noise transmitted from one device to another with no connection between the two devices.
Radio-frequency Interference (rfi): Occurs in the frequency band reserved for communications.
Random Noise: Irregular noise signal that has instantaneous amplitude occurring randomly in time.
Static: Radio interference detected as noise in the audio frequency (af) stage of a receiver.
Terrestrial Interference: Unwanted Earth-based communication signals.
Thermal Noise: Random noise generated through the thermal agitation of electrons in a resistor or semiconductor device.
White Noise: Electrical noise signal that has continuous and uniform power.

Antenna Specifications

Main and Side Lobes of an Antenna.
We can calculate beamwidth as a ratio of the main lobe of the dish to its side lobes. In terms of physical measurements, beamwidth is the width of the main lobe between points on the reflective surface where power has dropped by three decibels. *Figure 12-4* illustrates the difference between the main lobe and side lobes.

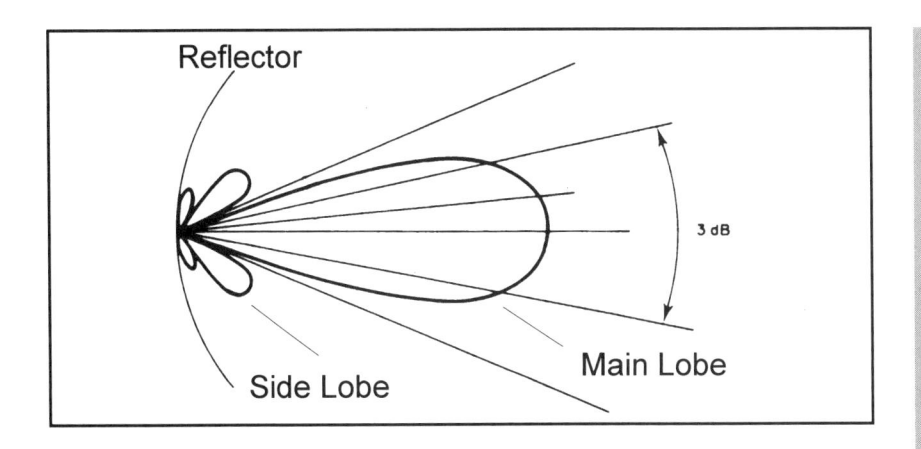

Chapter 12: Wireless Networks and Satellite Distribution

Figure 12-4.
The difference in gain between the main lobe and side lobes of a reflector antenna.

Looking at the figure, we see that the main lobe projects out toward the region that has the maximum amount of power. Side lobes project off to either side of the main lobe and indicate the capability of the dish antenna to pick up microwave signals from off-axis sources.

If the dish antenna has the proper design, the main lobe will appear much larger than the side lobes. Yet the width of the main lobe must remain consistent with the requirement to have a narrow bandwidth. Since commercial satellites have a 2° spacing between orbital slots, the magnitude of the side lobes becomes increasingly important. Side lobes located within this 2° range and with enough power would allow the dish antenna to receive signals from the adjacent as well as desired satellites simultaneously.

If we view the dish antenna in terms of a telescope pointed at a particular section of the sky, it becomes easier to say that the dish has an established field of vision. When we viewed the antenna gain equation, we calculated the gain for the main axis of the main lobe of the antenna. As shown in *Figure 12-5*, the gain of the antenna drops away from the main axis.

Beamwidth defines the preciseness of the region that the dish antenna can target. A dish antenna that has a narrow

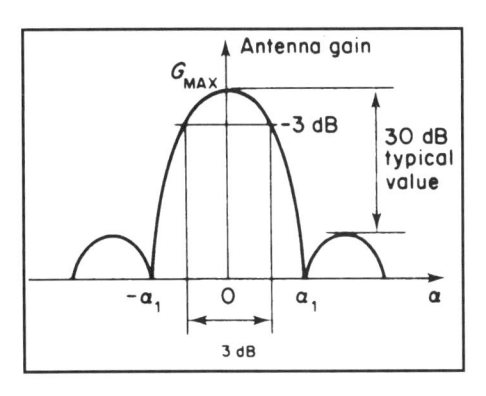

Figure 12-5.
Gain comparisons of a satellite antenna.

333

beamwidth becomes less susceptible to interference from adjacent satellites. A wide beamwidth allows the simultaneous reception of signals from the desired satellite and signals from the adjacent satellites. The signals from the adjacent satellites would appear as interference.

Referring back to Figures 12-4 and 12-5, we see that the equation for finding the half-power beamwidth of an antenna appears as:

$$\theta \text{ degrees} = k \lambda / D$$

where
θ degrees = 3 dB beamwidth (degrees)
λ = wavelength (meters)
D = Earth station diameter (meters)
k = the aperture illumination factor

When an antenna has a uniformly illuminated aperture, the aperture illumination factor has a range between 58 and 65. If the antenna has a tapered aperture, the illumination factor will have a value of 70.

Semiparabolic grid antennas work well for locations with the potential for high wind loading. When using a solid antenna, add a radome to reduce wind loading, protect against ice, and to prevent birds from roosting on the antenna feeds. For short U-NII links, panel, patch, or planar antennas may operate well. But these types of antennas do not offer good front-to-side, front-to-back, and cross-polarization response.

Broadband Wireless Transmission Frequency Bands

The type of receiving antenna and the resulting coverage—along with the type of transmitter—depend on the frequency band allocated for the system. *Table 12-3* lists descriptions and the frequencies of frequency bands used for wireless transmission. As the table shows, some frequency bands divide into channels, while others divide into blocks.

At the outset of wireless Internet access, many providers relied on MMDS, ITFS, and MDS analog television transmitters, because those bands remained underutilized for television signal transmission and reception.

Frequency Band Name	Frequency Range (MHz)	Basic Usage	Description
MMDS	2500-2686	31 individual 6 MHz television channels. Analog transmission to 35 miles and requires line-of-sight	Multichannel, multipoint distribution service
MDS1	2150-2156	Single channel 6 MHz. Analog transmission multipoint distribution service	Multipoint distribution service
MDS2	2156-2160	Same as MDS1. Single channel 6 MHz.	Multipoint distribution service
MDS2A	2156-2160	Same as MDS1. MDS2 truncated on one side to 4 MHz	Multipoint distribution service
WCS	2305-2320	5 MHz or 10 MHz blocks	Wireless communications service
WCS	2345-2360	5 MHz or 10 MHz blocks	Wireless communications service
ITFS	2500-2690	6 MHz channels shared with MMDS	Instructional television fixed service. Education service, includes Internet access
LMDS	27500-28350 31000-31300	Short range, 3 miles, 20 MHz channels.	Local multipoint distribution service
LMDS	31000-31300	Short range, 3 miles, 20 MHz channels. Propagation affected by rain	Local multipoint distribution service
ISM	902-928	0.5-mile range. Spread spectrum omni-directional.	Instructional scientific and medical. Unlicensed bands used for LANs and as a return path of 2-way modem systems.
ISM	2400-2483.5	0.5-mile range. Spread spectrum omni-directional. Can be engineered beyond 15 miles point-to-point as the return path for a cable modem system	Instructional scientific and medical. Unlicensed bands used for LANs and as a return path of 2-way modem systems.

Table 12-3.
Wireless transmission frequency bands.

In addition, the use of MMDS, ITFS, and MDS transmitters meant that two or three digital 2-MHz subchannels could work with adjustments to the analog transmitter diplexers and filters. The introduction of digital transmitters has allowed wireless networking to expand into the WCS, UHF, and VHF bands for Internet downstream access. Aside from the 5-MHz bandwidth found with the WCS systems, most wireless systems use a 6-MHz downstream bandwidth.

Telecommunication Technologies

> *MMDS/WCS transmitter systems for Internet access include channelized transmitters, channel combiners, automatic backup, and network management equipment.*

Each broadband fixed wireless system operates as a full-duplex system. Going back to the table, the use of two frequency bands allows two-way operation. While the higher frequency band establishes the high band in the link, the system considers the lower frequency set as the low band. During operation, the transmitter at one end of the link will use the high band; the transmitter at the other end will use the low band.

MMDS/WCS Internet Access

The efficiencies gained through digital video compression allow the dedication of a few RF channels per MMDS system to provide 10-Mbps broadband high-speed data service to Internet users. To meet FCC guidelines for cochannel and adjacent-channel interference, the downstream transmission must remain similar to one of the presently authorized digital video modulation formats. A single 6-MHz RF channel using 64 QAM can deliver a raw data rate of 30 Mbps or 27 Mbps after forward-error correction.

MMDS/WCS transmitter systems for Internet access include channelized transmitters, channel combiners, automatic backup, and network management equipment. Available with 5 to 100 watts of average output power, digital transmitters accept a 44-MHz centered QAM intermediate frequency then upconvert and amplify the signal. While crystal oscillators provide excellent phase noise, feedforward amplifiers and equalization techniques offer good gain, good phase linearity, and minimize power consumption. With WCS systems, the use of spectral shaping filters allows the system to meet stringent FCC specifications for out-of-band power. Channel combiners employ waveguide directional filters to combine either nonadjacent or adjacent microwave channels for transmission. Transmitter systems use the SNMP network management capability for remote configuration and fault management.

Licensed and Unlicensed Frequencies

The FCC offers licenses for the use of broadband wireless frequency bands through auctions. Current license holders, or those holding leases from these license holders, own the only permission rights to use radios on those frequencies. The MMDS band has 31 licensed 6-MHz-wide channels in the main part of the band. In addition, a license or lease agreement may combine the MMDS response channels that range from 2,686 to 2,690 GHz to create an additional 4 MHz of spectrum for data communication. Since the frequency band has channels or segments, installers can place multiple systems in a geographic area without interference.

An unlicensed band such as the U-NII band consists of a band of frequencies available for use by anyone and doesn't have license requirements. But users of an unlicensed band must use "type approved" radio equipment sanctioned by the FCC for use within the specific frequencies. Unlicensed band links offer much easier implementation and immunity to interference through the use of directional antennas.

Receiving the Signal

Equipment at the receiving nodes of a broadband wireless network includes an antenna, frequency downconverter, router, and wireless modem. As with the transmission site, the frequency band determines the type of receiving antenna and downconverter. The amplifier located at the antenna and the downconverter must provide sufficient amplification to maintain the signal at a level close to the high end of the range acceptable to the modem.

Reception of MMDS, MDS, and WCS signals requires a small antenna integrated with a downconverter. The antenna assembly may mount either on the roof, on a tower, or on the side of a building. Antennas used to receive MMDS, MDS, and WCS signals may include flat arrays that measure one foot square and produce 17 dB of gain, or dome-shaped designs with similar characteristics. Some manufacturers recommend the use of partial parabolas that measure two feet in diameter or two feet long cast Yagi antennas. Most downconverters supply internal gain as well as antenna gain so that output signal level remains close to the 0 dBmV needed for the modem.

Universal broadband routers produced by companies such as Cisco allow two-way transmission of digital data using either coaxial cable or broadband fixed wireless. The routers support IP routing with a wide variety of protocols and any combination of Ethernet, Fast Ethernet, High-Speed Serial Interface (HSSI), serial, and ATM media. Network interfaces reside on port adapters that provide the connection between the router and external networks. Cable or wireless interfaces reside on modem cards and provide the connection to cable or wireless networks.

Some broadband routers support the on-line insertion and removal of port adapters and modem cards. Depending on the router, the number of available slots for wireless modem cards varies from two to four. The router shown in *Figure 12-6* supports the installation of two port adapters, an input/output controller, and a network-processing engine. Some routers also use downloadable software and flash memory that allows a system upgrade without physically accessing the router.

The wireless receiver provides outdoor control and data interface to the indoor subsystems. In addition, the receiver operates as a frequency converter and supplies power amplification. Most receivers consist of an RF signal headend circuit, connector ports for IF signal input, control signals, and test signals along with an assembly for the connection of the antenna.

Figure 12-6.
A wireless broadband router, which supports the installation of two port adapters, an input/output controller, and a network-processing engine.

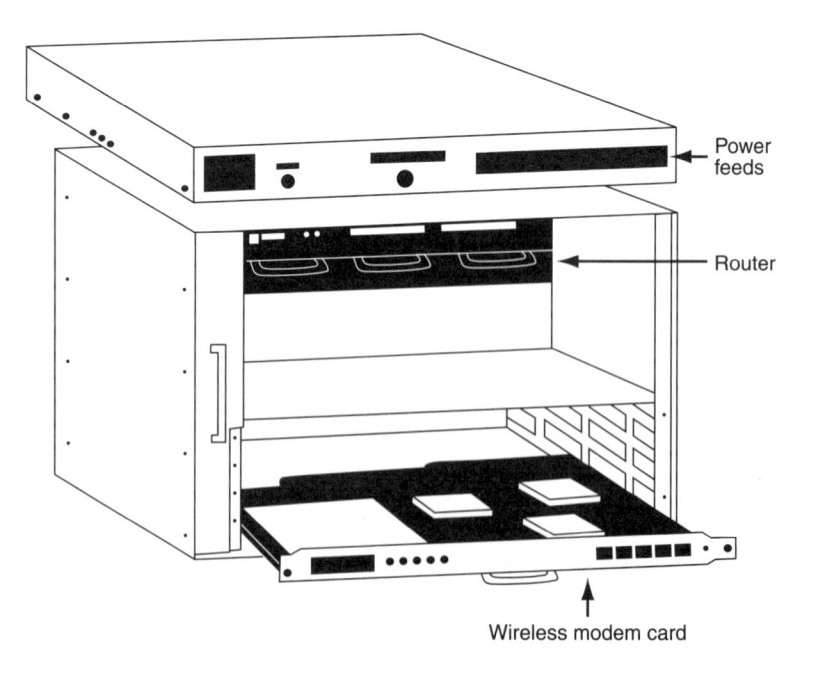

Wireless modem cards install in the router and become configured through the system console of the router. A wireless modem card provides the control and data interface to the system's digital motherboard and the radio frequency subsystem in the wireless transmitter and receiver. In addition, a wireless modem card provides the up/down conversion from baseband frequencies to an intermediate frequency.

Connecting to the Internet with a Satellite System

Since the DBS signals travel as digital packets, the signal providers can send video, audio, and computer data in any combination from the uplink center to IRDs. The large amount of bandwidth found with DBS transponders also allows DBS companies the option of providing data services such as Internet or interactive TV services.

The greatest drawback, however, occurs because of the asynchronous nature of DBS systems. Currently, DBS broadcasts occur only from the uplink center to the receiver and don't allow the sending of data from the receiver at the same speeds. Any return data from a DBS system travels through the attached telephone line.

The option to display the Internet or an interactive shopping channel on a television requires separate decoders or set-top boxes. Yet the use of special decoders provides several interesting options for viewers. For example, a decoder with built-in personal television features allows subscribers to set the decoder so that it watches for certain types of favorite channels. Storage devices within the decoder allow the saving of the favorite channels or Internet-based information for later viewing.

Satellite Distribution of Data Services

Satellite reception dramatically increases the available bandwidth for communications to the consumer and provides an instantly accessible communications infrastructure. For example, a satellite with 16 transponders has a downlink transfer capacity of 600 megabits per second. The asymmetric characteristics of satellite communications limit the transfer capacity to the downlink communications, however, due to the lack of bidirectional bandwidth capabilities. Upstream communications back to the satellite continue to require the use of a telephone system.

Satellite Networks

Networking and user services acquired through satellite technologies make use of:

- Low power and simple user terminals
- Interconnectivity between either the cellular network or the data services network and the PSTN
- Global coverage
- On-board signal processing, baseband amplification and switching
- Intersatellite links
- Data security and authentication,
- Dynamic channel access schemes to support a variety of applications and higher layer protocols

Although satellite data transmission has gained a reputation through bandwidth capabilities, future satellites will also incorporate intersatellite links, on-board switching, data buffering and signal processing. The ability to use on-board processing and multiple spot beams will enable future satellite systems to reuse the frequencies more than the current series of systems.

In addition, satellites launched for future applications will rely on dynamic channel allocation. Demand Assignment Multiple Access (DAMA) systems bases the number of connections and available channels on the number of requested accesses.

Satellite technologies have gained a foothold in networking and user services due to the capability to bridge large distances. As a result, both networking and user services influence the design of new satellite technologies. For example, the rapid growth of the cellular telephony market has pushed the telecommunication industry investment heavily into Mobile Satellite Services (MSS).

Satellites in Geosynchronous, Inclined, and Low Earth Orbits

Some of the most familiar communications satellites lie within geosynchronous orbits. Others use inclined orbits, while still others use low Earth orbits. Each type of satellite orbit yields different results and requires different types of receiving equipment. For example, most satellites used for television programming rely on geosynchronous orbits. Newer satellites used for cellular phone and pager communication have low Earth orbits.

Geosynchronous Orbits. In 1945, the scientist and science-fiction writer Arthur C. Clarke envisioned the placement of an artificial satellite in Earth orbit at 35,803 kilometers above the equator. In this geosynchronous orbit, a satellite will orbit the Earth at the same speed as the Earth's rotation and remain stationary with respect to any point on the Earth's surface. Due to Clarke's prediction, we now refer to the equatorial belt as the Clarke Belt.

Orbiting satellites occupy a specific orbital slot or subdivision. During transmission, satellite uplink stations transmit signals to the satellite. In turn, the electronic circuitry on the satellite processes the signals and then transmits the signals through a downlink to a receiving station on the Earth.

Inclined Orbits. We define the measure of inclination as any difference between the angle made by the plane of the spacecraft with respect to the plane of the equator. Geosynchronous satellites have a 0.1° inclination. Satellites in an inclined orbit, however, have inclinations that range up to three degrees. Rather than remain stationary in its assigned slot, a satellite placed in an inclined orbit follows a figure-eight pattern about its assigned location. The width of the figure eight increases as the inclination and north/south motion increase.

Because of this, a satellite reception dish receiving signals from an inclined orbit satellite must have the ability to track the satellite. This tracking ability occurs through adjustments in the polar mount elevation or the azimuth. Tracking outside a given range occurs through adjustments in both the elevation and the azimuth.

Each type of satellite orbit yields different results and requires different types of receiving equipment.

Figure 12-7.
A diagram of low Earth orbit (LEO) satellite orbital slots. Some LEO satellites operate as remote sensing and weather satellites.

Low Earth Orbit Satellites.
A low Earth orbit (LEO) satellite circles the Earth at an altitude of only 200 to 500 miles. Due to the proximity to the Earth's gravitational field, LEO satellites travel at 17,000 miles per hour and complete one orbit within 90 minutes. Depicted in *Figure 12-7*, some LEO satellites operate as remote sensing and weather satellites, because of the capability to capture highly detailed images of the Earth's surface from low altitudes.

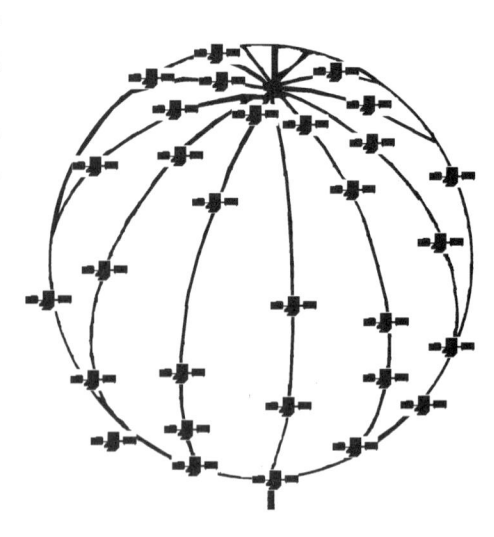

In addition, low Earth orbit satellites also serve one of the hottest trends in satellite technology by supporting global personal communications networks. The service offers global communications access through cellular phones or pagers.

Other telecommunications providers such as Teledesic, Globalstar, ICO, Ellipso, Astrolink, and Spaceway rely on low Earth satellites for not only cellular and pager communications, but also worldwide, high-speed access to computer networking, broadband Internet access, high-quality voice, and other digital data services.

Table 12-4 lists features of some of the modern satellite networks that use LEO, MEO as well as GEO satellite constellations.

MAC Protocols and Satellite Networks

Satellite channels have unique characteristics that require special considerations at the Data Link Control Layer of the OSI model. Since satellite links represent paths with high bandwidth and provide a broadcast channel, the transmission of data requires media sharing methods at the Media Access Control sublayer of the DLC. The traditional Carrier Sense

Chapter 12: Wireless Networks and Satellite Distribution

Network	Number of Satellites	Orbits	Orbit Type	MAC Method	Services
Inmarsat M	6-20	36000 km	GEO	FDMA	Voice, data (2.4 Kbps), FAX, Telex
GlobalStar	48 (8 spares)	1400 km, inclined	LEO	CDMA	Voice, data (9.6 Kbps), FAX location services
Teledisc	840	700 km	LEO	ATDMA, FDMA, SDMA	Voice, data up to 2 Mbps
Odyssey	12	10370 km	MEO	CDMA	Voice, data (9.6 Kbps), FAX, GPS
ICO	10 (2 spares)	8-10000 km	MEO	TDMA	Voice, data (2.4Kbps), FAX, GPS

Table 12-4.
Modern satellite networks and features.

Multiple Access/Collision Detection schemes typically used with local-area networks will not operate with satellite channels.

In addition, long round-trip delays for signals and high bandwidth-delay products complicate the design of DLC layers in satellite data networks. To isolate the effect of these undesirable features from higher layers and provide transparent interface with other terrestrial networks, most schemes use a combination of different protocols to achieve higher channel efficiency and highly dynamic channel allocation.

Due to the point-to-point characteristics of the link, earth stations don't sense the existence of a carrier on the uplink signal. With CSMA/CD at the downlink, a 270-microsecond delay would exist before an Earth station would become aware of potential collisions. Applications depending on CSMA/CD protocols cannot tolerate such delays. As a result, most satellite MAC schemes usually assign dedicated channels in time and/or frequency for each user.

Typical MAC schemes used for satellite data transmissions include ALOHA, FDMA, TDMA, and CDMA. With Pure ALOHA, competing stations can transmit anytime. But Pure ALOHA offers a very low efficiency of 18%. Slotted ALOHA (S-ALOHA) uses the satellite broadcasts to synchronize the ground station transmissions to the start of a slot time. With this approach, efficiency improves to approximately 36%.

Frequency Division Multiple Access (FDMA) remains as the oldest and most common method for channel allocation. With FDMA, the available satellite channel bandwidth divides into frequency bands for different Earth stations. As a result, FDMA requires the use of guard bands to provide separation between the frequency bands. In addition, Earth stations relying on FDMA must carefully control power dissipation to prevent the microwave power from interfering with frequency bands for the other channels.

With Time Division Multiple Access (TDMA), channels become time multiplexed in a sequential fashion. Since a TDMA satellite channel consists of multiple time slots in a framed structure, each Earth station can only transmit in a fixed time slot. But stations with more bandwidth requirements may receive additional time slots. Each time slot can carry data packets belonging to any one of the users.

At any instant, a TDMA frame consists of fixed numbers of reserved and empty slots. Since TDMA requires time synchronization between the Earth stations, one Earth station generates the broadcast signal through the satellite connection. Rather than use the fixed assignment of time slots to a channel, TDMA handles the task dynamically in real time. When an Earth station has reserved a time slot, the system recognizes the packets as safe packets. Each data packet carries a virtual circuit identifier (VCI) field that indicates the location of the receiving Earth station. Earth stations can recognize packets on the downlink broadcast by checking the VCI field in the packet.

Whenever a new Earth station tries to establish a channel, it sends an unsafe data packet in one of the available slots. If the packet reaches the destination without collision, the system reserves the particular time slot for the Earth station and allows the transmission of safe packets. If a collision occurs, collision avoidance schemes resolve the conflicts between the contending Earth stations. An Earth station with no data to send in a reserved time slot loses its reservation for that slot. If required, a user may reserve more than one time slot. *Figure 12-8* shows the TDMA method of channel allocation for simultaneous requests.

Code division multiple access (CDMA) uses a hybrid of time/frequency multiplexing and operates as a form of spread spectrum modulation. A

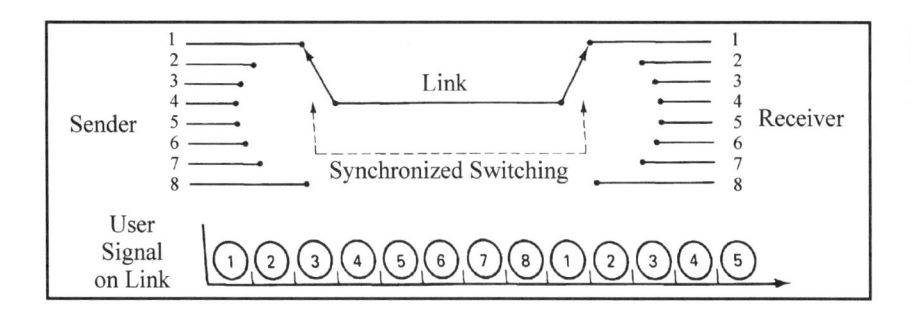

Figure 12-8.
TDMA method of channel allocation for simultaneous requests.

Spread Spectrum system spreads the transmitted signal over a frequency much wider than the minimum bandwidth required for sending the signal. CDMA provides a decentralized way of providing separate channels without relying on timing synchronization. As *Figure 12-9* shows, CDMA operation represents each binary transmission symbol with a spreading code that

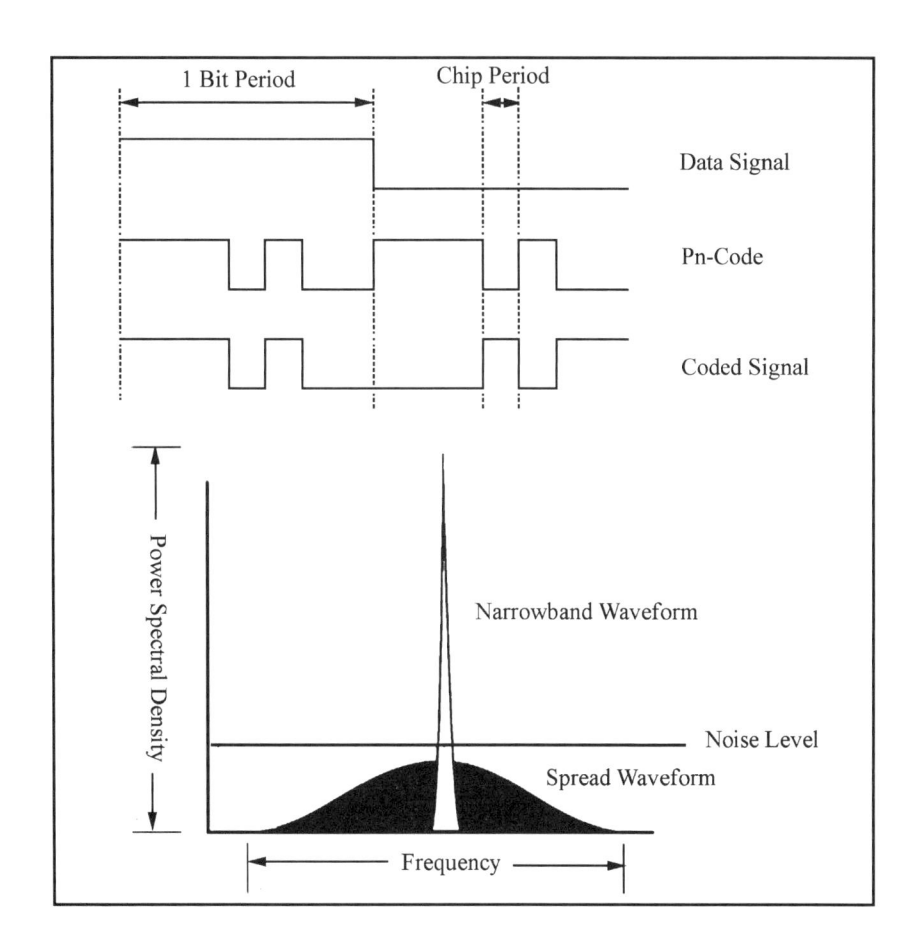

Figure 12-9.
CDMA operation, which represents each binary transmission symbol with a spreading code that consists of a zero-one sequence.

345

Telecommunication Technologies

consists of a zero-one sequence. The spreading code, or chip rate, has a higher bit rate than the symbol bit rate.

Each user has a unique orthogonal code and obtains the resulting signal as the product of the input data stream and the spreading code. As the receiver retrieves user data, it correlates the incoming bit stream with the spreading code of the receiver. If the data has a different spreading code and doesn't match the user, it appears as noise. CDMA provides several benefits for satellite data transmission in that:

- It solves the problem of multiple accesses without any coordination among the users
- The chip or spreading code provides a method to identify and authenticate the source transmitter without information in the packet
- This method provides high security against eavesdropping in satellite broadcast channels, because of the use of the unique spreading code for each user
- It allows the reuse of same frequencies in adjacent beams in a multiple spot beam satellite by assigning different spreading codes to each   user

Despite the benefits, CDMA remains limited in that the spread spectrum signal of N users increases the noise level.

Very Small Aperture Terminal Networks

A Very Small Aperture Terminal (VSAT) operates as a small satellite Earth station and features a reflector dish that measures 1.8 meters or less in diameter. The outdoor portion of a VSAT includes the antenna and equipment that performs RF transmission functions. Indoor VSAT equipment performs digital baseband signal processing functions.

VSAT networks provide a solution for centralized networks that have a central host and a number of geographically dispersed terminals. Typical examples of VSAT applications include small and medium businesses with a central office, banking institutions with nationwide branches, backbone links for an Internet service provider, and airline ticketing systems. Examples of common VSAT services include point-of-sale

VSAT networks provide a solution for centralized networks that have a central host and a number of geographically dispersed terminals.

applications and the distribution of management information within a corporation. VSAT solutions reduce the complexity of the system and, as a consequence, reduce transmission costs and the need for maintenance. Greater efficiency also occurs through end-to-end circuit control, network flexibility, and short installation time.

With VSAT networks, either the transmitter or the receiver antenna on a satellite link must have a larger size. Simplifying VSAT design involves utilizing a lower performance microwave transceiver and a smaller, lower gain dish antenna. The combination operates as a small, simple bidirectional Earths station that installs at the end user's premises.

VSAT networks rely on the S-ALOHA and TDMA MAC schemes. At the LLC sublayer, VSAT networks use a "look back N" protocol with a selective reject Automatic Repeat request (ARQ) retransmission strategy. The most common protocol implementation uses a transmission window with N=128 packets, where the receiver responds with retransmission requests for only erroneous or missing packets. This protocol, combined with forward-error correction, produces reliable data transfers while providing low average delays on satellite links. Most VSAT links use X.25 as a network protocol rather than TCP/IP.

VSAT networks use a star-based topology that supports each remote user with a VSAT. During operation, the Earth hub station acts as the central node and employs a large-size dish antenna with a high-quality transceiver. The satellite functions as the broadcast medium and acts as a common connection point for all remote VSAT Earth stations. In addition, the satellite provides immense bandwidth as well as amplification that compensates for the lower gain seen at the uplink. At the high-performance hub Earth Station, the received signal remains at a usable level.

This arrangement has a disadvantage in that communication between two VSATs requires two satellite transactions since all connections must pass through the hub ES node. Moreover, the data link supported from the hub to the VSAT has a slower 19.2 kbps data-transfer rate than the 512 kbps data rate seen from the VSAT to the hub. *Figure 12-10* shows two VSAT terminals communicating in a simple VSAT network.

Figure 12-10.
Two VSAT terminals communicating over a VSAT network.

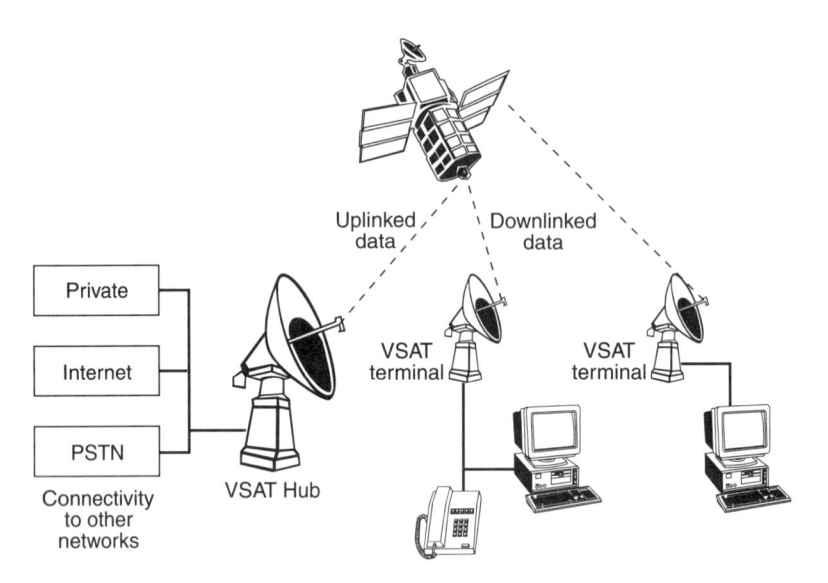

But modern satellite technologies have begun to override some of the early disadvantages seen with VSAT. Newer satellites employ on-board processing with switching functions. As a result, the satellite repeater can provide demodulation, amplification in the baseband, and retransmission at full power. Given this functionality at the satellite, the system can eliminate the basic function of the Hub Earth Station. In addition, the satellite can provide full point-to-point mesh connectivity between the VSAT Earth Stations along with larger bandwidth in both directions.

Receiving Data Services with DirecPC

With DirecPC, a service offered by Hughes Network Systems that provides Internet access through private satellite dishes, upstream requests for web pages travel through a normal modem connection to the DirecPC Network Operations Center. The installation of a DirecPC system requires the purchase of a receiving system, an adapter card for a personal computer, and a connection to an Internet service provider. DirecPC provides the optimum results when installed on a Pentium-class computer system with 32 megabytes of RAM and the Windows 95 or higher operating system. *Figure 12-11* shows a diagram of the DirecPC system.

DirecPC offers several options for consumers that include the Internet-only DirecPC installation kit and the DirecDuo Internet and television

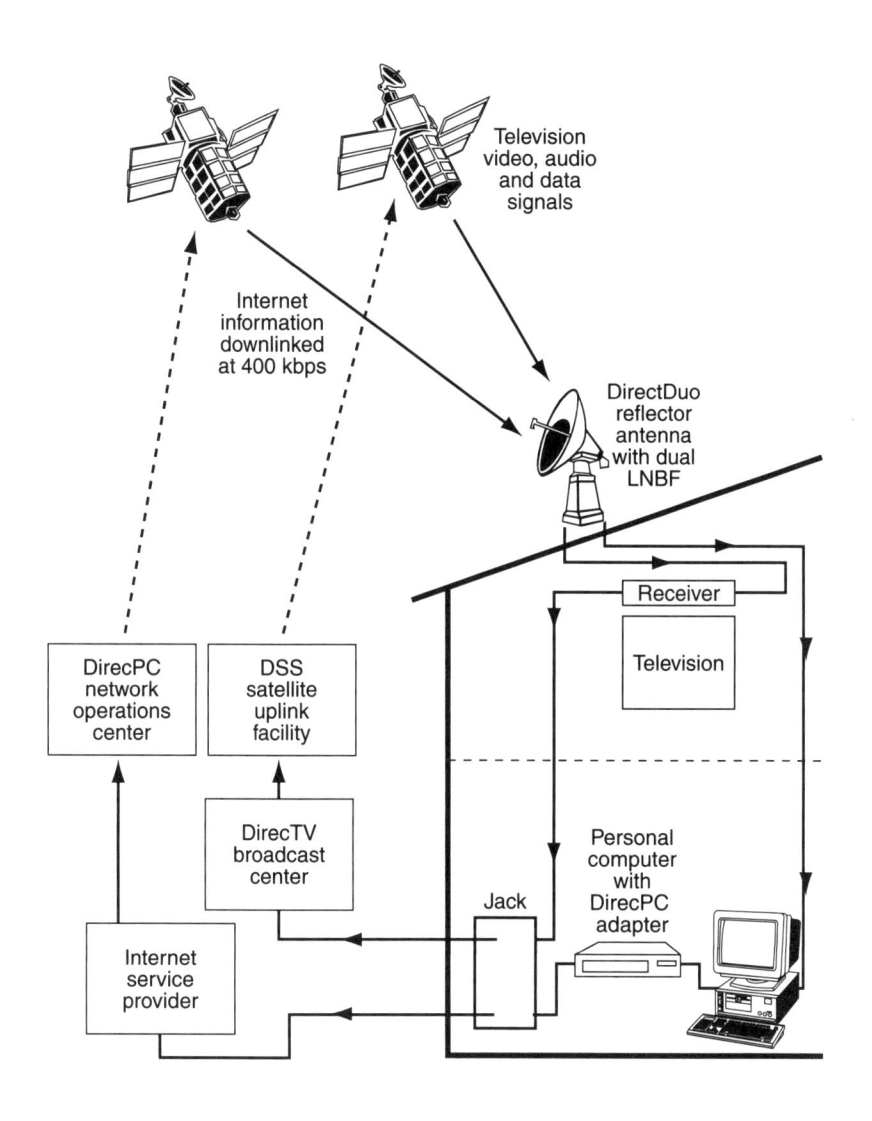

Chapter 12: Wireless Networks and Satellite Distribution

Figure 12-11.
A diagram of the DirecPC system. Upstream requests for web pages travel through a normal modem connection to the DirecPC Network Operations Center.

reception kit. While the Internet-only kits include a single-function DirecPC dish antenna, a DirecPC satellite modem, installation and access software, and a universal mount, the DirecDuo kit includes a slightly different dish antenna. The satellite modem connects either through an Industry Standard Architecture (ISA) connector, the Peripheral Connect Interface (PCI) connector, or the Universal Serial Bus (USB) connector found either within the computer or as a port.

DirecPC software installs with Windows 3.1, Windows 95, Windows NT, or Windows 98 and provides access to the Turbo Internet, Turbo Webcast,

Telecommunication Technologies

As the customer sends the address request for a web page, the DirecPC software attaches an electronic addressing mask called a tunneling code to the request.

Turbo Newscast services. The services provide a sports ticker that provides real-time delivery of scores of sporting events and a stock ticker that shows the status of all three stock exchanges. The Turbo Newscast service provides the real-time video stream delivery of news.

DirecPC Satellites

DirecPC relies on the GE-1 and GE-4 satellites placed in a geosynchronous orbit to carry and downlink the Internet data. Since GE-4 is an older satellite, systems aimed at GE-4 use the ISA rather than the PCI or USB satellite modems. In addition, the GE-1 Satellite relies on vertical polarization rather than the horizontal polarization seen with GE-4.

Along with the installation and applications software, the DirecPC also arrives with a satellite alignment software program that lists the elevation and azimuth for most cities in the United States. After the consumer selects the appropriate city, the program responds with the exact elevation, azimuth, magnetic azimuth, and polarization settings required for that location.

In addition, the software includes a signal strength meter that shows as a graphical output to the screen and has a maximum strength reading of 160.

DirecPC Operation. When a customer requests an address for a web page, the request travels from the customer's modem to the Internet service provider. As the customer sends the request, the DirecPC software attaches an electronic addressing mask called a tunneling code to the request. The tunneling code instructs the ISP to forward the request for the web page to the Network Operations Center instead of the ISP server.

After receiving the request, circuitry at the Network Operations Center strips the tunneling code from the request. Multiple T3 lines carry the web page request to the appropriate web site, where the Network Operations Center retrieves the desired content. The center uplinks the information to the DirecPC satellite. From there, the satellite downlinks the information at a rate of 400 kilobytes per second to the customer's system and computer. With the entire transfer from computer to the Network Operations Center, the satellite, and back to the DirecPC system and personal computer taking approximately 450 milliseconds, the downloading of a one-megabyte file takes approximately 30 seconds.

Installing and Running DirecPC with Windows 95 or 98

Since the DirecPC software originally installed with Windows 3.1, the installation process involved manually installing networking components, software drivers, and a terminate-stay resident program. Furthermore, the Windows 3.1 operating system had base memory limitations that complicated the DirecPC installation. The installation of the networking components required that the installer have knowledge about hardware requirements and IRQ conflicts that can occur in a personal computer.

An interrupt request (IRQ) line allows devices attached to the personal computer to send an interrupt signal to the microprocessor. The interrupt switches the microprocessor from one operation to another temporary operation. Adding a new device to a personal computer requires the setting of the device IRQ number. IRQ conflicts occur when two devices have the same IRQ number. The plug-and-play function of Windows 95 allows the system to automatically set IRQ numbers and accept the installation of new devices such as interface cards.

With the introduction of Windows 95, Windows 98, and Windows NT, DirecPC offers a new set of installation software that uses the Win95 Winsock and allows the installation of other network cards. In brief, Windows 95 supports 32-bit applications and removes the memory and networking limitations formerly seen with DOS-based operating systems. The Win95 Winsock functions as an application-programming interface (API) that allows communication between Windows programs and other machines through the Transmission Control Protocol/Internet Protocol (TCP/IP). TCP/IP represents a suite of communications protocols that connect host computers on the Internet. Network operating systems include protocols that support the TCP/IP standard.

Installing the DirecPC requires following a simple step-by-step process in Windows. After placing the DirecPC disk in the floppy disk drive of the computer:

- Select the Start button at the bottom left of the Windows desktop
- Select the Settings option from the list
- Select the Control Panel icon
- Select the Add/Remove Programs icon
- Highlight DirecPC in the selection window
- Select the Install button
- At the "Install by Floppy or CD screen?" select Next
- At the "Run Installation Program" screen, select Finish
- At the Setup prompt, select Yes
- Then highlight "US" for Antenna location and select Next
- At the Installation Path screen, select Next
- At the "Select folder" menu screen, highlight the "DPC" selection and select Next
- At the Choose Destination Location screen, select Next
- Select Finish to complete the installation and restart the computer

Gateways and DNS

Use of the DirecPC service requires that the personal computer communicate through either a Serial Line Internet Protocol (SLIP) or the newer Point-to-Point protocol (PPP) so that a connection can occur between the computer, an ISP, and the DirecPC Network Operations Center. In addition, the installer must designate a gateway address in the network setup portion of the Windows software.

A gateway functions as a combination of hardware and software and links two different types of networks. The gateway number shown in the DirecPC installation designates a specific DirecPC satellite.

Setting up a network connection in Windows also requires the designation of a domain name service (DNS) number. A DNS translates the domain names associated with web addresses into IP addresses. Each individual IP address contains a unique set of numbers that point toward a specific

Internet server. The installation of the DirecPC software specifies the DNS number (198.77.116.8) as a default setting.

Due to demands on the DirecPC server, however, using the DirecPC DNS number may slow the response of the system. Rather than use the DirecPC DNS, set the Windows designation for the DNS of the desired Internet service provider and set the DirecPC Navigator to use the modem for DNS lookups. The procedure includes the following steps.

- Locate the DNS designation for the ISP
- In Windows 95, select the Start function. Then select the Settings function and the Control Panel. Within the Control Panel, select the Network application
- Select TCP/IP from the listed choices. After selecting TCP/IP, select DirecPC Adapter. Open the properties selection for the DirecPC adapter and select the DNS tab
- Delete any previously set DNS numbers
- Set the ISP DNS number as the default and select OK
- Before restarting the computer, initialize the DirecPC Navigator application and select the Networking option
- Select the Terrestrial Tab and follow the instructions for setting up a port to access the terrestrial only page using the DNS setting

Restart the computer so that all changes record in the system. To run DirecPC, select the Internet icon. Then select properties and deselect auto dial. Starting the DirecPC TurboInternet application automatically dials the desired ISP. After logging onto the system, use an Internet browser to access the World Wide Web.

Using DirecPC on a Network

A computer system using DirecPC can function as a proxy server on a local-area network. Proxy server software acts as a gateway to the Internet for all client computers connected to the network, as well as enabling the DirecPC capabilities on the network. Since a proxy server requires a high-bandwidth connection, the server software resides on the computer using the DirecPC adapter card.

Proxy server software acts as a gateway to the Internet for all client computers connected to the network, as well as enabling the DirecPC capabilities on the network.

Before installing the proxy server software, attach the DirecPC computer to a functioning TCP/IP-based network. Install the DirecPC computer to the network and verify the compatibility of the DirecPC interface and the network interface card. Both cards must function at the same time on the network. The DirecPC computer will have TCP/IP settings configured for the DirecPC interface card. When configuring the TCP/IP settings on the DirecPC computer for network access, only change the TCP/IP settings for the network-interface card.

Summary

Chapter 12 described wireless technologies and satellite communication technologies. It emphasized planning for the wireless system and outlined potential problems that can prevent the system from operating normally. The planning portion of the chapter considered site selection and regulations concerning the placement of a tower and transmitter. Potential problems for wireless systems include multipath distortion, wind loading, obstructions, and interference.

Chapter 12 also provided information about antenna construction, polarization, and gain while describing key antenna specifications such as beamwidth and aperture illumination. The chapter moved from covering hardware issues to a description of broadband frequency bands and wireless Internet access. From there, the chapter changed tempo by considering the use of satellite technologies to access data networks and the Internet.

Before installing the proxy server software, attach the DirecPC computer to a functioning TCP/IP-based network.

Index

Symbols

10/100 switching 108
1000BaseCX 70, 121
1000BaseLX 121
1000BaseSX 121
1000BaseT 119, 121
100BaseT 115, 116, 118, 120
100BaseT2 116, 117
100BaseT4 117
100VG-AnyLAN 118, 119
10Base2 111, 195
10Base5 69, 112
10BaseT 14, 52, 100, 112, 115, 118, 120, 289
2B+D 197
2M-1 308

A

AAL 226, 236
AAL5 236
ABR 244
access concentrator 228, 255
access servers 145
access-control byte 126
accounting management 22
ACK 291
ACR 54
adapters 81
adaption, media-rate 106
ADC 167, 309
add/drop multiplexer 267, 271
address screening 231
address summarization 140
addressing 40
addressing, ATM 245
addressing, Internetwork 134
ADM 267, 268, 271. 272
adoptive equalization 192
ADSL 299, 303, 304, 307-309, 314, 315, 318, 319
ADSL Forum 302
ADSL Framing 310
ADSL modems 306-308
ADSL networks 305
AFE 185, 186, 309
AFI 245
alarms 288
ALOHA 345
Alta Vista 166

AM 168, 170
AM radio 3
AMD 100
America On-line 164
amplifiers 52, 292, 295
amplitude modulation 168, 284, 322
analog modems 189, 300
analog signals 4, 171
ANSI 186, 257, 273, 274, 302
antennas 322, 324-328, 330, 332-334, 336, 339
antennas, dish 330, 331, 332, 335, 349
antenna gain 331, 332
antennas, Yagi 339
API 353
Apple Macintosh 42
AppleTalk 39, 43, 44, 97, 138, 275
AppleTalk networks 35
AppleTalk Zone 137
Application Layer 31, 41, 42, 48, 291
applications software 15
Architecture, Ring 272
ARP 136, 137, 251
ARQ 349
ASICs 309
Astrolink 344
asymmetric switching 108
asynchronous communications 172
asynchronous transmissions 14
AT commands 177
AT&T 161
ATDM 145
ATM 63, 104, 140, 221, 225, 236, 240, 251, 253, 293, 301, 302, 305, 340
ATM addressing 245, 249
ATM ARP 245
ATM cells 222, 230, 255
ATM concentrators 228
ATM endpoints 236
ATM Forum 302
ATM interconnect 227
ATM Layers 223
ATM networks 221, 222, 225, 236, 241-243, 247, 253, 256, 273, 297
ATM services 242

ATM switches 236, 237, 240-242, 244
ATM WANs 242
ATP 43
attenuation 53, 54, 56, 84, 85, 300, 325
ATU 305, 306, 309, 310
audio library 243
audio signals 286
AUI 100, 111
autoreconfiguration 127
azimuth 343, 352

B

B/routers 142
backbone 141, 142, 226, 254, 255, 274, 286
backbone cabling 76, 82
balance transformers 57
baluns 57, 81
band 52
band, voice 159
bandwidth 8, 19, 51, 52, 107, 122, 302, 303, 306, 319, 337, 342
bandwidth, dedicated 301
bandwidth, guaranteed 316
Banyan VINES 44
barcode readers 11
baseband 12, 112, 117
baseband transmissions 12
Basic Link 86, 87
Basic-rate Access 303
baud rate 187
BCD 7
beaconing 127
beam switching 321
beamwidth 330, 334, 336
bearer services 196, 197
BECN 212
Bell Operating Companies 220
best-effort service 19
binary numbers 6
binary-coded decimal 7
bit 7, 33
bit errors 10
bit, parity 182
bit, start 173, 180
bit, stop 173, 180
BITS 260
BNC connectors 70
BOCs 161
Boole, George 7
BPI 290
BRA 303
branch office 210, 302
BRI 196, 197, 200, 209
bridge/routers 97
bridges 36, 37, 96, 103-105, 139, 142, 145, 148

bridging 104, 141, 274
broadband 12, 19, 221, 235, 268, 269, 289, 294, 300, 303, 321, 322, 336, 339, 344
broadband ISDN 196, 235
broadband transmission 12
broadband wireless 322
broadcast 89, 90, 109, 110, 141, 276, 294, 297, 321
broadcast radio 15
browser 29, 165
buffer 114
buffering 40, 321
burstiness 235
BUS 248, 250, 251
bus topology 90
byte 7
bytes, overhead 259, 262
bytes, frame-control 126

C

cable 286
cable balance 55
cable, coaxial 15, 25, 31, 61, 68-70, 292-294, 311
cable, copper 8, 73, 154
cable, dual 13
cable, feeder 295
cable, fiber-optic 25, 31, 33, 56, 73, 74, 82, 254, 292-295, 311
cable headend 286, 288
cable modems 283, 284, 286, 287, 289-291, 294, 315
cable networks 286
cable, RS-232C 174
cable, single 13
cable television 12, 283, 293
cable testers 82-87
cable toners 84
cable tray 76, 78, 82
cable, twisted-pair 33, 58, 92, 112, 116
cable-data networks 286, 288, 289, 290
cable-delivery methods 51, 77
cables, jumper 82
cables, patch 82, 102
cabling 33, 51
cabling, backbone 76, 77, 82
cabling, copper 73
cabling interfaces 33
cabling, structured 76
cabling, twisted-pair 31, 33, 54-57, 112, 116, 117
CAC 239
call blocking 159
call control 183
call forwarding 159
call waiting 143

Index

call establishment 207
call termination 207
callback 217
caller-ID 205, 216
calls, inter-LATA 161
calls, intra-LATA 162
calls, long-distance 162
capacitance 83, 160
carrier detect 175, 190
carrier, loop 269
carrier signal 284
carrier waves 168-171, 190
Category 3 61, 63, 81, 116
Category 4 61, 63, 81
Category 5 61, 63, 81, 116
Cat 5E 61
Category 6 61
Category 7 61
CBR 243
CCITT 14, 273
CCP 210
CCS 159, 160
CD 175, 190
CD-ROM 312
CDMA 345, 346, 347, 348
CDV 239, 240, 241, 244
cells 316
cell delineation 226
cell jitter 241
cell switching 222
cellular phones 343, 344
central office 153-155, 157-160, 201,
    202, 252, 269, 270, 300, 301, 305,
    308, 309, 311, 317, 319
CES 252, 253, 254
channels 12, 14, 86, 159, 183, 222,
    307, 308, 310, 312, 316, 336,
channel testing 86, 87
character codes 42
chat 291
chip sets 309, 310
CIR 253
circuit group 207
circuit switching 150, 151, 207, 222
circuit-switched network 40, 147, 150
Cisco Systems 213, 339
Clarke Belt 343
CLECs 303
clients 15, 76, 94, 95, 165
client, thin 128
client/server 16, 93, 94
clocking 259
closet, telecommunications 76, 77
cloud, network 92
CLP 242, 256
CLR 239, 240, 242
CMOS 100
CMTS 287
CODEC 236

COM ports 176
common-channel signaling 203, 204
communications, asynchronous 172
communications, synchronous 172
compressed video 313
compression 10, 180,
    184, 186, 210, 338
CompuServe 164
concatenated signals 266
concentrator 267, 269, 277, 280
conditioned line 153
conduit 76, 82
configuration management 21
congestion 36, 47, 139,
    242, 243, 287
connection management 40
connectionless networks 148
connectivity 20, 95, 195,
    197, 207, 219
connectors 173, 174
connectors, BNC 70
control, flow 182, 183
control information 27, 28
couplers, hybrid 81
CPE 200, 229, 230, 235,
    254, 313-316
CRC 106, 111, 213, 236,
    275, 309, 310
cross-connect 268, 269, 272
cross-polarization 333, 336
crosstalk 54-56, 318, 319
CS 227, 236
CSMA/CD 89, 110, 118, 121,
    122, 125, 247, 345
CSPDN 218
CSU 142, 143, 148
CSU/DSUs 139
CTD 240, 241, 243
CTMS 290
CTS 175
CW 169

D

DAC 167, 277
DAMA 342
DAS 277, 278
data 1, 14, 15, 17, 47, 68,
    76, 78, 144, 149, 151, 153,
    228, 253, 277, 308, 312
Data Link Layer 33-35, 37, 38,
    44, 96, 104, 107, 109, 131,
    135, 139, 145, 223, 289,
    290, 344
data transfer 27
data transmission 42, 89, 122, 192
data-flow control 40
data-transfer rate 187, 192
datagrams 33, 39, 149, 150, 207, 209

357

daughter board 309
DAVIC 302
DBS 341
DCE 143, 173, 174, 182, 183, 211-214
DCLI 212
DDR 146, 208, 210, 216, 217
DDS 142, 150
DE 212, 213, 256
DEC LAT 275
DECnet 137, 275
decompression 184
demodulation 171, 288, 290
demultiplexing 15, 268, 271
detectors 171
deterministic networks 125
deviation 170
device drivers 44
DFT 314
DHCP 289
DIA 309
dial backup 146, 215, 216
dialing 158, 159, 176
differentiated service 20
digital equipment 109, 213
digital information 6
digital transmission 15
diplex filters 295
DirecDuo 350
DirecPC 350, 352-356
discard eligibility 213
discriminator, frequency 171
dish antenna 330-333, 335, 349
display 1
distance learning 243
distortion 10, 15, 322
distribution 129
distribution hubs 288
DLC 270, 309, 344
DLCI 211, 212, 213, 256
DMT 307, 308, 310, 314, 315
DNIS 205, 216
DNS 48, 289, 354, 355
DOCSIS 289
DOS 353
downstream channel 307, 308, 312, 316
downtime 22
DQDB 230, 235
DS-0 294
DS-1 231, 258, 260, 264, 265, 266, 268
DS-3 231, 264, 265, 269
DS0 205, 251, 252
DSL 299, 303, 318
DSLAMs 306
DSP 245, 300, 301, 309
DSR 175
DSS-1 204
DSU 142, 143, 148, 236

DTE 142, 173, 174, 182, 183, 211, 212, 213, 214
DTMF 158
DTR 175
dual-homing 279, 280
duplex channel 307, 308
DWDM 258
DWMT 307, 308

E

e-mail 17, 291
E.164 245
E1 306, 312, 318, 319
ECSA 257
EHF band 3
EIA/TIA 51, 54, 60, 77, 78, 85
ELAN 248, 249, 250, 251
element management 288
Ellipso 344
EMI 73
emissions 60
EMS 288
encapsulation 142, 207, 256
encoding 185
End Delimiter 126, 277
end-to-end 195, 207, 244, 270, 300
energy management 76
enterprise network 17, 133
envelope delay 153, 155
EOC 310
EPD 240
EPD/PPD 240
equalization, adoptive 192
equalizer, frequency-domain 314
equipment rack 78
error correction 27, 35, 180, 183, 184, 191, 213, 226, 308
ESS 159
Ethernet 33, 35, 44, 52, 63, 89, 97, 101, 104, 107, 109-111, 114-116, 119-123, 139, 148, 195, 241, 247, 267, 274, 277, 286, 289, 340
Ethernet LANs 105, 241
Ethernet monitors 84
EtherTalk 35
ETSI 302
Excite 166
extenders 95, 100

F

FAA 322, 328
Fall Back 192
Fall Forward 192
Fast EtherChannel 123, 124

Index

fault locators 84
fault management 22
fault tolerance 279, 280
FAX machines 63, 177, 196
FCC 188, 322, 328, 338, 339
FCS 111, 127, 214, 277
FDDI 63, 89, 107, 229, 257, 273-275, 277-279
FDM 15, 145, 307, 308, 313
FDMA 345, 346
FECN 212, 213, 256
feeder cable 295
FEQ 314
fiber-optic cable 8, 16, 25, 31, 33, 56, 73, 74, 82, 154, 292, 293, 294, 295, 308, 311, 322
fiber/coax 293-295, 297, 304
Fibre Channel 122
file control 183
filtering 106, 109, 138
filters, Diplex 295
filters, media 57
firewalls 100
FISU 206
flow control 182, 183
FM 168
FM broadcast 3
format 1
forward-error correction 308, 313, 338
forwarding 106
fractional DS-3 225
fractional T1 147, 162
fractional T3 222
fragmentation 109
frame bursting 122
frame control 276
Frame Relay 140, 148, 152, 210-213, 216, 219, 225, 244, 251-253, 255, 256, 293
Frame Status 127
frame switch 107
frame-control bytes 126
frames 33, 34
framing 262, 290, 310
frequency 1, 153, 155, 168, 170, 314, 321, 330
frequency discriminator 171
frequency downconverter 339
frequency modulation 168, 169, 322
frequency response 183
frequency-division multiplexing 15
frequency-domain equalizer 314
Fresnel Zone 329
FTP 48, 59, 164, 291
FTTC 299, 300, 311
full-duplex 109, 121, 122, 205, 338

G

gain 330-332, 335
gateways 145, 354
GBIC 121
GE-1 352
GE-4 352
GEO 344
geosynchronous orbits 343
GFC 223
GIF 186
Gigabit EtherChannel 123
Gigabit Ethernet 119-122
global networks 16
Globalstar 344
GND 175
ground waves 2, 3
grounding 58-60, 324
GTE 161
guaranteed service 20

H

H2TU-C 319
H2TU-R 319
half-duplex 121, 122
handshaking 34, 191
hanging up 176
HDSL 299, 311, 317, 318
HDSL2 318, 319
HDTV 286, 312
headend 283, 286, 287, 291-295, 297, 322
headend, cable 286, 288
header 27, 29, 30
HEC 226
Hello Protocol 136, 137
hertz 8
heterodyne 168
HF band 3
HFC 293-295, 297
home office 215
Home-Termination Units 294
Host-to-Network Layer 46, 47
HSSI 340
HTML 165, 186
HTTP 48, 164, 291
HTU 294
hubs 33, 76, 78, 95, 101, 102, 114, 118, 286, 288, 292, 316
hub network 272
hubs, super 286
hubs, switching 114
hubs, wiring 16
hub-to-hub 113
hubs, 100VG-AnyLAN 119
hubs, distribution 288
hubs, stackable 102

359

Hughes Network 350
hybrid couplers 81
hybrid fiber/coax 51, 293, 294, 297

I

IA/TIA 77
IANA 134
IBM 42, 45, 95, 124
ICA 129
IDFT 314
IDI 245
IDUs 28
IE 205
IEC 161, 162
IEEE 14, 100, 115, 122, 124, 135, 229, 231, 235, 275
IF 340
ILEC 303
impedance 53, 83
in-band signaling 203
in-channel signaling 203, 204
inclined orbits 343
indexer 166
integration 90
Intel 109
inter-digit time 158
inter-LATA calls 161
interactive television 294, 341
interactive videotex 196
interference 60, 184, 300, 333, 338, 339
Internet 19, 29, 45, 47, 48, 147, 163, 164, 190, 251, 283, 286, 288, 291, 297, 299, 304, 310, 312, 317, 318, 321, 322, 337, 338, 341, 344, 348, 350, 355
Internet Explorer 165
Internet Layer 47, 48
Internet Protocol 47, 165
internetwork 17, 18, 131, 132, 134, 135, 137, 138-140, 142, 196, 208, 210, 213, 218, 251, 254-256, 273, 305, 306, 312
interoperability 90, 255, 305
Interrupts 176
intra-LATA calls 162
IP 47, 165, 208, 209, 245, 339, 354
IPCP 209
IPLs 162
IPX 39, 43, 137, 208, 245, 275
IRDs 341
IRQ 176, 353
ISA 351, 352
ISDN 139, 150, 195-199, 200-205, 207, 208, 210, 214, 215, 219, 283, 304, 313, 315
ISO 13
ISO/IEC 51, 77
isotropic antenna 332
ISP 19, 163, 164, 190, 251, 352
ISUP 206
ITFS 337
ITU 188, 219
ITU-T 245
IXCs 258

J

jack, modular 62, 63
jitter 259
jumper cables 82

K

Kflex 188

L

LANs 12, 14, 16, 17, 33, 36, 73, 89, 95, 100, 101, 104, 106, 115, 124, 132, 133, 274, 286, 304, 310, 312, 317
LANs, Ethernet 105
LANs, Virtual 129
LAN Manager 97
LAN switches 104, 106, 107, 108, 236
LANE 236, 247
LAPB 219
LAPD 203, 219
lasers 295
LATAs 161, 162
Layer, Application 31, 41, 42, 48, 291
Layer, Host-to-Network 46, 47
Layer, Internet 47, 48
Layer, IPX 43
Layer, MAC 290, 291
Layer, Network 33, 34, 37-39, 44, 131, 135, 137, 140, 289, 291
Layer, Physical 31, 33, 35, 44, 95, 98, 131, 223, 230, 275, 289, 301
Layer, Presentation 41-43
Layer, Session 41-44
Layer, SPX 43
Layer, Transport 39-41, 43, 44, 47, 291
layers 26-28, 30, 51, 93
LCL 56
LCP 208
LCTL 56
LDN 198
leased lines 146, 148, 150, 198, 252, 254
LEC 161, 162
LEDs 75
LEO 344
LF band 3

lightning protection 327
line equalization 184
line overhead 262, 263, 267
line utilization 21
Link, Basic testing 86
Link, Data Layer 33-35, 37, 38, 44, 96, 104, 107, 109, 131, 135, 139, 145, 223, 289, 290
LLC 96, 104, 349
LMI 213, 214
local loop 153, 155-157, 159, 300
local-area networks 12, 14, 16, 17, 33, 36, 73, 89, 95, 100, 101, 104, 106, 115, 124,132, 133, 274, 286, 304, 310, 312, 317
LocalTalk 35
long-distance calls 155, 162
long-distance carriers 162, 163
long-distance networks 75
loop carrier 269
LSSU 206

M

MAC 96, 100, 104, 109, 115-117, 121, 135, 136, 230, 247, 274, 275, 289-291, 295, 313, 316, 344, 345, 349
Macintosh 45, 94, 95
MacTCP 45
magnetic azimuth 352
mainframe computers 11
maintenance channels 308
MAN 17
management, element 288
management, network 107, 260, 289, 305
management proxies 21
management, traffic 252, 268
MANs 12, 17
master/slave 93, 95, 260
MAU 111
maxCTD 239
MBS 239, 243, 244
MCI 161
MCR 239, 244
MDS 337, 339
media filters 57
media, transmission 15, 17, 57
media-rate adaption 106
medium 28
medium, transmission 16
MEO 344
messaging 30
MF band 3
MIBs
microcomputers 11, 15
microcontrollers 30

microprocessors 30
microsegmentation 107
Microsoft 44, 165
microwave radio 9, 329, 330, 346
microwave relays 11
microwave signals 324
MLP 209
MMDS 337-339
MNP 184, 186
modal dispersion 75
mode specifications 75
modems 16, 97, 139, 146, 148, 160, 167, 171-174, 176, 177, 180-182, 184, 188-190, 192, 199, 283, 284, 286, 287, 289-291, 294, 300, 301, 303, 306-308, 315, 318, 319, 339
modular jacks 62, 63
modulation 167, 168, 169, 170, 171, 290, 313, 322
monitoring, performance 262
motherboard 97
MP 209
MPEG 304
MSAUs 124, 127
MSS 342
MSU 206
MTS 306
multicast 89, 90, 109, 110, 213, 214, 276
multifrequency dialing 158
multilayer switch 109
Multilink 208, 209, 210
multimode 75
multipath distortion 322
multiplexers 144, 152, 259, 267, 269, 270-272, 303, 306
multiplexing 14, 15, 40, 222, 259, 260, 266, 268, 273, 307, 308, 313, 316
multipoint-to-multipoint 253
multiport repeater 101
multitasking software 15

N

N-connectors 72
N-ISDN 196
NAPs 164
narrowband 221
NAT 214, 291
Navigator 355
NCPs 43, 209
NDCs 287
Near-end crosstalk 318
Netscape 165
network cloud 92
network-interface cards 15, 16, 17, 30
network internetworking 254-256

Network Layer 33, 34, 37-39,
    44, 131, 135-137,
    140, 206, 289, 291
network management 16, 20, 107,
    260, 289, 305
network standards 13
network switch 104
network terminations 315
network throughput 21
network topology 76, 90
networks, ADSL 305
networks, ATM 253, 256, 273, 297
networks, cable-data 289
networks, circuit-switched 40, 147, 150
networks, client/server 16
networks, connectionless 148
networks, enterprise 17, 133
networks, FDDI 278
networks, global 16
networks, local-area 12, 16,
    17, 33, 36, 89, 286
networks, long-distance 75
networks, metropolitan-area 12, 17
networks, optical 146, 257, 273
networks, packet-switched
    40, 147, 148, 151, 207
networks, private/public 11
networks, satellite 342, 344
networks, SONET 267
networks, telephone 73, 195
networks, voice 151
networks, VSAT 349
networks, wide-area 12, 16, 17,
    39, 131, 286, 289
networks, wireless 321, 340
NEXT 54, 55, 85, 86
NFS 44
NI-1 220
NICs 16, 96, 97, 165, 247
NIF 255
NMSs 21
NNI 223, 237
nodes 11, 31, 34, 38, 41, 89,
    90, 95, 106, 145, 289
noise 10, 53-55, 57, 58, 82, 84,
    155, 160, 184, 285, 295, 297,
    319, 333
non-ISDN terminals 198, 199
Northern Telecom 213
Novell 35, 43, 44, 97, 137
NT1 199, 200, 202, 220
NTs 315
NVP 83, 84

O

OA&M 206
OAM 256
OAM&P 262, 269
OC 258, 270
on-line service 164
ONU 311
open standards 13
optical channel 222
optical networks 146, 257, 273
optical receiver 270
OPTIS 319
orbits 343
oscillator 168, 170, 181
OSI 13, 26, 30, 31, 33, 35, 37, 39, 40,
    42, 44, 45, 48, 51, 89, 93, 131,
    135, 138, 140, 205, 206,
    210, 223, 231, 274, 275, 289,
    344
out-of-band signaling 160
overhead 259, 261, 262, 263,
    264, 265, 266, 267, 269

P

PABX 143
packets 30, 33, 34, 38, 102, 138, 141,
    209, 240, 275, 291, 346, 349
packet carriers 162, 163
packet switching 147, 148, 222
pagers 343, 344
pair changers 81
pair skew 56
PAM5 120
PAM5x5 117
parabola 331
parity bit 182
parity check 181
passband 153
patch cables 82, 102
patch cords 63
patch panels 78
path overhead 264-266
pay-per-view 243, 286
PBXs 318
PC+R 239
PCI 351, 352
PCR 243, 244
PDUs 30, 227, 230-232, 236, 275
PECL 123
peer-to-peer 93
performance management 21
performance monitoring 262
peripherals 89
personal computers 94
phase modulation 168
phones, cellular 343, 344
phones, ISDN 199
PHY 274, 275
Physical Layer 31, 33, 35, 44, 95,
    98, 131, 205, 223, 230,
    275, 289, 301
pinout configurations 63

Index

PLCP 230
plug-and-play 309, 353
PMD 225, 274, 275
POH 261, 264, 265, 267
point-of-sale devices 11
Point-to-Multipoint 271
point-to-multipoint 236, 253, 271, 272
Point-to-Point 270
Point-to-point 150, 236
point-to-point 76, 122, 146, 208, 236, 253, 271, 272, 299, 300, 301, 315
point-to-point services 146
Pointers 261
pointers 261, 262
Polarization 330
polarization 330, 352
Polling 21
POP 162
ports, serial 180
POTS 157, 302, 306, 309, 318
POTs 300
PPD 240
PPP 208, 209, 216, 217, 354
PRC 259
Preamble 110, 276
preamble 316
preambles 290
Presentation Layer 41, 42, 43
PRI 196, 197, 198, 205
Printers 44
printers 15, 16, 97
Private networks 11
Propagation delay 56
propagation delay 56
Proprietary standards 13
Protocol 230, 250, 291
protocol 94, 110, 136, 137, 140, 183, 238, 255, 256, 274, 291, 295, 304, 349, 354
protocol converter 145
Protocol, Hello 136, 137
Protocol, Internet 165
protocol, signaling 151
protocol stack 28
protocol, subnet 32, 33
protocol suite 28, 29
protocols 25, 26, 27, 29, 30, 44, 45, 104, 138, 140, 141, 160, 163, 184, 188, 191, 208, 209, 213, 245, 274, 275, 305, 313, 321, 340, 344, 345
protocols, LAN 89
protocols, modem 188
protocols, peer 30
provisioning 262
Proxy 355
PS EL-FEXT 55
PS NEXT 55
PSD 319

PSTN 216, 219, 301, 342
PSTNs 151
PT 236
Public networks 11
Pulse dialing 158
pulse dialing 158
pulse modulation 170
pulses 5, 116, 259
pulses, rectangular 4
punchdown block 67
punchdown panel 77
punchdown tool 67
PVC 152, 212, 242, 255, 256
PVCs 241

Q

QAM 283, 284, 285, 288, 289, 290, 293, 314, 338
QoS 19, 239, 240, 241, 251, 254
QPSK 283, 284, 290
QPSK waveform 284
Quality of Service 252
Quality-of-Service 238, 239
Quality-of-service 19
quality-of-service 19, 237

R

raceways 76
radio 8, 154, 168, 327
radio, broadcast 15
Radio frequency 2
radio, Microwave 9
radio signal 329, 330
radio waves 168
radio, wireless 31
RADSL 299, 316, 317
ramp waveform 5
range 1
RBOCs 258, 303
RDI 266
receive data 175
rectangular pulses 4
rectangular wave 4
rectangular waveform 4
redirectors 44
Reed-Solomon 290, 309
regenerator 267
REI 266
relays 33
reliability 295
remote office 210
repeater 33, 95, 98, 102, 118
repeater, multiport 101
resistance 83
resisters, terminating 81
RF 2, 168, 171, 283, 321, 331, 333, 340, 348

RFI 73, 266
RI 175
ring architecture 268, 272
ring indicator 175
ring topology 91
RISC 228
RJ-11 62
RJ-45 62, 112
RLE 186
rooms, equipment 77
routers 30, 33, 36-38, 78, 95, 96, 129,
       136, 138-141, 148, 200, 208,
       217, 236, 238, 245, 280, 286, 288,
       306, 339, 340
RS-232 33, 174, 199
RSMI 290
RTS 175
RX 175

S

S-ALOHA 345, 349
safety alarms 76
SAPI 203
SAPs 28
SAR 227, 236
SAS 277
satellite 11, 15, 31, 154, 321, 330,
       332, 336, 341-344, 349, 350, 352
sawtooth waveform 4, 5
scalability 141
SCCP 206
SCR 239, 243, 244
SCS 76
ScTP 59
SDH 273
SDM 15
SDS 162
SDSL 299, 311
SDUs 28, 230
SEAL 236
search engine 166
section overhead 262, 267
security 22, 109, 141, 208, 216,
       289, 290
segmentation, broadcast 141
serial communications 172
serial ports 180
servers 15, 29, 31, 76, 94, 95, 129,
       139, 145, 165, 286, 289, 297
service, best-effort 19
service, differentiated 20
service, guaranteed 20
service internetworking 254, 256
session layer 41-44
set-top box 286, 297
SHF band 3
shielding 58, 59
sidebands 170

signal attenuation 153, 160
signal, carrier 284
signal distortion 10
signal ground 175
signal headroom 54
signal, radio 329, 330
signal skewing 56
signal, video 285, 308
signaling 181-183, 160, 195, 203, 204
signaling protocol 151
signals 2, 4, 6, 56, 57, 151, 156, 157,
       159, 160, 171, 176, 259, 260, 266,
       267, 269, 284, 286, 301, 319, 322,
       324, 329-331, 333
sine wave 4, 170
SIP 230-232
SIWF 256
sky wave 2, 3
sliding window 291
SLIP 354
Slotted ALOHA 345
small office 215
SMDS 140, 226, 228-
       231, 235, 242, 293
SMT 274, 275
SMTP 48
SNA 163, 275
SNI 229, 231
SNMP 218, 291, 338
SNR 314
sockets 39
SOF 110
software 15, 16
SOHO 215
SONET 225, 226, 257-260,
       262, 264, 266-
       269, 272, 273, 293, 311
source address 231, 277
source-route bridging 104
space-division multiplexing 15
SPE 261, 262, 266
speed dialing 159
SPID 197, 198
spider 166
splitters 286, 306, 313
Sprint 161
SPX 43, 44
square wave 5
SRL 53
SS7 204, 205, 206
SSB 170
SSI 290
stackable hubs 102
standards 1, 13, 14, 16, 51,
       77, 187, 188, 191
star topology 76, 92, 112
start bit 173, 180
Start Delimiter 126, 276
stop bit 173, 180

364

Index

storage devices 16
STP 57, 58, 59, 60
StrataCom 213
structured cabling 76
STS 258, 260-262, 264, 266-269
subnet protocol 32, 33
super heterodyne 168
super hub 286, 289, 290
supertrunk 155, 292
SVC 151, 152
switch-to-router 122
switch-to-server 122
switch-to-switch 122
switched internetworks 138
switched services 146, 147, 152
switches, tandem 155, 160
switches, ATM 236, 240-242, 244
switches, frame 107
switches, LAN 104, 107, 108
switches, multilayer 109
switches, network 104
switches, solid-state 159
switches, symmetric 107
switches, WAN 139, 140
switching, 10/100 108
switching, asymmetric 108
switching, beam 321
switching, circuit 150, 151, 222
switching, hub 114
switching, packet 147, 148, 222
switching systems 159
symbol 284, 310
synchronization 262, 273, 275, 290
synchronous communications 172
synchronous transmissions 14

T

T1 63, 144, 146, 147, 149, 162, 198, 222, 225, 228, 303, 306, 309, 312, 313, 317-319
T3 147, 162, 352
tandem switches 155, 160
TC 225, 226
TCP 47, 48
TCP/IP 44-48, 51, 138, 240, 275, 291, 349, 353, 355, 356
TDI 44
TDM 145
TDMA 290, 345, 346, 349
TE1 198
TEI 203
telecommunications closet 76, 78
teleconferencing 310
Teledesic 344
telemetry 15
telephone 15, 73, 153, 155, 156, 157, 158, 159, 160, 195, 294, 342
telephony 196, 243, 293, 294, 312, 318, 321
television 12, 15, 283, 286, 292, 293, 294, 312, 330, 341
terminal adapter 198, 199, 208
terminal multiplexer 267, 271
terminating resisters 81
testers, cable 82, 83
testing, channel 86, 87
Thicknet 69, 112
thin client 128
Thinnet 69, 111
throughput 97, 106, 120, 134, 208
TIA/EIA 81
time-delay 153
time-division multiplexing 15
timing 290
token 125
Token Ring 33, 35, 44, 63, 89, 101, 104, 107, 124-127, 139, 247, 272, 274, 277
tone dialing 158
topology 16, 25, 76, 90
topology, bus 90
topology, ring 91
topology, star 92, 112
towers 324, 327, 328
TP-PMD 275
traffic management 252, 268
trailer 27, 29, 30
transceiver 112, 118, 122, 123
transformers, balance 57
translational bridging 104
transmission, baseband 12
transmission, broadband 12
transmission, data 42, 89, 122, 191, 192
transmission, digital 15
transmission, full-duplex 14
transmission, half-duplex 14
transmission links 11
transmission media 15, 16, 17, 57
transmission, satellite 15, 31
transmission, wireless 15
transmission convergence 225
transmissions, asynchronous 14
transmissions, synchronous 14
Transmit Data 175
transmitter 322, 327
transparent bridging 104
Transport Layer 39, 40, 41, 43, 44, 47, 291
transport overhead 262
trellis 290
triangle waveform 4
triangular waveform 5
tributary signals 267
trunks 154, 155, 156, 157, 159, 160, 198, 291, 292, 293, 295, 297

365

TSB-67 81, 85-87
tunneling code 352
Turbo Internet 351, 352, 355
twinaxial 70
twisted-pair cable 25, 31, 33, 54, 56-58, 92, 112, 116, 117
twisted-pair ports 100
TX 175

U

U-NII 339
U.S. Robotics 188
UART 180, 181
UBR 244
UDP 48, 214
UHF 3, 337
UNI 223, 237, 239, 241, 245, 256
unicast 89, 90, 110, 276
UNIX 40, 94
UPC 240
URL 164, 165
USB 351, 352
user interface 94
user-to-user 207
UTP 57, 58, 60, 63, 67, 112, 275, 304

V

V.90 187-189
VCC 250
VCI 242, 244, 245, 346
VDSL 299, 311-316
VHF 3, 337
video 1, 14, 15, 17, 47, 68, 76, 78, 221, 228, 235, 253, 285, 286, 293, 294, 308, 310, 312, 313, 338
video-on-demand 243, 304, 312, 317
videoconferencing 17, 138, 243, 311, 312
videophone 196
videotex, interactive 196
VINEs 97
virtual channel 240, 243
virtual circuit 151, 152, 211, 212, 242
virtual tributary 260, 265, 266
visible light 8
VLAN 129, 247
VLSI 100
voice 1, 14, 15, 17, 47, 68, 76, 78, 144, 149, 151, 153, 157, 159, 221, 228, 235, 243, 253, 312
VPI 242-245
VPN 162
VSAT 348, 349, 350
VT 260-262, 265, 266, 268

W

WAN 17, 73, 104, 139, 140
waveform, QPSK 284
waveforms, 4, 5
waves, carrier 168-171, 190
waves, ground 2, 3
waves, radio 168
waves, rectangular 4
waves, direct 2
waves, modulated 168
waves, sine 4, 170
waves, sky 2, 3
waves, square 5
WCS 337, 338, 339
WDM 258
Web 29, 48, 164, 165, 166, 186, 189, 219, 286, 350
WebCrawler 166
wide-area network 12, 16, 17, 39, 131, 286, 289, 312
wideband 268
Win95 Winsock 353
wind loading 325
Windows 44, 94, 97, 128, 129, 350, 351, 353, 354, 355
wire-management system 78
wireless 15, 16, 31, 151, 321, 322, 324, 327, 329, 321, 327, 330, 336, 339, 340
wiremap 85
wiring closet 77, 78, 101, 120
wiring hub 16
workstation 25, 89, 236

X

X.25 140, 148, 163, 197, 207, 211, 219, 251, 253, 349
xDSL 225, 252, 299, 300-302, 318
Xerox 109, 137
XNS 137

Y

Yagi antennas 339
Yahoo 166
Ys 81

PROMPT PUBLICATIONS

Exploring the World of SCSI
Louis Columbus

Focusing on the needs of the hobbyist, PC enthusiast, as well as system administrator, *The World of SCSI* is a comprehensive book for anyone interested in learning the hands-on aspects of SCSI. It includes how to work with the Logical Unit Numbers (LUNs) within SCSI, how termination works, bus mastering, caching, and how the various levels of RAID provide varying levels of performance and reliability. This book provides the functionality that intermediate and advanced system users need for configuring SCSI on their systems, while at the same time providing the experienced professional with the necessary diagrams, descriptions, information sources, and guidance on how to implement SCSI-based solutions. Chapters include: How SCSI Works; Connecting with SCSI Devices; and many more.

Communication
500 pages • paperback • 7-3/8" x 9-1/4"
ISBN 0-7906-1210-0 • Sams 61210
$34.95

Dictionary of Modern Electronics Technology
Andrew Singmin

New technology overpowers the old everyday. One minute you're working with the quickest and most sophisticated electronic equipment, and the next minute you're working with a museum piece. The words that support your equipment change just as fast.

If you're looking for a dictionary that thoroughly defines the ever-changing and advancing world of electronic terminology, look no further than the Modern Dictionary of Electronics Technology. With up-to-date definitions and explanations, this dictionary sheds insightful light on words and terms used at the forefront of today's integrated circuit industry and surrounding electronic sectors.

Whether you're a device engineer, a specialist working in the semiconductor industry, or simply an electronics enthusiast, this dictionary is a necessary guide for your electronic endeavors.

Electronics Technology
220 pages • paperback • 7 3/8 x 9 1/4"
ISBN: 0-7906-1164-4 • Sams 61164
$34.95

To order today or locate your nearest Prompt® Publications distributor at 1-800-428-7267 or www.samswebsite.com

Prices subject to change.

PROMPT PUBLICATIONS

Handbook for Parallel Port Design
James J. Barbarello

This book "demystifies" the parallel port. First addressing the basic tools of inputting to and outputting from the port, it logically progresses from the simple to more complex, showing how to use display devices (LEDs) and input sensing devices (such as light-sensitive resistors, IR LEDs, phototransistors, and rotary encoders). Then it walks you through the design process until you have an ample understanding of the parallel port and the skills to use it effectively.

A companion diskette is included, containing 52 support files, viewable in DOS, Windows® or Qbasic. These data and executable files will help you better understand and master the concepts presented, such as: Keycard Circuit Alignment; Rotary Encoder Demonstration; ADC Application; Computer Based Logic Probe Application; Executive Decision Maker Application; and others.

Electronics Technology
344 pages • paperback • 7-3/8 x 9-1/4"
ISBN: 0-7906-1177-5 • Sams: 61177
$29.95

Administrator's Guide to E-Commerce
Louis Columbus

Electronic commerce is here to stay, and its impact is growing daily. The multitude of books on the subject reinforce that fact, but the majority of them focus on theory, not practice. *The Administrator's Guide to E-Commerce* is a comprehensive hands-on guide to creating and managing websites focused on commerce using the Microsoft BackOffice product suite. From taking an existing web server and creating an electronic storefront, this book also explains what e-commerce really is, the role of networking technologies contributing to the growth of this industry, issues of privacy and security, electronic payment systems and client-based software.

Specifically, you will learn how to use Microsoft Windows NT Workstation and Server 5.0 and products from Microsoft's BackOffice product suite, SQL Server, Transaction Server, Internet Information Server and Commerce Server, including StoreBuilder Wizard. Examples of how companies have created customized applications are included, as well how to create links to content-based databases, ensuring scalability and performance consistency as traffic grows on your web site.

Business Technology
416 pages • paperback • 7-3/8" x 9-1/4"
ISBN 0-7906-1187-2 • Sams 61187
$34.95

To order today or locate your nearest Prompt® Publications distributor at 1-800-428-7267 or www.samswebsite.com

Prices subject to change.